P9-CKX-149
3 1404 00760 4025

MAY 24 2004

Ozone and Plant Cell

Ozone and Plant Cell

by

Victoria V. Roshchina

and

Valentina D. Roshchina

Russian Academy of Sciences,
Institute of Cell Biophysics, Russia

KLUWER ACADEMIC PUBLISHERS
DORDRECHT / BOSTON / LONDON

A C.I.P. Catalogue record for this book is available from the Library of Congress.

ISBN 1-4020-1420-1

Published by Kluwer Academic Publishers,
P.O. Box 17, 3300 AA Dordrecht, The Netherlands.

Sold and distributed in North, Central and South America
by Kluwer Academic Publishers,
101 Philip Drive, Norwell, MA 02061, U.S.A.

In all other countries, sold and distributed
by Kluwer Academic Publishers,
P.O. Box 322, 3300 AH Dordrecht, The Netherlands.

Printed on acid-free paper

Printed in the Netherlands.

CONTENTS

PREFACE

The problem of the influence that ozone can cause in living organisms have now become blatantly apparent due to the depletion of the Earth's ozone layer and the increase of such anthropogenic sources of the gas as motor vehicle exhaust and modern devices, including computers, that emit ultraviolet radiation.

Plants play significant roles as both indicators and origins of oxygen and ozone in Nature. Therefore, the interaction between plants and ozone is important for plant biophysics, physiology, biochemistry and ecology. Since the issue now needs to be considered on the cellular and molecular levels, the aim of this book is to present a real picture of the functioning of living plant cells upon ozone to a wide group of specialists and naturalists.

In the first chapter, the properties of ozone and its transformations in both the atmosphere and water are described. The second chapter is devoted to the ways in which ozone penetrates into living plant cells, the subsequent reactions with cell components and the role of cellular cover and membranes in these processes. Information about such ozone effects as membrane damage, changes in cell permeability and membrane malfunctions is also provided. Molecular mechanisms characteristic for the interaction of ozone with cell components are described in chapter 3. To facilitate an understanding of these mechanisms, modern conceptions about free radicals, peroxides and protective reactions of cells against ozone are also considered. The fourth chapter is devoted to steady-state protectory systems that counteract ozone and other reactive oxygen species. Adaptive changes in cellular metabolism to O_3 exposure are considered in chapter 5. The information in chapter 6 focuses on the cellular monitoring of ozone, and deals with such issues as cellular sensitivity to this substance and the possibility of devising plant cellular models for use as ozone biosensors and indicators. Possible responses to ozone indicated by the various plant reactions are also described. In the conclusion, the importance and possible mechanisms of ozone action when large and small doses display diametrically opposite effects on cells are considered.

This monograph will be useful for ecologists, plant physiologists, biochemists, biophysists and botanists, including lecturers and students, as well as for specialists in the field of environmental management.

ACKNOWLEDGMENTS.

The authors are extremely grateful to two corresponding members of the Russian Academy of Sciences, Professor Alexander Kuzin and Professor Arkadii Budantsev, for their useful comments and careful editing of the manuscript. In addition, we wish to thank Dr. Valerii N.Karnaukhov, Head of the Laboratory of Microspectral Analysis and Cellular Monitoring of Environment and Vice-Director of the Institute of Cell Biophysics at the Russian Academy of Sciences, for his constant support, his willing assistance with the work and his eagerness to provide any necessary discussion. We also thank Ludmila I. Mit' kovskaya, Larisa F. Kun'eva and Andrei V.Rodionov for their help in producing the computer version of the manuscript.

INTRODUCTION

In 1785, the Dutch scientist Martinus van Marum noticed that air exposed to electrical sparks had a characteristic smell and demonstrated redox properties. This phenomenon was explained in 1840 by Christian Schönbein, the Swiss researcher, as the formation of a special gas, which was named ozone, based on the Greek word "ωοξ" or "ozo", which meant "smell". Later it was shown, that ozone is an

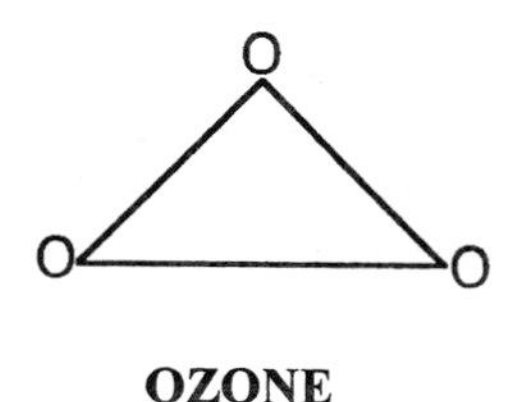

OZONE

allotropic modification of oxygen trioxide and is constantly present in air. Its concentration increases during thunder storms or in predominantly coniferous forest areas. Recently, anthropogenic sources of O_3 are also known: exhaust from automobiles and wastes of manufacturing plants, as well as any technology that produces ultraviolet irradiation, from photocopiers to computers. Below is an indication of the main sources of ozone divided into outdoor and indoor sources of the gas:

OUTDOOR SOURCES
NATURAL
Ultraviolet irradiation of air
Thunder storms,
Forest excretions
ANTHROPOGENIC
Exhaust from automobiles and
wastes from manufacturing

INDOOR SOURCES
Ultraviolet - producing technique

In the stratosphere, ultraviolet irradiation and thunder storms are the main contributors to ozone accumulation. However, in the lower layer of the atmosphere known as the troposphere, automobile exhaust and industrial wastes, both containing nitrogen oxides and carbon monooxide (Sanderman et al., 1997; Guderian, 1999), along with the monoterpenes of woody plants (Finlayson-Pitts and Pitts, 1986; Roshchina and Roshchina, 1993; Roshchina, 1996) are the primary anthropogenic and biogenic sources of ozone.

Ozone is a small part of the atmosphere—accounting for 0.64×10^{-6} of the total air mass. The Earth-encompassing layer of ozone is estimated around 0.4-0.6 cm thick.

Distribution of ozone in the atmosphere is, however, non-uniform. The greatest density (number of molecules per unit of the volume) is observed at the height of 24-26 kms in tropical areas and 18-20 km in the polar zones. In other areas of the atmosphere, its amount is, as a rule, somewhat less. At the surface of the Earth, the ozone concentration is about 10^{-6} - 10^{-4} %. Like other gases, ozone participates in the general circulation of the atmosphere. It is heated up in the subequator zone, rises upwards, and moves along meridians to the southern and northern poles, where it is cooled, descends, and moves back towards the equator. The amount of ozone overhead therefore depends on geographical latitude and season. It is higher in high latitudes, and is at a maximum in spring and a minimum in autumn. As the ozone concentration in the air of any region depends on climatic and many other factors, the degree of the atmosphere saturation with ozone constantly changes and only regional averages can be provided.

Ozone contained in the upper layers of the atmosphere wholly absorbs short-wave ultra-violet rays (Baird, 1995; Ryerson et al., 2001). Due to this quality, the ultraviolet radiation (shorter than 295 nm) physiologically dangerous for life does not reach the earth's surface. The ozone layer is the earth's natural "sunscreen" (Baird, 1995). This layer began to deteriorate in the 70's. And in the 80's, the ozone layer in the stratosphere began to decrease constantly (Executive Summary. Special of Atmospheric Science Panel.1998). By the 90's, the ozone layer around the Earth had become, on average, 3-4% thinner than it was in the previous decade. In 1992, the level of ozone had fallen by another 3% (Gleason et al. 1993). An especially strong reduction (the ozone hole) is observed annually in Antarctica (Executive Summary. Special of Atmospheric Science Panel. 1998). There the decrease in ozone quantities reaches 50%. Apparently, there is a large air convection flow to the industrial regions, where increased destruction of ozone due to the emission of harmful substances into atmosphere is occurring. The depletion of the ozone is also observed in the Northern Hemisphere, but in smaller percentages (22%) than in the Southern Hemisphere. In the 90's, a huge ozone anomaly was formed over territory stretching from the Arctic Ocean to the Crimea, and ranging from west to east over an area that included both the USA and Europe. In February–March 1995, the quantity of ozone above Siberia twice decreased (Bridges and Bridges, 1996). In the Northern Hemisphere, all observed ozone anomalies ("holes") are primarily attributable to dynamics (changes) in the atmosphere arising from termobaric and circulation changes, or from increases in the amounts of CO_2, NO_2, and water vapor (the so-called major "green house" gases), as well as other atmospheric components (Rousseaux et al., 1999).

The main effect of ozone-layer depletion in the stratosphere is an increase in ultraviolet radiation in ranges dangerous to life (250–320 nm) and normally absorbed by ozone (Mandronich et al., 1998). The ultraviolet radiation strengthened by the reduction of the ozone layer around the Earth causes chromophores to be activated, then photoproducts to be subsequently formed, and biochemical changes in cells to occur. Often, the result is the formation of tumors, skin cancer, cataracts, immune-system problems and other potential injuries to human beings (Longstreth et al., 1998). Appreciable changes in common ecosystem conditions can also be observed, especially those involving ground vegetation (Caldwell et al., 1998), phytoplankton (Hader et al., 1998), and the realization of biogeochemical cycles (Zepp et al., 1998). Thus, absorption of ultraviolet radiation by the ozone layer protects living organisms. More specifically, it is possible to indicate cellular components important for life as nucleic acids and proteins that are needed to be protected first. Damage of these substances can result in mutations or even the destruction of cells (Blum, 1969; Aleksandrov and Sedunov, 1979; Mizun, 1993).

Ozone also helps to define the thermal status of the atmosphere. In the infrared region of the spectrum, ozone has a line of absorption at about 960 nm. Thus, ozone helps to contain energy in the infrared range radiated by the Earth and keeps it within the atmosphere, so that the thermal balance of our planet is kept at a steady state. Otherwise our planet would quickly cool.

If stratospheric ozone is decreasing, its concentration in the ground-level layer of the air, the troposphere, is increasing and, in a number of instances, reaching toxic levels (Tang et al., 1998). Systematic measurements of the ozone concentration confirm that the greatest quantitative changes in ozone occur in the troposphere, where all organisms live (Karol', 1992 ; Semenov et al., 1999). In Russia, data on the harmful pollutants emitted to the atmosphere in cities and industrial centers, complete with discussions of the possible effects involving ozone, are issued annually in special collective publications. According to the data in the 1994 yearbook, 0.000906 thousand tons of ozone per year are added to the atmosphere of St.-Petersburg, and 0.000212 thousand tons to the air in the Voronezh area. Both amounts exceed the ozone content in non-polluted agricultural regions ("Emission of harmful substances" Yearbook, ed., M.E.Berlyand, 1994). The changes of ozone concentration around the Earth is analyzed every year by World Meteorological Organization and published as Annual Global Ozone Research and Monitoring Reports,Geneva, and Air Quality Criteria for Ozone and Related Photochemical Oxidants, National Center for Environmental Assessment, U.S. Environmental Protection Agency.

In high concentrations, ozone is a dangerous toxicant. In many countries, various resolutions have been passed in order to limit the concentration of ozone in the atmosphere. More often than not, such concentrations of ozone are estimated at 0.0001mg/l (10^{-5} M). Despite the smallness of the number, the special properties of ozone mean that it has an influence on all the life on the surface of the Earth. A 10% reduction in the amount of ozone has already had a global effect on living organisms—some crop yields have diminished and various kinds of pathologies in both animals and humans have been reported. In many cities, a recurring springtime disturbance of the respiratory function (Hazucha, 1992) is now being commonly observed. In particular, there are increased complaints about chronic lung diseases, nervous and immune system disorders, skin cancers and an eye-retina damage (Aleksandrov and Sedunov, 1979; Karol', 1993). Such problems can especially be linked to seasonal changes in the ozone concentration of the atmosphere.

Higher than usual oxidant concentrations in the troposphere are, furthermore, related to air pollution. In the troposphere, ozone is a component of "photochemical smog", which resembles a fog and which is becoming heavier and darker due to constant emissions of industrial wastes and the byproducts from the combustion of coal and petroleum. The smog problems that were first described in Los Angeles are now observed over many of the world's cities.

The decline of the ozone concentration in the stratosphere is connected to various chemical compounds, penetrating into the upper atmosphere as a result of nuclear explosions, space flights, flights of supersonic aircraft, applications of clororganic fertilizers in agriculture, as well as by other means.

Dangerous effects of the ozone layer depletion or tropospheric ozone accumulation for living organisms are studied by specialists of Environmental Medicine and Plant Biology, according to the programs " Ozone Depletion and Human Health Effects", "Stratospheric Ozone and Human Health " and "Ambient Ozone and Plant Health" (see Internet sites: http://sedac.ciesin.org/ozone or http://www.epa.gov or http://www.nps.ars.usda.gov).

Ozone does, nevertheless, have some beneficial uses. Rather small doses of ozone can appear to be useful in stimulating plant development, as O_3 can modify or increase the efficiency of the regulatory processes in living organisms. The oxidizing properties of ozone can also be used as a disinfectant for such purposes as the purification of drinking water or the modification of the ways in which microorganisms exchange substance so that more valuable metabolites are obtained. Ozonated water, for instance, is applied as a pesticide in order to eliminate some plant parasites.

In ecological studies of the environment and habitat of organisms, it is necessary to consider both the negative and positive effects of ozone on living cells. The threat of ozone depletion in the stratosphere, the commonly excessive increase of its concentration in the ground-level air, and the possible salutary influence of small concentrations of the compound means that research into the interaction between life and ozone needs to be given the highest scientific priority.

Ozone is formed not only from abiogenic sources, but also from plants (Bauer et al., 1979; Heath and Taylor, 1997). As producers of oxygen via photosynthesis, plants play a key role in maintaining the Earth's ozone balance. Plants can also serve as indicators of ozone damage, when the gas occurs as a pollutant. Consequently, the interrelationship between plants and ozone has planetary significance. Previous and most contemporary scientific endeavors mainly dealt with the compound's meteorological or pathologic (pollutant) profiles. Significant findings are astutely analyzed and described in such monographs as those written by Feder and Manning (1979) or Treshow and Anderson (1989), or Semenov et al.(1999) as well as in books edited by Treshow et al. (1984); Guderian (1985; 1995; 1999); Sandermann et al. (1997), De Kok (1998), Fuhrer and Achermann (1999), and Weber et al. (1999 . Wide-ranging studies into the influence of ozone on plants in general and the effects of ozone pollution on plant life in particular have been conducted in many laboratories and have been summarized in several specialist publications, including some appearing on the Internet (see appendix 1). Readers should consult them if they require background information on ozone-related plant activity, the ways in which plants can function as indicators of O_3-induced damage, or as monitors of the ozone level in industrial and urban regions. We intend to examine the problem of ozone not only by investigating its function as a screen for the Earth's biosphere or its properties as an air pollutant, but by studying the activities of ozone and the agents of ozonolysis within plant cells and even on the cellular surface. It is this part of the picture that has, up to now, been insufficiently studied.

In this book, the main focus falls on the plant cell as a receiver and possible source of ozone. Specific attention will be given to the cellular mechanisms involving ozone, the effects of O_3 on cellular components and the cellular monitoring of ozone presence in the air.

CHAPTER 1

ATMOSPHERIC OZONE

1.1. PHYSICO-CHEMICAL PROPERTIES OF OZONE.

Ozone is an allosteric modification of oxygen. At normal temperature and pressure, it is gas of dark blue color with a characteristic sharp smell, discernible at concentrations of 10^{-4}%. At temperatures below - 111,5 ^{o} C, ozone condenses and turns to an unstable, dark-blue colored liquid. Ozone is condensed at higher temperature than oxygen (-182.97 ^{o} C), a factor that makes it possible to separate these gases. O_3 is, however, not termostabile, and transforms into oxygen at temperatures of 100 ^{o} C. In pure water, ozone decomposes faster than it does when in its gaseous phase. The ozone solubility in water 0.394 g/L (at 0° C), which makes it 15 times more soluble than oxygen. In saline water solutions, O_3 dissolves more poorly than it does in water alone.

Ozone – an optically active gas—has spectral properties that greatly distinguish it from oxygen (Fig.1). Ozone mainly absorbs ultraviolet radiation from the sun in the 200-360 nm range of wavelengths, whereas oxygen absorbs UV radiation at wavelengths <200 nm. Thus in the region of the spectrum shorter than 220 nm, there are bands of absorption characteristic of both oxygen and ozone. In the 200-320 nm, band most characteristic of ozone, O_3 has maximum absorption at 255 nm. In the visible part of the spectrum, beams from 450 to 650 nm (Fig.1, right scale) are also absorbed. The maximum value of the absorption coefficient in the visible area of the spectrum is two thousand times less than in the ultraviolet area. Ozone also has a band of absorption in the infrared part of spectrum at 650-1000 nm, which is especially intensive at 957 nm. Due to these absorptive properties of O_3, our planet is protected against thermal disturbance. Without an no ozone layer, the earth's surface temperature would decrease by 1.5-3 ^{o}C (Gilbert, 1981) and the biosphere would be gravely endangered.

Due to O_3's absorption of sun light at wavelengths less than 290 nm, the most biologically active part of solar radiation does not reach the earth's surface. As shown on Fig.1, the spectral region of ozone absorption is the same one as that in which "substances of life" - DNA and proteins - also absorb ultraviolet light. Therefore, the ozone layer protects all living organisms against hazardous UV-radiation <300 nm.

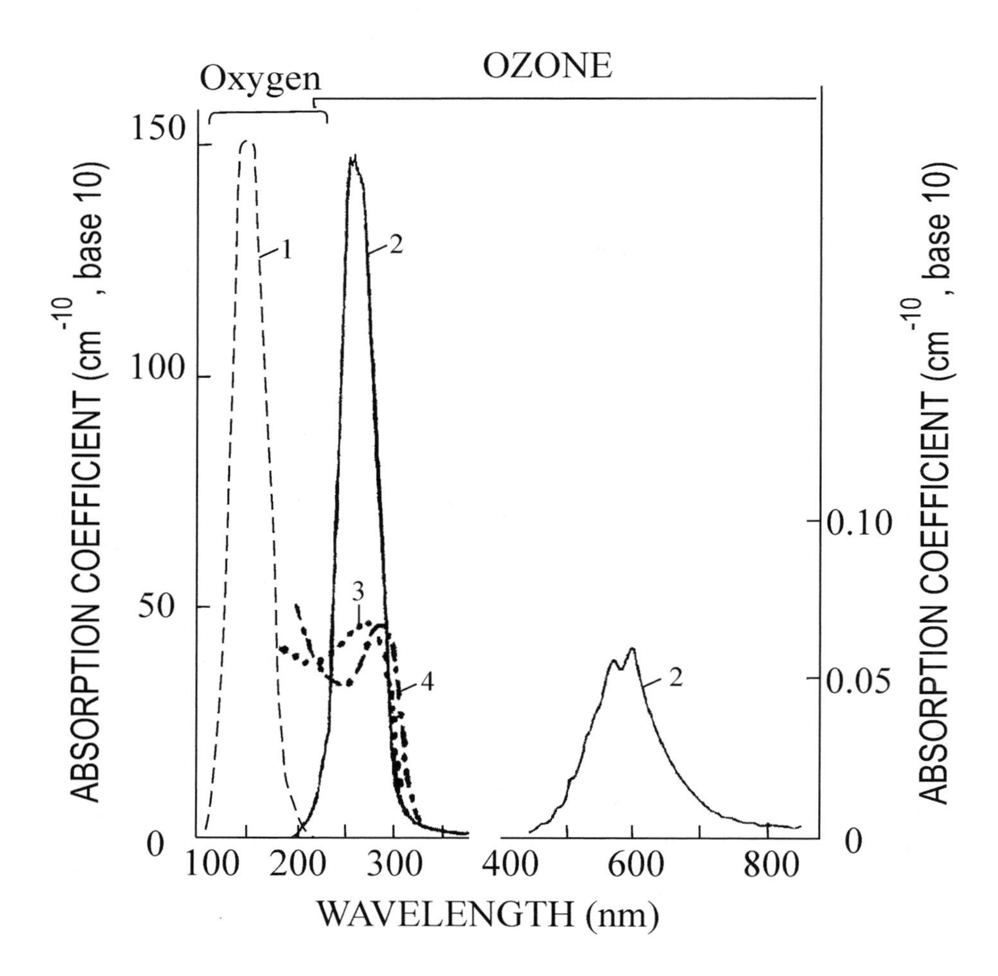

Figure 1. The absorption spectra of ozone (1), oxygen (2), DNA (3) and protein (4). (Sources: Prokof'eva ,1951 , Razumovskii, 1979; Perov and Khrgian, 1980; Gilbert,1981, Becker et al.,1985; Guderian et al., 1985;Finlayson-Pitts and Pitts, 1986)

Among other differences between ozone and oxygen, the following are worth noting. Ozone is colored, diamagnetic, explosive and poisonous. Its reactivity is extremely high. It oxidizes all metals, except for silver and platinum. Interacting with some inorganic and organic compounds, ozone will form ozonides—one kind of peroxide. Ozonides will be formed when ozone reacts with the hydroxides of alkaline metals, for instance

$$2\ KOH + 2\ O_3 \rightarrow 2\ KO_3 + H_2O + 1/2\ O_2$$

Inorganic ozonides of these compounds are characterized by the presence of an O_3 ion, which causes red coloration. Oxides of nitrogen also react easily with ozone, and form corresponding ozonides (Razumovskii, 1979).

$$NO + O_3 \rightarrow (NO_2)^x + O_2 \rightarrow NO_2 + O_2 + h\nu$$

$$NO_2 + O_3 \rightarrow NO_3 + O_2$$

The intermediate product in this case is nitrogen dioxide, and the reaction is accompanied by a chemiluminescence. In organic ozonides the O_3 group is covalently bound to two radicals

R_1-C - O - CH-R_2
⌊ O...O ⌋

Given these properties, chemical, optical and other methods can be used to detect the presence of ozone in the air. The chemical methods frequently involve iodometry (Ehmert 1959, Byers and Saltzman 1959; Ingolds et al. 1959; Boyd et al. 1970), in which potassium iodide oxidized by ozone, as O_2 does not react with this reagent. The reaction occurs as follows:

$$O_3 + 2\,KI + H_2O = I_2 + O_2 + 2\,KOH$$

The optical methods involve observations of the following: the ozone-induced chemiluminescence of such dyes as rhodamine (Bersis and Vassilliou, 1966; Guderian, 1985), the ozone-induced fluorescence resulting from the oxidation of non-luminescent dihydroacridine to fluorescent acridine (Prokof'eva, 1951), or the quenching dark blue fluorescence of luminol (Peregud et al., 1962; Peregud, 1976; Peregud and Gorelik, 1981; Grushko, 1987). The procedure involving chemiluminescence is most frequently used. The detection limit of chemiluminescence is less than 1 ppb of ozone and is not subject to interference by other atmospheric constituents (Guderian, 1985). Besides rhodamine (Regener, 1964) and luminol, chemiluminescence of 40 other dyes (e.g. eosin, riboflavin, etc) have been found to react only when ozone is present in the air (Ray et al., 1986). Consequently, they can also be used to indicate ozone levels. Additionally, electrolytic (galvanic) methods of detecting ozone in the air are also frequently used (Hersch and Deuringer, 1963).

Ozone has a strong toxic effect on living organisms. It is 3 times more toxic than hydrocyanic acid, 5 times more than phosgen(e), 10-15 time more than chlorine, and 50 times more toxic than phenol (Razumovskii, 1979).

Useful information about ozone's physico-chemical properties can be found' in a number of the monographs, reviews and manuals (Prokof'eva, 1951; Razumovskii and Zaikov, 1974; Perov and Khrgian, 1980; Razumovskii, 1979; the Chemical encyclopedia, 1964, т.3, Gilbert, 1981; Becker et al., 1985; Finlayson-Pitts and Pitts, 1986; Le Bras (ed) 1997, Hov (ed), 1998).

1.2. FORMATION OF OZONE

The formation of ozone occurs both in the upper layers of the atmosphere and at ground level . In the upper layers, the greatest quantity of ozone will be formed photochemically, in reactions involving the sun's short-wave radiation. This process is dependent on the occurrence of atomic oxygen. The first theory of photochemical formation of ozone was advanced by Chapman in 1930 (cit. Noyes and Leighton, 1941). According to this theory, oxygen dissociates to atoms when exposed to intense ultraviolet solar radiation at wavelengths less than 242 nm,:

$$O_2 + h\nu \rightarrow O + O$$

The reaction occurs in the atmosphere at heights greater than 20 kms above the earth's surface, since radiation with $\lambda < 240$ nm hardly penetrates below this level. The resulting atoms of oxygen join with oxygen molecules in the presence of a third component—particle M—which consumes a part of the co-striking energy. Any molecule or atom can serve this role as M.

$$O + O_2 + M \rightarrow O_3 + M$$

Atomic oxygen has a very short lifetime in the atmosphere and will convert to ozone almost at once (see the book edited by Hov, 1997). At a level higher than 30 kms, ozone exists much longer (hours and days). In 1965 Hunt in USA (cit. Perov and Khrgian, 1980) specified that, in reactions formative of ozone, the significant role belongs to a water vapor. Radicals $H^\bullet$ and $OH^\bullet$ are formed from water when ultraviolet light acts. Atomic oxygen then reacts with the $H^\bullet$ radical to produce a perhydroxy radical, which ensuingly participates in the formation of ozone (see 1.3.3.1).

The destruction of ozone in the top layers of atmosphere occurs when this gas is exposed to solar radiation at 200-320 nm, so that the following reaction occurs:

$$O_3 + h\nu \rightarrow O_2 + O$$

The loss of ozone also occurs in its reactions with atomic oxygen, or between two molecules of O_3:

$$O_3 + O \rightarrow O_2 + O_2$$

$$O_3 + O_3 \rightarrow 3\ O_2$$

Moreover, various amounts of ozone break down in reactions with chemical compounds formed in the ground-level atmospheric layer (in particular involving anthropogenic pollutants) and then lifted into the stratosphere by air circulation. Except for photochemical reactions, the concentration of ozone in the stratosphere is only affected by dynamic processes, such as air movements in vertical or horizontal directions. All of this leads to the rather non-homogenous distribution of O_3 in the atmosphere.

Troposphere—the bottom layer of the atmosphere—extends up to height of 8-15 kms to see level. (Stratosphere is the portion of the atmosphere from 15 to 50 km in altitude immediately above the troposphere). All living organisms are further concentrated in the lower 30-50 m thick layer of troposphere. Here, 70% of the ozone concentration is provided by a diffusion of the gas from the stratosphere, and 30% through O_3 formation from reactions occurring in the same ground-level layer.

Usually, the amount of ozone in the troposphere is not high. However, owing to human industrial, it is in the process of accruing. Especially in the ground-level layer above large cities (Tokyo, Los Angeles and others), a 40-50-fold increase in the concentration of atmospheric ozone can be observed. In the last decade, its level has also increased in agricultural areas because air pollutants can be transported hundreds of kilometers from their source (mainly industrial enterprises and vehicle exhaust). In many agricultural regions, the concentration of ozone is 2-3 times higher than it would be without human influence (Heagle, 1989). Ozone is also found among the gases emitted by volcanoes (Kerr, 1993). The importance of ozone for any life on other planets has also recently been underlined when ozone was detected on the satellites of Saturn (Noll et al. 1997).

The last decade of the 20th century was characterized by the global deterioration of atmospheric quality. Insofar as this issue is concerned, the significant role is played by increasing concentrations tropospheric pollutants, including ozone (Liu and Liptak, 2000). For the last 30 years, the amount of O_3 in troposphere has increased on average by 60%, and over the entire century as much as 5 times in the area above Europe alone (Zvyagintsev and Kruchenitskii, 1997; Friedrich and Reis, 2000). In effect, ozone is a secondary product produced by the pollution of the atmosphere, as it is formed when the primary polluting substances (carbon monoxide, hydrocarbons, nitrogen oxide, and hydrogen) are exposed to solar light. We shall therefore consider the distribution of these primary pollutants and the role that they play in the formation of ozone.

Carbon monoxide (CO) is formed by combusting any kind of hydrocarbon-based fuel in ways which result in the incomplete oxidation of the carbon. In the Northern Hemisphere, atmospheric CO content is increasing at a rate of 2% a year (Dimitriades, 1981). Vegetation also contributes somewhat to this increase. The emission of carbon monoxide by plants was investigated in 1970 (Delwiche, 1970), and it is now assumed that light causes polyphenols to disintegrate and to produce CO. According to the calculations of Bauer et al. (1979), world production of carbon monoxide from plants is about 0.7×10^{14} g/year. Other research seems to confirm this data (Zimmerman et al., 1978). Greater amounts of CO are also formed in the air when isoprene and terpenoids released by plants are oxidized (Zimmerman et al., 1978). The isoprene oxidation will produce formaldehyde, which decomposes upon light to yield carbon monoxide as one of its products. The oxidation of terpenoids is a process that has been given less attention, though it is known that either ozone or free radicals attack its double bonds. As a consequence, the terpene rings can break down into segments, including short-chain aldehydes, which will subsequently decompose to form carbon monoxide and hydrogen (Zimmerman et al., 1978) (see section).

The oxides of nitrogen, which also contribute to ozone formation, are found in oil-refinery emissions, but they are also formed from nitrification of the soil, as well as in other processes. The largest emissions of CO and NOx come from motor-vehicle traffic, as these two compounds constitute about 94% of the gas content in automobile exhaust (Heagle, 1989). The various oxides of nitrogen are also released by plants, so that these life forms are also potential participants in ozone synthesis (Brown and Roberts, 1988; Ryerson et al., 2001). Both hydroxyl radical ($\dot{O}H$) and perhydroxyl radical ($H\dot{O}_2$) participate in the formation of O_3. The reactions proceed as follows: in the beginning CO_2 and hydrogen will be formed from CO and hydroxyl radical. Then hydrogen, in a reaction with oxygen, produces perhydroxyl radical, which interacts with nitrogen monoxide, forming nitrogen dioxide:

$$CO + \dot{O}H \rightarrow CO_2 + H$$

$$H + O_2 + M \rightarrow H\dot{O}_2 + M$$

$$H\dot{O}_2 + NO \rightarrow NO_2 + OH$$

M - is any particle -molecule or atom.

Further photolysis of nitrogen dioxide occurs, followed by the interaction of the oxygen atom with a molecule of oxygen. This last step is the basic mechanism of ozone formation.

$$NO_2 + h\nu \rightarrow NO + O$$

$$O + O_2 + M \rightarrow O_3 + M$$

In plants, nitrogen oxides are released by some of the processes involved in the nitrogen metabolism (Roshchina and Roshchina, 1993; Wojtaszek,2000). Moreover, NO can participate in the complex mechanisms dealt with the secondary messengers in all living organisms (Alberts et al., 1994) including plants. Gas nitric oxide is considered to be a signaling molecule in Vertebrates, where it accumulates in blood vessels, serves as a local mediator helping activated macrophages and neutrophils to kill invading microorganisms, and is used by many types of nervous cells to signal neighboring cells (Alberts et al., 1994; Wojtaczek,2000). It is formed by deamination of the amino acid arginine in the presence of NO synthase, and its secretion is stimulated by acetylcholine. Nitric oxide produced in this way may act only locally because it has a short half-life (5-10 seconds) in extracellular spaces before it is converted to nitrates and nitrites by oxygen and water. Furthermore, NO reacts with iron in the active site of guanylate cyclase (guanylyl cyclase), stimulating it to produce the secondary messenger cyclic GMP. Due to the formation of nitrogen oxides in plant cells, they may also serve as local sources of ozone for both the plant cell involved and its surrounding spaces. Although this subject is new, it has recently been recognized as important for understanding the local formation of O_3 and its action in living organisms.

The increase in the ozone concentration in the atmosphere is promoted by the increase in the quantity of methane (CH_4). Global increase in CH_4 concentration occurs at a rate of about 1% a year. Methane is formed in microbiological processes of the soil, as a byproduct of coal mining and as a result of fires. Total production of methane in the world amounts to $3\ x10^{-8}$ tons per year (Koyama, 1963; Altshuller, 1983). Depositions of lakes and flooded soils are major sources of methane emission to the atmosphere (Koyama, 1963). In particular, much methane is produced by tundra soils, which alone yield 10% of global atmospheric methane (Whalen and Reeburgh, 1990; Khalil, 1999). Today, more than 60% of the CH_4 in the air comes from human activity, including rice cultivation, coal mining, natural gas usage, biomass incineration, and the raising of cattle (Khalil, 1999).

The key role in methane oxidation is played by hydroxyl radical ($\dot{O}H$), which participates in the following O_3-producing series of reactions. In the beginning methane reacts with hydroxyl radical, forming ($\dot{C}H_3$) radical

$$CH_4 + \dot{O}H \rightarrow \dot{C}H_3 + H_2O$$

Then the ($\dot{C}H_3$) -radical reacts with oxygen to produce alkylperoxide

$$\dot{C}H_3 + O_2 + M \rightarrow CH_3O_2 + M$$

where M is the energy generated by the collision of molecules.
Subsequent reaction with nitrogen results in the formation of a $CH_3\dot{O}$ radical

$$CH_3O_2 + NO \rightarrow CH_3\dot{O} + NO_2$$

which then reacts with oxygen to generate perhydroxyl ($H\dot{O}_2$)

$$CH_3\dot{O} + O_2 \rightarrow CH_2O + H\dot{O}_2$$

The interaction of ($H\dot{O}_2$) with a molecule of nitrogen oxide produces a hydroxyl ion and nitrogen dioxide

$$H\dot{O}_2 + NO \rightarrow OH^{\cdot} + NO_2$$

Exposed to light, the nitrogen dioxide reacts with oxygen, forming ozone.

$$NO_2 + h\nu + O_2 \rightarrow NO + O_3$$

The formation of ozone in processes involving carbon monoxide and methane has been described in several papers (e.g. Dimitriades, 1981; Larson, 1978). Other known compounds participating in the formation of ozone include hydrogen and hydrogen peroxide (Noyes and Leighton, 1941). Although O_3 can be formed in many ways, reactions involving such unsaturated hydrocarbons as pentene and hexene are noteworthy because they are found in automobile exhaust and, to a lesser degree, in a composition of gases released by vegetation. For a long time, it was felt that the contribution of natural organic compounds to the formation of ozone was small. However, data published by Trainer et al. (1987) indicate that, in agricultural areas, this contribution can be significant and even to exceed the contribution of hydrocarbons from anthropogenic origins. Volatile organic compounds can be sources of O_3 which is usually formed in reactions of hydrocarbons with NO_x (Dimitriades, 1981; see also section 1.3.2 below). In addition, Loyd and coworkers(1983) have determined the influence of natural hydrocarbons on the accumulation of ozone in air. On the basis of these data, it was assumed that there was an optimal ratio of natural hydrocarbons to oxides of nitrogen in which ozone production could best proceed. In agricultural regions, where hydrocarbon/NO_x ratios were typically high, it was thought that small amount of ozone is formed because the concentration of NO_x is not enough for the O_3 production . In regions where low ratios existed, there are also no best conditions for ozone formation because the abundant NO_x caused ozone to break down (see section 1.3.1.). Atmospheric ozone is now considered to be a relatively stable product of the photo-oxidation of nitric oxide and hydrocarbons (Sandermann, 1996; Sandermann et al., 1997).

Recently one of modern hypothesis is considered that O_3 is formed in all processes in which atomic oxygen arises — such as at ultraviolet radiation, electric discharges, decomposition of peroxides, the oxidation of phosphorus, etc (Rosolovskii, 1992). From this point of view, biological objects that generate atomic oxygen at electric processes, formation and destruction of peroxides, various oxidative reactions on membranes (including photosynthesis and respiration in plants), both within and out living cells , could form ozone. Besides generating atomic oxygen by means of the above-mentioned processes, NO, CO and other compounds are also produced within all living cells (Roshchina and Roshchina, 1993; Alberts et al., 1994), and these substances may also be used for cellular ozone formation. In the presence of an additional local energy source, they can also participate in the formation of intracellular ozone. Such views must still be verified by the scientific research, but it seems a constant production of atomic oxygen in both non-organic and organic Nature leads, subsequently, to ozone formation.

1.3. DECOMPOSITION OF OZONE

Ozone is decomposed by means of a mechanism peculiar to chain reactions (Razumovskii, 1979) involving M—a particle that absorbs a part of the energy produced by the collision (co-striking) of molecules

$$O_3 + M \rightarrow O + O_2 + M$$

$$O + O_3 \rightarrow O_2 + O_2^x$$

$$O_2^x + O_3 \rightarrow O + 2\ O_2$$

The initiators of the chain disintegration could be small quantities of various impurities (NO_x, HCl, HBr, Cl_2, hydrocarbons and other), the majority of which are of anthropogenic origin (for more detail, see below). Ozone also interacts with organic substances and gases excreted by plants. Such reactions involving ozone occur at all altitudes, right down to the surface of the earth.

1.3.1. Reaction of ozone with oxides of nitrogen and halogen atoms

The reactions of ozone with oxides of nitrogen occur as the chain processes. Nitrogen monoxide is formed in internal combustion chambers at high temperatures and emitted as a constituent of the exhaust gases

$$N_2 + O_2 \rightarrow 2NO$$

Nitrogen monoxide is easily oxidized to form nitrogen dioxide

$$2\ NO + O_2 \rightarrow 2NO_2$$

The oxides of nitrogen react with ozone, destroying it :

$$NO + O_3 \rightarrow NO_2 + O_2$$

$$NO_2 + O_3 \rightarrow NO_3 + O_2$$

Nitric oxide also occurs in plants (Wojtaszek, 2000) and animals (Vanin,2001). Therefore, it is likely that similar interactions with ozone occur in living cells.

Chlorine and bromine interact in a similar fashion with ozone as the oxides of nitrogen do. Collecting in the bottom layers of the atmosphere, they are, ultimately, transferred to the stratosphere, where they participate in cycles of O_3 destruction. In general, the reactions can be formulated as follows:

$$X + O_3 \rightarrow XO + O_2$$

$$XO + O \rightarrow X + O_2 \qquad \text{(X - chlorine or bromine)}$$

The oxides of nitrogen, chlorine and bromine work as catalysts, each of which can interact with thousands molecules of ozone. The same processes are characteristic of freones (composed of chlorine and fatty-line fluorohydrocarbons), which are used in a refrigerating industry and other manufacturing products, such as of deodorants, varnishes, insecticides ,etc. Freones are considered as a major source of chlorine in the atmosphere. These molecules are very stable. Once released into the air, they remain there for many years. Exposed ultraviolet radiation, a molecule of freone decomposes, liberating atomic chlorine. The atoms of chlorine or fluorine react with ozone and transform it into molecular oxygen. As already mentioned, the research on the air above Antarctica has shown that the concentration of chlorine oxide inside the "ozone hole " is very high. The places where the amount of ozone is the smallest are the same ones where many molecules of chlorine oxide can be found. Thus, the concentration of O_3 decreases with chlorine accumulation.

Under light ($\leq$ 320 nm) irradiation carbon monoxide and dioxide can also induce ozone decomposition:

$$CO + 2H_2O + O_3 + h\nu \leq 320 \text{ nm} \rightarrow CO_2 + 2H_2 + 2O_2$$

Ozone is produced not only from abiogenic sources, but also by plants (Bauer et al.,1979). Moreover, spectral analysis indicates that O_3 as well as oxides of carbon and nitrogen are present in the atmosphere of the satellites of Saturn (Noll, 1997). It is therefore likely that the formation and destruction of ozone occurs on other planets of our Galaxy as well.

It is necessary to note that, depending on the conditions in which a reaction occurs, the same components of air can be either absorbers or sources of ozone. For instance, the oxides of nitrogen can participate both in a formation of O_3 (in reactions with organic compounds—see section 1.2.2.) and in its destruction (see section 1.3.1.). Hydrocarbons may serve as a source of ozone (by means of the hydrocarbon -NO_x. reaction—see section 1.2.) or as a contributor to ozone depletion (see section 1.3.2.). However, if air contains significant quantities of organic substances, these compounds more frequently function as absorbers than as sources of O_3 (Finlayson Pitts and Pitts, 1986).

1.3.2. Reaction of ozone with organic components in the atmosphere

Ozone can virtually interact with all types of organic compounds (Pryor, 1979). Rates and mechanism of these reactions depend on the nature of the substance with involved. In reactions with saturated hydrocarbons, alcohols, ketones and acids are included among the final products (Bailey, 1958). The reactions involve the free radical mechanism. Saturated organic compounds, containing oxygen, react with ozone so that, as a rule, a C-H bond is broken. Below are examples involving saturated hydrocarbons and aldehydes (Fig. 2) .

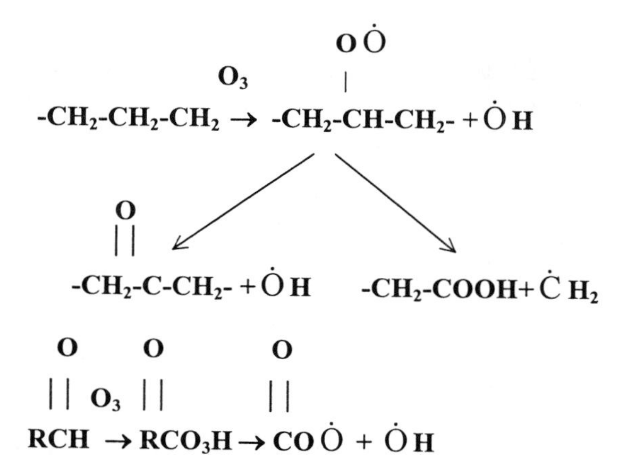

Figure 2. Reaction of ozone with saturated organic compounds

Ozone reacts very quickly with unsaturated substances to form ozonides. This gas links with their double bonds and breaks them, forming peroxide or ester compounds (Fig. 3).

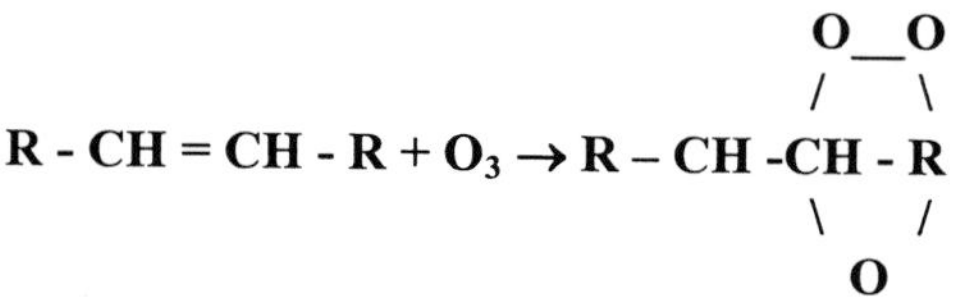

O__O

/ \

$$R - CH = CH - R + O_3 \rightarrow R - CH - CH - R$$

\ /

O

Figure 3. Reaction of ozone with unsaturated organic compounds

Tropospheric ozone actively interacts with olefines and other more oxidized derivatives excreted by plants. Among these substances are ethylene, isoprene, terpenoids (α-pinene and β-pinene, limonene, cymene, myrcene, Δ^3-carene, menthol, citral, geraniol and other). As a result, larger amounts of oxidized organic (aldehydes, ketones and acids) and inorganic (CO_2; H_2O; H_2) compounds will be formed.

Among plant excretions, there are many compounds with aromatic rings. Such substances are also capable to react with ozone. Originally, this reaction proceeds very slowly because additional energy and time are required for an ozone molecule to weaken a bond in the aromatic system (Razumovskii, 1979). After this has occurred, weakly linked C=C bonds are, as a rule, broken down by ozone. As a final result, unsaturated ozonides will be formed (Fig.4). The reaction with aromatic alcohols and phenols occurs mainly by means of the alienation (estrangement) of hydrogen from the hydroxyl group and the opening of the aromatic cycle. Polycyclic aromatic hydrocarbons react with ozone easier than monocyclic ones. The inclusion of ozone in one of the aromatic cycles induces a protective action, retarding the O_3 interaction with other cycles (Razumovskii,1979).

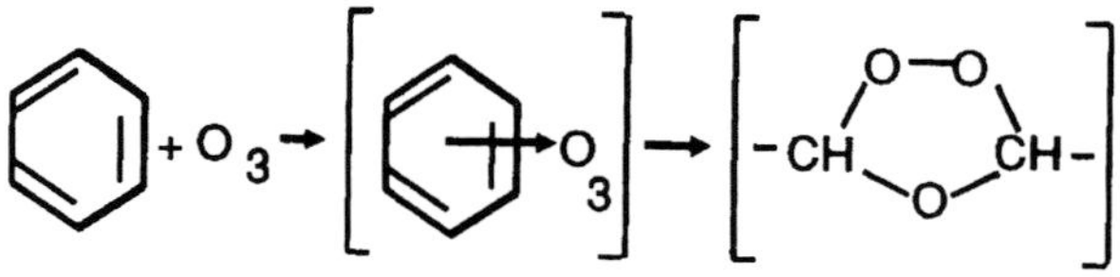

Figure 4.Reaction of ozone with aromatic compounds

The interaction of gaseous plant excretions with ozone occurs constantly in nature. It is an important process for plant and human life, as such reactions alter the air environment.

1.4. CHEMICAL REACTIVITY OF OZONE IN WATER

The reactions of ozone in water and in water solutions are of special interest, as the reactions with O_3 in living cells occur in a water environment. Interacting with pure water, ozone decomposes to form radicals and peroxides. Stachelin and Hoigne

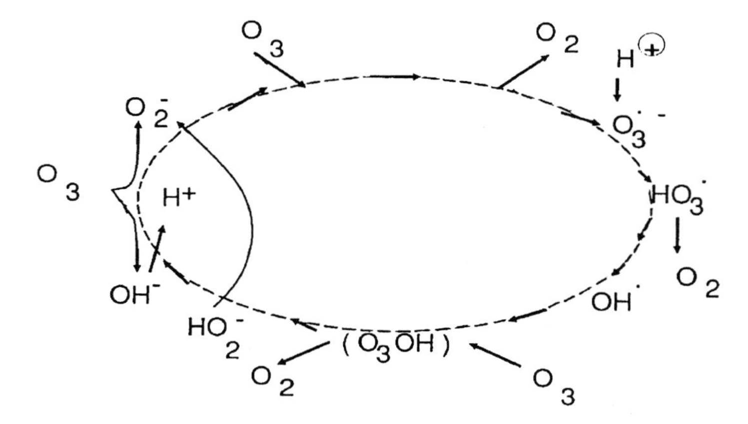

Figure 5. The cycle of chain reactions occurs at the solution of ozone in water

(1982; 1985) have described current views on the chain reactions arising when ozone is dissolved in H_2O. Evidently these reactions can be represented as a cycle (Fig.5). The reactions occur in the following sequence:

1. Water always contains H^+ and OH^- ions, and ozone reacts with hydroxylion to produce superoxide radical ($\dot{O}_2^-$) .

$$O_3 + OH^- \rightarrow 2\dot{O}_2^- + H^+$$

2. Superoxide radical in turn interacts with a new molecule of ozone; as a result trioxide radical (ozonide) is formed

$$\dot{O}_2^- + O_3 \rightarrow O_2 + \dot{O}_3^-$$

3. Trioxide radical reacts with an ion of hydrogen, and hydrotrioxide ion radical is the result:

$$\dot{O}_3^- + H^+ \rightarrow H\dot{O}_3$$

4. Hydrotrioxide ion radical is broken with a formation of hydroxyl radical and molecular oxygen:

$$\dot{O}_3 \rightarrow \dot{O}H + O_2$$

5. Interaction of the hydroxyl radical with ozone produces hydroxyl trioxide:

$$\dot{O}H + O_3 \rightarrow O_3OH$$

6. Hydroxyl trioxide decomposes to form hydrodioxyion and oxygen:

$$O_3OH \rightarrow O_2 + HO_2^- .$$

7. Hydrodioxyion can interact with hydroxylion, forming a superoxide anion radical:

$$H\dot{O}_2 + OH^- \rightarrow \dot{O}_2^- + H_2O$$

The cycle is repeatable.

It is necessary to note that, in a water solution, ozone dissociates faster than it does in air. Ozone breaks down very quickly in alkaline solutions and remains largely stable in acidic ones. According to Hoigne and Bader (1983), the reactions of ozone with non-dissociated organic compounds in water can proceed in three ways:

$$
\begin{array}{c}
\text{Scavenger} \leftarrow O_3 \rightarrow +M \Rightarrow M_{ox} \\
\downarrow \\
+OH^- \downarrow \\
\dot{O}H \rightarrow +M \Rightarrow R \\
\downarrow \quad \downarrow \\
P \leftarrow \text{organic compounds} \rightarrow P
\end{array}
$$

(1) Scavenging of ozone by a scavenger; (2) Direct reaction with dissolved substance (M), an organic non-dissociated compound, which will form an oxidized substance (Mox); (3) Radical type of reaction in which ozone reacts with OH^- ions to form an $\dot{O}H$ radical, which, in turn, reacts with molecules of organic substance (M) to produce radicals (R). All this chain is broken in new stabile products (P).

1.5. PROPERTIES OF THE ACTIVE OXYGEN SPECIES FORMED IN WATER AS A RESULT OF OZONOLYSIS

Reactions of ozone in water occur in all biological systems and, as seen in section

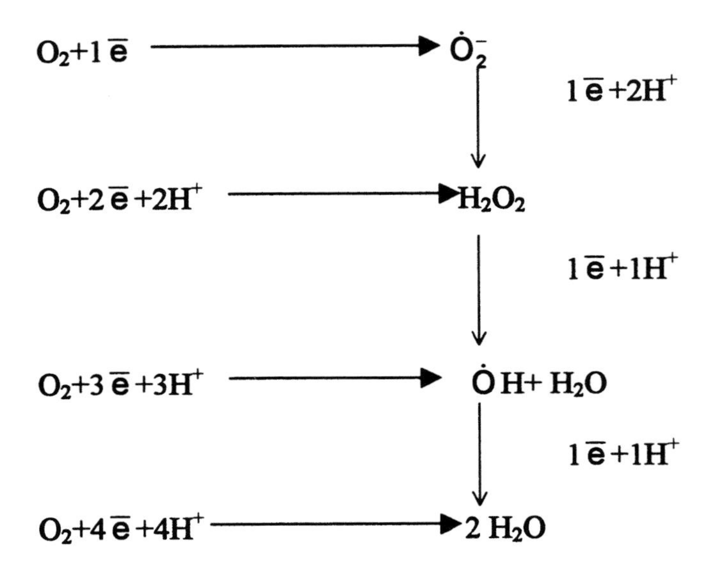

Figure 6. The sequential reduction of an oxygen molecule (Salin, 1987). The intermediate active products formed are: superoxide anion radical $\dot{O}_2^-$, *hydrogen peroxide* H_2O_2 *and hydroxyl radical* $\dot{O}H$.

1.4., involve ozonolysis. The oxygen radicals produced, $\dot{O}_2^-$, $\dot{O}H$, and $H\dot{O}_2$, are, along with H_2O_2 and singlet oxygen (also known as the "active [reactive] species of oxygen), the most important forms of oxygen for biological processes in water. Hydrogen peroxide and singlet oxygen (O_2^1), though not radicals, play significant roles in oxidizing reactions that form or transform free radicals. The latter active species of oxygen is formed by a sequential reduction of molecular oxygen, as shown is the diagram of Fig.6. The connection of one electron to oxygen molecule results in the formation of a superoxide anion radical. The linkage of two electrons and two protons to a molecule of oxygen or one electron and two protons to a superoxide anion ($\dot{O}_2^-$) produces H_2O_2. Hydrogen peroxide serves as a source of hydroxyl radical $\dot{O}H$ (by attaching 1 electron and 1 proton). The same occurs by linking three electrons and three protons to O_2. Water, most reduced form of oxygen, will be formed when 4 electrons and 4 protons are connected to oxygen, or 1 electron and 1 proton - to a hydroxyl radical. In the step reduction of oxygen, the intermediate forms of restored oxygen can play a certain role in oxidizing processes. Capacity of the oxygen radicals ito interact with biological systems may differ widely from each other (see table 5 in section 3.2). Below we separately consider each of the most important active oxygen species.

1.5.1. Superoxide anion radical

Superoxide anion radical is a primary product of the reduction of molecular oxygen (Fig.6). Upon protonation, it is transformed into hydroperoxy radical ($H\dot{O}_2$), which is a weak acid:

$$O_2 \xrightarrow{+e} \dot{O}_2^- \xrightarrow{+H^+} H\dot{O}_2$$

At one-electronic reduction of superoxide anion radical, peroxidion will be formed, while protonation produces hydrogen peroxide

$$\dot{O}_2^- \xrightarrow{+\bar{e}} \dot{O}_2^{2-} \xrightarrow{+2H} H_2O_2$$

Superoxide anion radical will also be formed upon one-electron oxidation of the peroxide :

$$H_2O_2 \xrightarrow{-H^+ + \bar{e}} H\dot{O}_2 \xrightarrow{-H^+} \dot{O}_2^-$$

Thus, superoxide anion radical can be formed both from the reduction of molecular oxygen and from the one-electron oxidation of hydrogen peroxide. Probably the interaction between $\dot{O}_2^-$ and hydrogen peroxide (the Haber-Weiss reaction, see section 3.2.) leads to hydroxyl radical formation in living cells.

$$\dot{O}_2^- + H_2O \longrightarrow O_2 + HO^- + H\dot{O}$$

The chemical properties of superoxide anion radical have been considered in a number of reviews and research papers (Gilbert, 1981; Halliwell, Gutteridge, 1984; 1989; Aver'yanov and Lapikova, 1988; Afanas'ev, 1979; 1989; Vladimirov et al., 1991; Zenkov and Men'shchikova, 1993; Zenkov et al., 1993).

1.5.2 Hydroxyl radical

The precursors of hydroxyl radical are the superoxide anion radical and hydrogen peroxide. Hydrogen peroxide participates in the Fenton reaction (also see section 3.3.). It is the main source of the hydroxyl radical. The reaction occurs according to the following equation (Halliwell and Gutteridge, 1985; 1986):

$$H_2O_2 + Fe^{2+} \rightarrow Fe^{3+} + \dot{O}H + OH^-$$

Hydroxyl radical is also formed in the Haber and Weiss reaction, (1932), which produces both hydroxylanion (OH^-) and the hydroxyl radical ($H\dot{O}$):

$$H_2O_2 + \dot{O}_2^- \Leftrightarrow O_2 + H\dot{O} + OH^-$$

The lifetime of a hydroxyl radical in a cell is about 100 μs, and the distance over which it can travel from its place of formation to a target molecule does not exceed 100 nm (Slater, 1976). As a consequence, the place of formation and the place of reaction of an $\dot{O}H$ radical must be in direct affinity with each other. Hydroxyl radicals react with all compounds present in biological systems. It is the strongest of all known oxidizers and can break down C-H or C-C bonds. $\dot{O}H$ radicals can participate in three basic types of reactions (Larson, 1978):

1. Detachment of hydrogen atoms

$$RH_2 + \dot{O}H \rightarrow (\dot{R}H) + H_2O$$

2. Linkage with double bonds

$$H\dot{O} + R \rightarrow (\dot{R}OH)$$

3. Electron transfer

$$H\dot{O} + R \rightarrow \dot{R} + OH^-$$

1.5.3.Hydrogen peroxide

Hydrogen peroxide, the most stable among reactive oxygen species (see also section 3.3.), is not a radical, i.e. has no unpaired electrons. We are considering the species here because its formation and reactions are closely connected with the free radical oxygen chain. Hydrogen peroxide will be formed in any water systems where superoxide radicals appear. As a result of the dismutation of $\dot{O}_2^-$, hydrogen peroxide and singlet oxygen occur.

$$\dot{O}_2^- + \dot{O}_2^- + 2H^+ \rightarrow H_2O_2 + O_2$$

In a molecule of hydrogen peroxide, the -O-O- bond is fragile; therefore the molecule easily converts in hydroxyl radical (see section 3.2.2.). Reaction between a superoxide radical and hydrogen peroxide also produces hydroxyl radical:

$$\dot{O}_2^- + H_2O_2 \rightarrow \dot{O}H + OH^- + O_2$$

Like the superoxide radical, hydrogen peroxide can work both as an oxidizer and a reducer. The presence of hydrogen peroxide as a generator of hydroxyl radicals in a medium can be determined by spectrophotometric method in the ultraviolet range of the spectrum or by the using of various colorimetric reactions, controllable with catalase or peroxidase (Hildebrandt et al., 1978). The chemical method of H_2O_2 detection can also be based on its ability to oxidize epinephrine (adrenaline) in adrenochrome, a characteristic that is measurable by means of the fluorimetric method (Fridovich, 1979).

1.5.4. Singlet oxygen (O_2^1)

Singlet oxygen is a species of oxygen which has one excited electron. This is due to one of the outside electrons has assumed a higher orbit that changes its spin. The usual form of oxygen (O_2) is in a basic triplet state, and can convert to a singlet state if provided with additional energy. The return to the basic state can proceed by transferring the excited electron of singlet oxygen to another molecule or by radiation of the excessive energy. The formation of singlet oxygen occurs in many chemical, photochemical and biochemical reactions, including those involving free radicals, peroxides of lipids or products of photooxidation (Krinsky, 1979; Murray, 1979). It also occurs as a byproduct in many reactions with a participation of reactive species of oxygen. Singlet oxygen reacts with alkenes to form hydroperoxides. Like ozone, O_2^1 can be a source of other compounds that generate radicals in vivo (Larson, 1978).

1.6. DOSES OF OZONE REQUIRED TO INDUCE BIOLOGICAL EFFECTS IN LIVING ORGANISMS

The concentrations of ozone in atmosphere (Wang and Isaksen, 1995) and the amounts of the gas that induce effects on biological systems (Sandermann et al., 1997) need to be determined. Table 1 represents most significant values of ozone in the air. The increased concentrations of ozone in the surface air are observed only after thunder storms or in areas of high pollution. Ozone's characteristic odor is an indicator of enhanced amounts of O_3 in air. Quantities higher than this minimum detectable by the nose are dangerous for the biosphere as a whole and for plants in particular. Ozone pollution causes visible symptoms of leaf injury. It is well-seen when the damage is more than 5% of the total leaf surface (Heck and Brandt, 1977; Guderian et al., 1985). Special journal issues have considered this question in detail (Feder, 1978; Guderian et al., 1985; Sandermann, 1996; Sandermann et al., 1997; Günthardt-Goerg et al., 1999), so here we shall only give it a little attention. More data will be provided in chapter 5.

The effects of ozone on living organisms can be sharp i.e. acute (exposures at a high concentration of the gas over a short time interval) or chronic (resulting from exposure to low concentrations over a long period of time). Thus, various parameters of plant damage can be observed on the levels of the individual cell, tissue or entire organism. Since low quantities of ozone constantly exist in a nature, only when amounts exceed a certain limit it should be considered as a pollutant (Treshow, 1984; Günthardt-Goerg, 2001). To estimate the ozone concentration and the duration (extent) of its effects on living organisms , notion of the dose is used. The minimal dose required to induce symptoms of their damages is called as the threshold. If a number of research plants belonging to the same species is injured by smaller dose, the species is considered to manifest a sensitivity to ozone.

Accordingly, especially sensitive, moderately sensitive and tolerant types of organisms can be distinguished.

Table 1. Concentrations of ozone in atmosphere

Location	*Concentration*			
	%	*Moles/L*	*μg/m³*	*μl/L or ppm*
Stratosphere	2.3×10^{-5}	10^{-8}	4.3×10^{-16}	9.3×10^{-14}
Troposphere under normal conditions	10^{-7}-10^{-6}	42.8×10^{-12}-10^{-11}	2.1 - 21	0.001 - 0.01
Troposphere after a thunderstorm	10^{-6}-10^{-4}	42.8×10^{-11}-10^{-9}	21-2100	0.01--1,0
Troposphere in areas of heavy air pollution	$> 10^{-5}$	$>42.8\times10^{-10}$	>210	>0.1
Discernible ozone smell	$>10^{-6}$	42.8×10^{-11}	214	0.01
Threshold of the air concentrations harmful for living organisms	10^{-5}	42.8×10^{-10}	210	0.1

Sources : Short Chemical Encyclopedia M: Khimiya, 1964; Gilbert, 1981; Finlayson-Pitts and Pitts, 1986; Chemical Encyclopedia M: Major Soviet Encyclopedia ,1992, Vol.3.pp. 332-333

The interrelation between ozone concentration and time required to effect depends not only a single threshold dose. It is frequently the case that the same degree of the ozone damage can be caused by acute dose (higher, than 0.1 ppm [196 μg/м³] for one hour) and a chronic exposure (smaller, than 0.05 ppm [98 μg/м³] for 10 hours). Reader can see "Diagnosing Vegetation Injury Caused by Air Pollution, 1978" for more detail. The question of the threshold dose required to cause visible damage of sensitive plant species, is a very important one (Feder, 1978). As a whole, it reachs at the exposure to the ozone concentration of around 0.05 ppm for 4 h for more sensitive species (Heggelstad and Menser, 1962). Although this threshold can vary depending on environmental conditions, it is accepted as a basic threshold value (Heyden van der et al.,1999; 2001).

Various means of measuring ozone doses.

1 ppm = part per million = 1 μl/L = 1000 parts per billion = 1960 μg/m³ = $0{,}41\times10^{-7}$ mol/L = $9{,}3\times10^{-5}$% ~
1 ppb = part per billion = 1nl/L = 0.001 ppm =0.0011 μl/L =1,96μg/m³ = $0{,}41\times10^{-11}$mol/L
1 mol/L (M))= $2{,}4\times10^{7}$ ppm = $2{,}4\times10^{7}$ μl/L = 2,4 x 10^{4} ppb =48×10^{9}μg/m³

1 μg/m^3= 0,5 ppb at 20-25 °C = 0.0005 ppm =0.0005 μl/L = 2,08×10^{-11} mol/L
volume% =1,07×10^4 ppm = 1,07×10^{4}μl/L = 1,07×10^7 ppb = 21,4×10^6 μg/m^3= 42,8×10^{-5} mol/L

For additional means, see (Stockwell et al., 1997)

SUMO - "total dose" or sum of all hourly mean ozone concentrations without regard to threshold concentrations
AOT 30, 40 ...- sum of 1-h mean ozone concentrations (whole day or day light hours) above a threshold of 30, 40 ...nl/L, respectively, over a defined time period (e.g., 1 year, a number of months). The unit is nl/L per h. The threshold and the time period should also be indicated

CONCLUSION

Atmospheric ozone arises mainly from the photodissociation of O_2 upon an exposure to the hard ultraviolet radiation of Sun < 242 nm. It forms the ozone layer in stratosphere. In troposphere the O_3 formation occurs as a result of reactions involving CO, NO, hydrocarbons, hydrogen, etc. Ozone is especially accumulated due to air pollution when these substances occur in high concentrations. Certain amounts of O_3 may be also produced indoors by ozone-emitting techniques. In living cells, there are also conditions leading to ozone formation that could be a problem of the future studies. Atmospheric ozone interacts with air components and water, forming ozonides and active oxygen species, both of which can influence biological systems. O_3 concentrations in the ground-level layer of the Earth's atmosphere represent the biologically active doses for biosphere in general and for plant cells in particular.

CHAPTER 2

TRANSPORT OF OZONE IN PLANT CELLS AND CELLULAR REACTIONS

This chapter is devoted to the ozone movement from the surface to the interior of plant cells and its subsequent action on complex cellular systems. On cellular level, ozone effects depend on many factors: the structural organization of plant cells and tissues, characteristics of the linkage with various surfaces, the possibilities of ozone penetration, the rate of its transport, etc. At this level, there is little difference in the ozone effects on different groups of organisms having no coating in addition to the plasmalemma. Distinctions can be made between the representatives of different realms of living organisms, which have different cellular ultrastructures, chemical compositions and metabolisms. In the present chapter, it is important to emphasize two structural peculiarities that distinguish plant cells from animal ones and determine the movement of O_3 along cellular surface and from the surface to the interior of a cell: the presence of a pectocellulose cover (cell wall) and of secretory structures produced a large variety of protective organic substances. The separate stages of the ozone transport in plant cells, the interactions resulting from this movement, and the primary cellular responses will each be given consideration in their turn.

2.1. MOVEMENT OF OZONE FROM SURFACE INTO PLANT TISSUE

The picture and the extent of biological effects caused by ozone depends on the degree, how far O_3 penetrates the living cell. On its way into the cell, the gas must overcome the resistance of some compartments and structures, where it reacts with biologically important cellular components and modifies them. O_3 must transgress a cell wall, an extracellular space between the cell wall and plasmic membrane, and then plasmalemma itself before it enters cytoplasm.

If the primary site of ozone damage is the plasmalemma, all other symptoms of cellular damage observed under the microscope as well as any other metabolic disturbances are not necessarily a direct effect of this gas. Some researchers even believe that ozone does not penetrate plasmalemma at all (Urbach et al., 1989;

Lange et al.1989). Such a possibility would be explained by the high reactivity of O_3, contacting with components of plasmatic membrane. Therefore, it would seem improbable that ozone might sometimes reach a central vacuole in the completely differentiated leaf cell. Urbach et al.(1989) estimated that only a small percent of the ozone, reaching plasmalemma, would diffuse into cytozole (cytoplasm), as most of it would link double bonds in membranous components in a series of reactions that cause damage to plasmatic membrane. Similar effects are also possible for intracellular membranes, if a small amount of ozone should happen to penetrate into cytoplasm.

We shall consider the means of transporting ozone from the surrounding air into the plant, concentrating on the movement into the cell itself, and the cellular reactions involving the gas and various cellular components that occur along the way. Figures 7 and 8 show the main mechanisms of ozone transport into the following types of plant cells: (1) cells of leaf and stem tissues; (2) secretory cells; (3) cells of microspores (vegetative and generative).

2.1.1.Cells of leaf and stem

The significant role in the flow of O_3 into a leaf and stem is played by the morphological peculiarities on a plant's surface: density and arrangement of the trichomes and the properties of cuticle covering the epidermal cells. Bennet et al. (1973) have shown that downy or pubescent leaf surfaces absorb more ozone than smooth ones. The resistance of *Petunia hybrida* to ozone is, for example, related to the trichome density on the surface of its leaves (Elkiey and Ormrod, 1979;1980). Moreover, secretions from any secretory hairs covering any plant tissue may construct a primary antioxidant barrier (see section 2.1.2). Fig.7 (left) shows how various excretions from secretory structures on the plant surface, be they volatile or liquid, can interact with ozone in the air.

In leaves and shouts, the penetration occurs in either of two ways: (1) through open stomata into intercellular spaces to cells of mesophyll and/or (2) through the epidermal layer of cuticle-covered cells to the underlying cells (Fig. 7). In every cell, ozone must pass through a cell wall, an extracellular space between cell wall and plasmic membrane, and then through plasmalemma into cytoplasm:

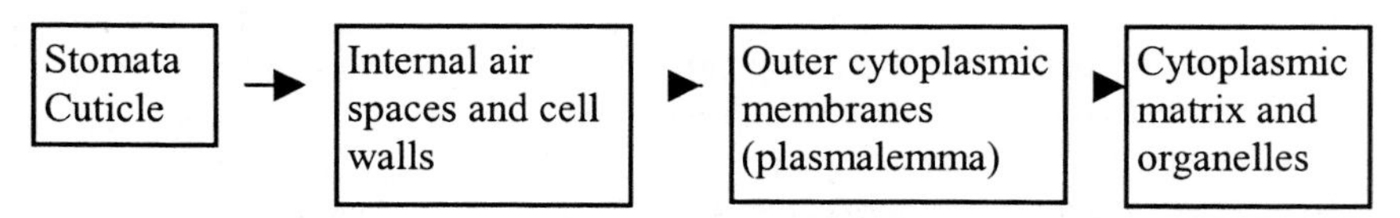

The first cellular barrier that ozone needs to overcome consists of stomata and cuticle through which ozone must find a way in order to enter apoplast (internal air space) or to penetrate the cellular wall. Thus, it is these structures that serve as the primary targets or primary receptors for ozone in plants (Treshow, 1984).

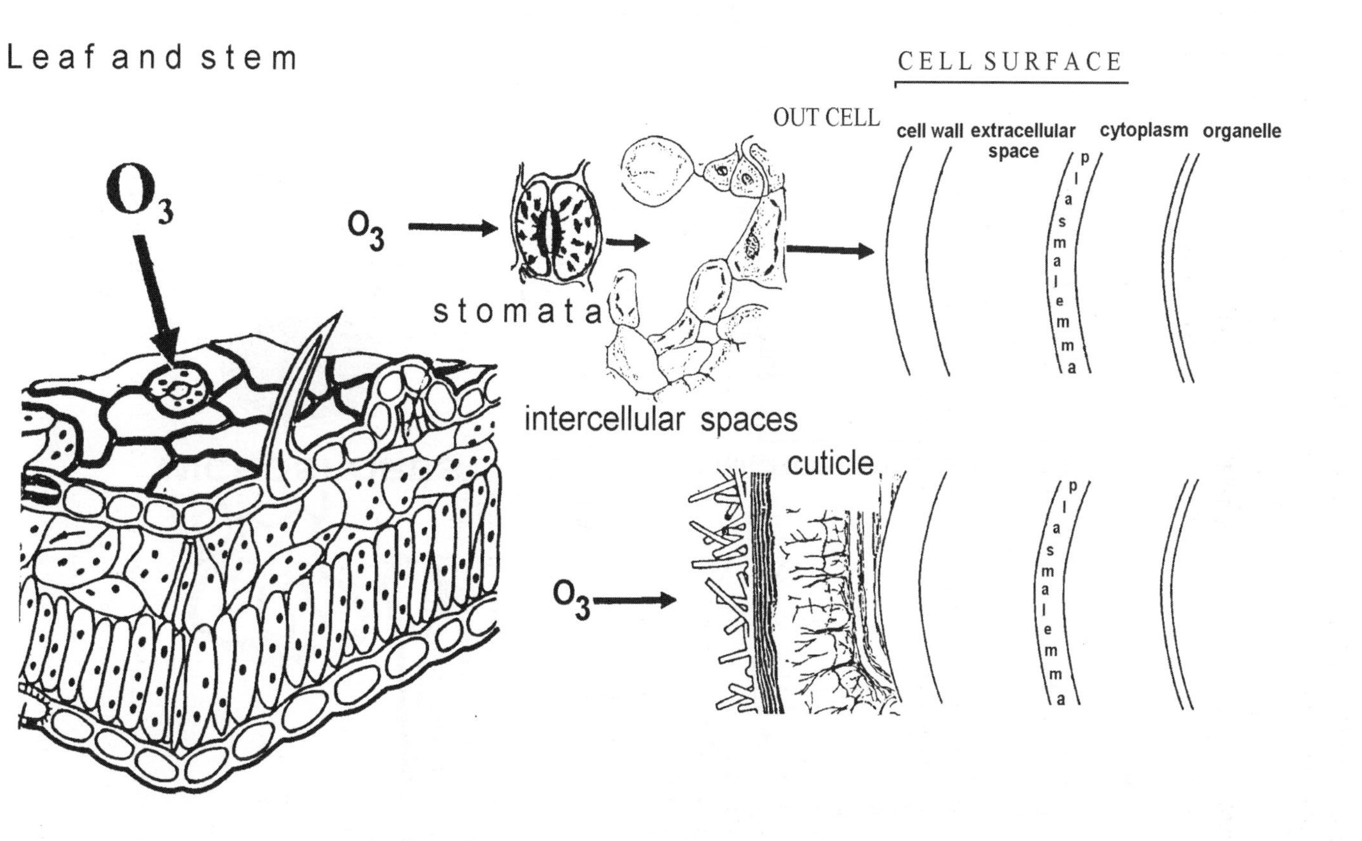

Figure 7. Transport of ozone into tissue and cells of leaf and stem.

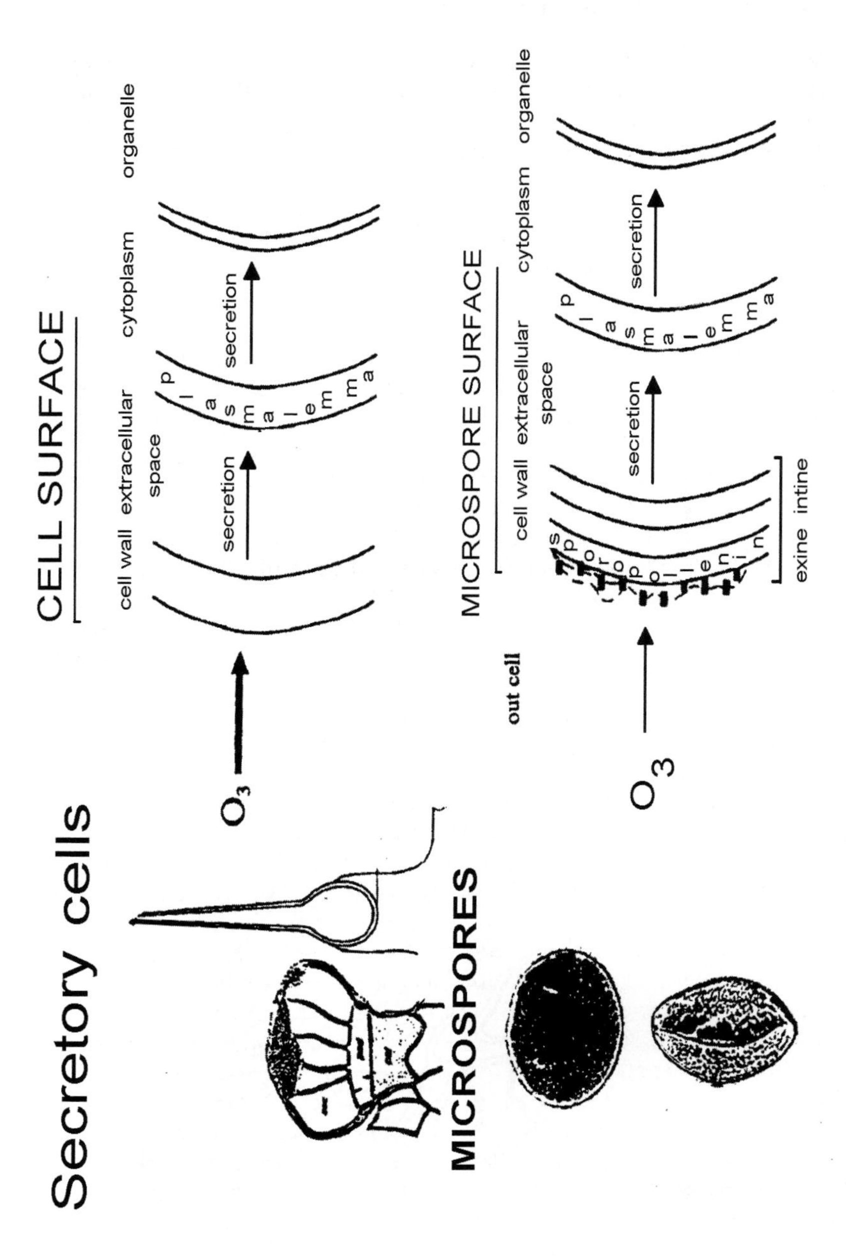

Figure 8. Transport of ozone into secretory cells and microspores

2.1.1.1. Pathways through stomata and cuticle.

Stomata. Ozone enters a leaf (as well as CO_2), passing through the stomata by a diffusion according to a gradient of concentrations (Heath, 1975). It then flows through the intercellular spaces into the parenchyma cells (Fig.7). Therefore, factors, reducing stomata conductivity, would also reduce the negative effects of O_3 (Fingey and Hogsett, 1985).

Upon fumigation with ozone, a sharp decrease in stomata conductivity is observed. However, the effect largely depends on a number of the factors: first of all the ozone concentration in the atmosphere, then, the exposure time; and finally, the peculiarities of the plant species. Consequently, the absorption of ozone by Lucerne *Medicago* occurs in rather low amounts, as compared to its concentration in air (Guderian et al., 1985). In the presence of higher concentrations, the stomata close, causing a proportional reduction in ozone absorption. When exposed to a range of the O_3 concentrations of 0.05; 0.075 and 0.1 μl/L, the stomata opening was approximately 0.5; 0.33 and 0.25,relatively, of those seen in a control without any treatment (Mudd, 1994). Gerini et al. (1990) have shown that leaves of string bean *Phaseolus vulgaris* to be fumigated with ozone induced a sharp decrease in the permeability of the affected stomata within 12 minutes. When presented with high concentrations of O_3, the stomata apparatus appears to be damaged (Salisbury and Marinson, 1985).

Many external factors influence the ozone effect. O_3 damage in leaves of tobacco *Nicotiana tabacum* and string bean *Phaseolus vulgaris* increases when there is increased humidity (Otto and Daines, 1969). Under water deficit the stomata are closed, and the ozone effect is, consequently, minimized (Treshow, 1984).

According to data provided by parametrical measurements, damage levels correlate with the size of the stomata apertures, which grow larger with an increase in humidity. (Otto and Daines, 1969). On the contrary, the addition of abscisic acid, which induces stomata closure, reduces ozone damage (Fletcher et al., 1972). Upon ozone fumigation of rice *Oryza sativa*, the content of abscisic acid is increased. Moreover, the concentration of the acid is higher in plants that are ozone-tolerant (Jeong et al., 1980).

As mentioned above, stomata are considered as one of the main barriers preventing ozone to enter a leaf. Approximate accounts of the gas flows into leaves have been provided by Laisk et al. (1989). The rates of transpiration and absorption of ozone by the foliage of sunflower *Helianthus annuus* L. were simultaneously measured for different stomata openings and various concentrations of O_3. On the basis of these experiments and the theoretical studies, information was acquired about the rates of ozone exchange and of the diffusion resistance of water vapor in *Helianthus annuus* leaves (Laisk et al., 1989). Ozone was added to the bottom of a leaf and its output measured on the top. It was concluded that ozone enters a leaf by diffusion through stomata and is quickly destroyed in the cellular wall and plasmalemma. Under the influence of ozone, the functioning of the stomata complex of cells is impeded. The extent of this influence on stomata conductivity depends on many factors, including the age and level of light exposure. Young leaves have higher stomata conductivity than old ones do (Reich, 1987). Fumigation of 6-9 and

19-21 day old leaves with ozone increases stomata conductivity, whereas similar treatment of 12-14-day-old leaves causes a reduction of such activity. The effect also depends on light exposure. Under low light conditions, the conductivity of old leaves varied more greatly when treated with ozone than they did when exposed to a high light intensity. On the top of a leaf, the decrease in the stomata conductivity had a monotonous character for all ages, while the conductivity in the bottom of s leaf at first increased in leaves aged from 4 h to 10 days, and then decreased in older leaves. In conditions of water stress (deficiency or surplus of moisture), no distinctions in the stomata conductivity on the top or bottom of a leaf were observed.

There is some information that a stomata conductivity does not depend directly on O_3 concentration but may be connected to a change in the intensity of photosynthesis and to the intercellular concentration of CO_2. Such was shown to be the case for a Norwegian fir tree, *Picea abies* (Wallin et al., 1990). More than likely, ozone causes toxic free radicals to be formed. In any case, the peak of sensitivity to ozone is unrelated to the number of stomata on both surfaces of the leaf or to their resisting mechanisms. A sharp decrease in the stomata conductivity can directly result from ozone action on proteins in the membranes that regulate the permeability of plasmalemma and the output(efflux) of ions in the guard cells. This point of view has been verified in an experiment on string bean *Phaseolus vulgaris*. In 3-4 hours of exposure to a rather high ozone concentration of ozone (0.37-0.60 x 10^{-6} M or 9-15 ppm), the ability of the guard cells and the mesophyll cells to restore water volume after the turgor decreased markedly (Sober and Anu, 1992). The reduced capacity was probably caused by the increased rigidity of the cellular covers (cell walls) and/or increased plasmalemma permeability. In any case, there was an appreciable influence of smaller ozone concentrations (0.16-030 x10^{-6} M or 3.9-7.3 ppm) during the long-time (for 1-2 days) fumigations. The disturbances in the regulatory properties of the stomata under such ozone treatment can serve as a model for an indication of ozone stress long before any visible symptoms of leaf damage appear (Shabala and Voinov, 1994).

Through the stomata, ozone penetrates into the air-filled portions of the intercellular spaces and then flows into the mesophyll cell walls. Another way in which O_3 enters the leaf is via the cuticle-covered epidermal surface cells. If the stomata are closed, then the main flow of ozone passes through the cuticle along a pathway that also should be given some consideration.

Cuticle. The cuticle absorbs part of the ozone, contacting the surface of a vegetative cell. The rest of O_3 transgresses the cuticle to enter the interior structure. Ozone absorption and its transmission across the cuticle are important areas of research for analysis of the ozone effects in vivo (Ledzian and Kerstiens, 1988; 1991). The physiological and chemical structures of cuticle play an important role in both the absorptive and transmitting processes.

The cuticle cover on the surface of a leaf consists of the cutin heteropolymers submerged in wax. There are, in fact, more than 20 monomers of cutin, representing a number of C16 and C18 fatty acids, most of which are usually hydroxylated (Kolattukudy, 1987; Airansinen and Paaso, 1990). Three layers of cuticle can be distinguished. The first is the outside superficial layer, consisting of wax (epicuticular layer). Under it, there is a layer immersed in a wax. The final internal layer represents a mixture of cutin, wax and polysaccharides, belonging to the cell wall. Plant cuticles contain various types of waxes, the structure of which depend on the components involved (Percy et al., 1994). There is a lamellar type of structure, which consists of carbohydrates and primary alcohols; a tubular type occurs when secondary alcohols, ketones and diketones are present; a rod-like type indicates the presence of aldehydes; a ribbon-like structure is connected to diols; and an amorphous type is associated with alkyl ethers and triterpenoid- ectolids. It is necessary to note that the cutin structure can widely differ between various plant species and plant structures. However, the physical and chemical properties of cutins are identical. They are resistant to chemical destroyers, acidic hydrolysis and oxidants. Recently, the effects of ozone on the epicuticular waxes of pine needles have been investigated, and it was shown that these were appreciably decomposed when subjected to chronic ozonation (Bytnerowicz and Turunen, 1994).

Ozone damage of the cuticular layer occurs by means of its reaction with lipids (peroxidation) and unsaturated carbohydrogens (Garrec, 1994). At an ozone concentration of more than 70 ppb (0.07 ppm), secondary alcohols, diols, and fatty acids disappear (Percy et al., 1994). This phenomenon is observed, for instance, in the needles of red fir tree *Picea rubens*. Exposure to a fog of increased acidity, only causes secondary alcohols to be altered. In the ozone-sensitive species *Pinus strobus*, ozone accelerates wax degradation (Huttunen, 1994). In addition to this, it should be noted that, due to ozonolysis, high concentrations of monoterpenes can accumulate in the cuticle of such coniferous plants as the genera *Abies*, *Picea*, and *Citrus*, (Shmid and Ziegler, 1991), and it also contributes damage to the cuticular layer.

The rates of ozone destruction and cuticular permeability have been determined for various cuticles isolated from some plants (Kerstiens and Lendzian, 1989). Due to the destruction of ozone in the cuticle, ozone migration through this layer is lower in thick cuticles than in thin ones. However, even in the most permeable cuticles, total plant absorption of ozone under natural conditions occurs much less through the cuticular layer (about 1/10000) than it does through open stomata.

Isolated cuticles react ambiguously to high concentrations of ozone. Depending on the plant species and the duration of ozone exposure, hydraulic permeability was either reduced by up to 79 % or increased by 295 % (Kerstiens and Lendzian, 1989; Kerstiens, 1994). The permeability to water (P) of external cuticles is less than 0.7 ms^{-1}, that peculiar to very poor permeable membranes (Kerstiens, 1994).

A degree of ozone damage in the O_3 fumigated cuticles depends on their capacity to absorb ozone. To illustrate this point, three *Petunia hybrida* Vil clones were enclosed in a chamber and exposed to 40 parts of ozone per 100 thousands parts of air (i.e. a 400 ppm ozone concentration). The absorption of the gas through both the stomata and the leaf surface was observed (Elkiey and Ormrod, 1980). “Capri,” the variety the least damaged by ozone, absorbed more total O_3, but the largest part of this absorption occurred on the leaf surface. “White Cascade,” an ozone-sensitive clone, demonstrated the opposite results. The clone "White Magic" with intermediate sensitivity to ozone was characterized by intermediary results (Elkiey and Ormrod, 1980).

The absorption of ozone by the cuticle occurs better if rain or dew moistens this epidermal substance, and it is much worse on a dry waxy surface (Fuentes et al., 1994a, Jagels, 1994). Acid fog (pH 3,0) in combination with a high concentration of ozone (2.5 % or 2.68 [104 ppm for 1-2 hours]) reduces the permeability of isolated leaf cuticles extracted from *Ficus elastica* Roxb and *Hedera helix* L. by 20 %. In contrast, the same conditions increases ozone's ability to migrate across cuticles extracted from the leaves of *Citrus aurantium* L., *Ilex aquifolium* L., *Prunus laurocerasus* L, as well as those from the fruits of *Capsicum annuum* L. and *Lycopersicon esculentum* Mill. (Kerstiens and Lendzian, 1989d). In fact, the increased permeability in a variety of *Capsicum annuum* reached as much as 200 %. Consequently, a certain amount of ozone can penetrate into a leaf through a wet cuticle, especially in places where the surface is damaged. Despite the fact that ozone solubility in water is rather low (0.052 g in 100 g of water at 20 ^{o} C), ozone is easily diffused through fine apertures. Accordingly, the basic way in which ozone penetrates into living leaf cells are through stomata and/or cracks in the epidermis.

The fast reactions involved in the interaction of the cell cover with ozone could cause changes of the fluorescence spectra of individual cells (Roshchina and Melnikova, 2000;2001). Significant green-yellow fluorescence is developed in some of the cells in the topside of leaves of *Plantago major*, *Raphanus sativus* and *Hippeastrum hybridum* (Roshchina and Melnikova, 2000). Fig. 9 shows the fluorescence changes that may occur. This phenomenon appears to take place in the fluorescing products of lipid peroxidation under oxidative stress (see section 3.1). The fluorescence spectra of pots (green - yellow fluorescence at wavelengths of 480-550 nm), arising on the leaf surface of *Plantago major* after fumigation with ozone (0.05 ppm h $^{-1}$ during 3-9 hours), are illustrated in Fig. 9. Similar fluorescing stains are observed after ozonolysis in leaves of the corn *Zea mays* and, in a smaller degree, in the skin-like leaves of the hippeastrum *Hippeastrum hybridum*, which is more tolerant of ozone (Roshchina and Melnikova, 2000). Upon longer expositions to ozone, the number of maxima (480-550 nm) and the area of a fluorescing pot surface, especially on the top surface of a leaf, are increased. If the exposure to the gas is, in total, about 3-9 hours and if there is not any subsequent treatment, the leaves develop normally, and the fluorescing pots disappear after 2-3 weeks of growth. Apparently, there are active reparation processes that take place in such cases. In conditions, involving chronic fumigation with ozone > 12-15 hours, the sensitive plants of *Plantago major* develop necrotic pots or other symptoms in the sites on a leaf where fluorescing pots appear. As illustrated by the spectra of leaf fluorescing pots in *Raphanus sativus*, the composition of the products formed in response to ozonolysis are rather similar to the lipophilic content of secretory hairs (Roshchina and Melnikova, 2000).

2.1.1.2. Pathways through cell wall and across extracellular space

Penetrating the stomata and entering the leaf, ozone passes through air-filled portions of intercellular spaces and is absorbed by the damp cell walls of the mesophyll. Along this route, a number of reactions may occur involving: the decomposing of ozone in the water phase of cellular medium, the reaction of ozone with olefins excreted by plant cells, and the reaction of ozone with antioxidants.

In interacting with water, ozone decomposes into radicals and peroxides. In alkaline solutions, this process occurs very quickly, and O_3 concentration is therefore reduced (Gurol and Singer, 1982). It is possible that ozone induces the cell-wall solutions to change pH. It has been established that a change of pH from 6 to 8 does not manifest itself in any symptoms of cellular damage because this reaction is carried out in the apoplast (intercellular spaces and cell walls) outside the main metabolic zone. Destruction of ozone in a water phase or solution (see section 1.4) is accompanied by the formation of superoxide radical ($\dot{O}_2^-$), which will, in turn, form such radicals as hydroxyl radical ($\dot{O}H$) , singlet oxygen (O_2^1), and H_2O_2 (Asada, 1980). Although the rates of formation of the hydroxyl radical are very low, accounts of buffer capacity (Nieboer and Farlane, 1984) show that a change in pH from 6.5 up to 9.5 can be achieved by the release of 2 (10-17 equivalents) hydroxyl groups. This can, in fact, happen when ozone breaks down in water.

In the air cavities and cell walls of leaves, ozone enters into reactions with the olefins produced by plants, mainly ethylene and isoprene. The mechanism of these reactions is discussed in sections 3.1; 3.2; and 3.3. In a layer of a liquid surrounding the mesophyll cells of plants, concentrations of oxy radicals and peroxides are increased at the expense of the ozone concentration.

Chameides (1989) has calculated the concentration of isoprene necessary for a 50 % loss of the ozone, migrating from a cell's environment to its plasmalemma. If the distance (length) of its diffusion is extended, the necessary concentration of isoprene (for 50 % ozone loss) can be reduced 2-3 times. It is not known exactly if such concentrations in the apoplast of living cells are possible. Certainly, they would remain higher than the amount of isoprene in the air above a forested area (10^{-7}M). In the cell wall, the concentration of isoprene can reach 10^{-5}M. Measurements of the internal concentration of isoprene in aspen *Populus tremula*, oak *Quercus* L. and poplar *Populus* L. have shown it to be 1-2 ppm. Thus, it seems quite probable that the ozone concentration in the air contained in both the intercellular spaces of a leaf and the cellular environment may decrease due to reactions with endogenous hydrocarbons. However, Chameides (1989) nevertheless suggests that the reactions of ozone with the water contained it the cell wall and with olefins are not so important for ozone absorption.

Most of the ozone that diffuses into a leaf reacts with the ascorbic acid contained in the cell wall. Accounts provided by Chameides (1989) indicate that the plants can protect themselves by concentrating ascorbate in the cover. Reacting with ozone, ascorbic acid limits the amounts of O_3 that can penetrate the cell wall and reach a more vulnerable part of the cell located inside cellular cover— plasmalemma.

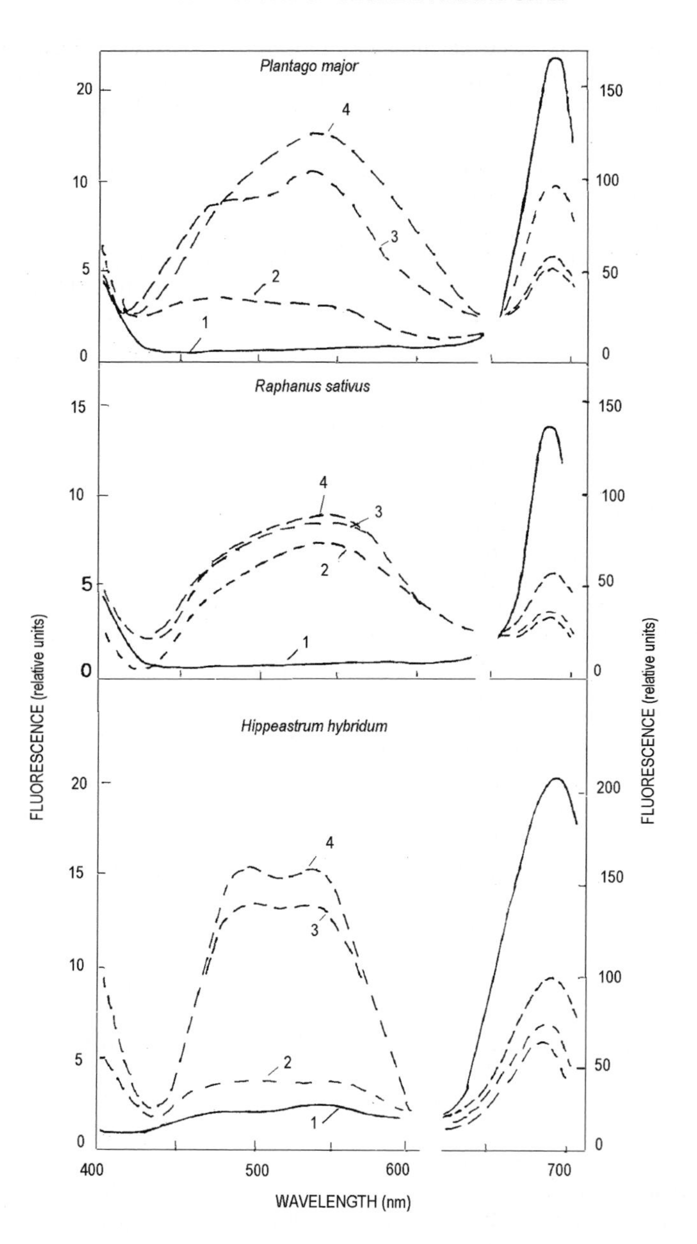

Figure 9. The fluorescence spectra of a leaf cells from 14 day-old seedlings with (broken line) and without (unbroken line) treatment by ozone. 1- control (without ozone); 2-4 - with ozone treatment of 3,12, and 15 hours with doses of 0.15; 0.6; 0. 75 ppm (adopted from Roshchina and Melnikova, 2000 and unpublished data). Right Y scale for red region.

Both reduced and oxidized ascorbate is present in the water phase (solution) existing in the extracellular space. According to data supplied by Luwe et al. (1993), the ratio (reduced ascorbate : reduced + oxidized ascorbate) in the apoplast of spinach *Spinacea oleracea* L. leaves falls from 0.9 to 0.1 after 6 hours of the plant fumigation with ozone 0.3μl/L or 0.3 ppm. Up to 0.9 picomoles of ozone could be absorbed by 1 cm^2 of leaf surface/second. The reduced ascorbate present in the apoplast is slowly oxidized by the ozone. On terminating the fumigation, the quantity of oxidized ascorbate in the apoplast slowly decreases, reaching the control level after 70 hours. However, oxidized ascorbate cannot be reduced in apoplast. It is transported into the cytozol and reduced there (Luwe et al., 1993). The reduced ascorbate is then slowly transported back to the apoplast. After 48 hours of fumigation with ozone, the total level of ascorbate falls and, simultaneously, necrotic pots (stains) appear on leaves. At his point of view, the antioxidizing defense is exhausted, and ozone has reached the plasmalemma.

Beside ascorbic acid, another antioxidants in the apoplast appear to be phenols, amino acids containing sulfhydryl groups, and other compounds. Redox enzymes (superoxide dismutase, peroxidase and catalase) can also execute the detoxication of ozone in a plant cell, as they protect living cells against free radicals and peroxides formed in reactions between O_3 and unsaturated hydrocarbons. Research undertaken by Bennet et al. (1979a) with use of a preparation of EDU (N(2-2*-oxo-1-imidazole indinidyl)ethyl) N' – phenylurea), which induced a resistance to ozone, is of a special interest. It has shown that O_3 tolerance of the ozone-sensitive leaves of the string bean *Phaseolus vulgaris* was closely connected with the presence of superoxide dismutase, of which the content can be increased with application of the EDU preparation (Bennet et al., 1979a). The ozone tolerance of plants growing in soils to which EDU has been added is also enhanced due to the increase in the dismutase activity. The same effect is observed when a solution of the preparation was directly applied to the leaves (Heagle, 1989). Thus, it is likely that detoxification of ozone in the apoplast is one of the primary means employed by cells to protect themselves against ozone.

2.1.2. Secretory cells

The secretory cells of glands, glandular trichomes and other structures located on the surface of a leaf, stem or leaf-derived parts of flower (petal, sepals, anthers, stamens), also interact with ozone (Fig.8). Secretions are stored within the cell and may be excreted either into the extracellular space (between cell wall and plasmalemma) or outside of the cell. They can interact with ozone, causing their composition and the properties of their individual constituents to change (Roshchina and Roshchina, 1993; Roshchina, 1996). The secretions contain many secondary compounds that can serve as antioxidants (Roshchina, 1996). Thus, a secretion evacuated either into the extracellular space or out of the cell completely may also interact with ozone, leading to changes in the fluorescence of the secreting cell (Figs.10,11). For instance, the secretion fluorescence of the secretory hair belonging

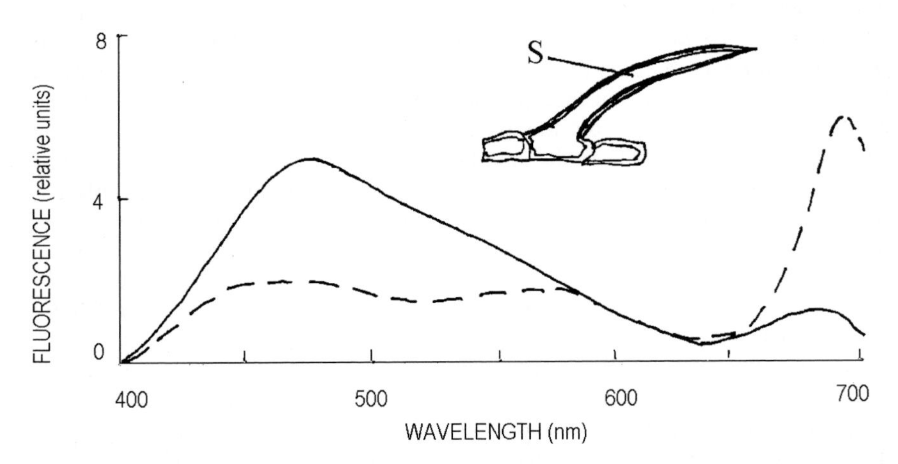

Figure 10. The fluorescence spectra of a leaf secretory hair belonging to the tomato Lycopersicon esculentum with (broken line) and without (unbroken line) the treatment by ozone 0.05 ppm for 1 hour.

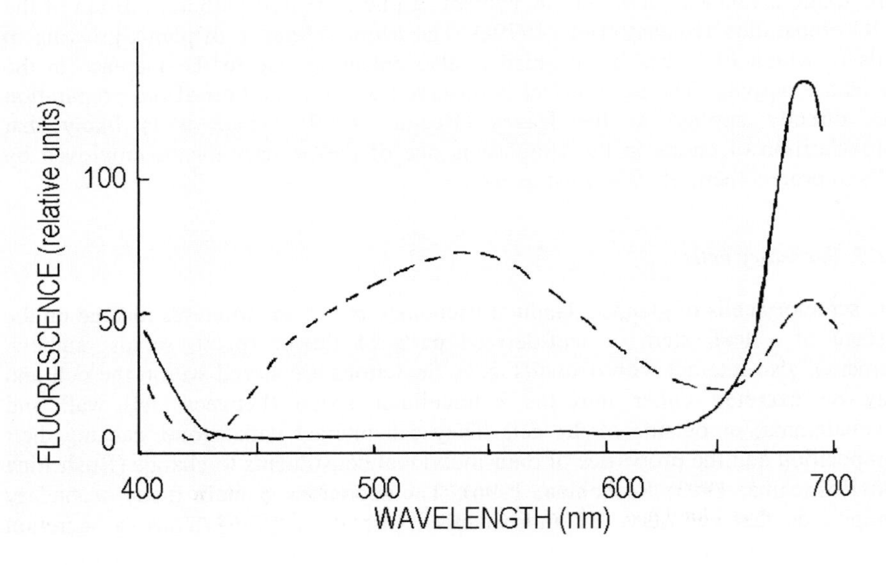

Figure 11. The fluorescence spectra of a secretory hair on a leaf of a 14-day-old Raphanus sativus seedling with (broken line) and without (unbroken line) exposure to ozone at 0.15 ppm for 3 h.

to the tomato *Lycopersicon esculentum* decreases in the blue region of the spectra, whereas the chlorophyll fluorescence at 680 nm increases (Fig.10).

As illustrated by the spectra of leaf fluorescing pots in *Raphanus sativus* (Fig.9), the composition of the products formed by ozonolysis are rather similar to the lipophilic content of the secretory hairs (Fig.11). Fluorescing pots will be often formed on the basis of such hairs (Roshchina and Melnikova, 2000). The fluorescence spectra of the secretory hairs is significantly changed by the ozone treatment, so that green-yellow light emission (maximum 540-550 nm) arises. Consequently, the influence of ozone causes changes inside the hair to occur that are observable as a change in the fluorescing products.

2.1.3. Cells of microspores

As known for both generative (pollen, male gametophyte) and vegetative microspores ozone penetrates, first of all, through the cover structured into an exine (outer layer) and intine (internal layer). Exine consists of a polymerized material called sporopollenin. After penetrating this cover, the ozone must still pass through cellular wall to plasmic membrane (Fig.8 below). Ozone often induces the green-yellow fluorescence of the dry vegetative microspores of *Equisetum arvense* (Roshchina and Melnikova, 2001). Fig. 12 shows the characteristic band at 475 nm and a maximum band at 540 nm. This green-yellow fluorescence is developed in a

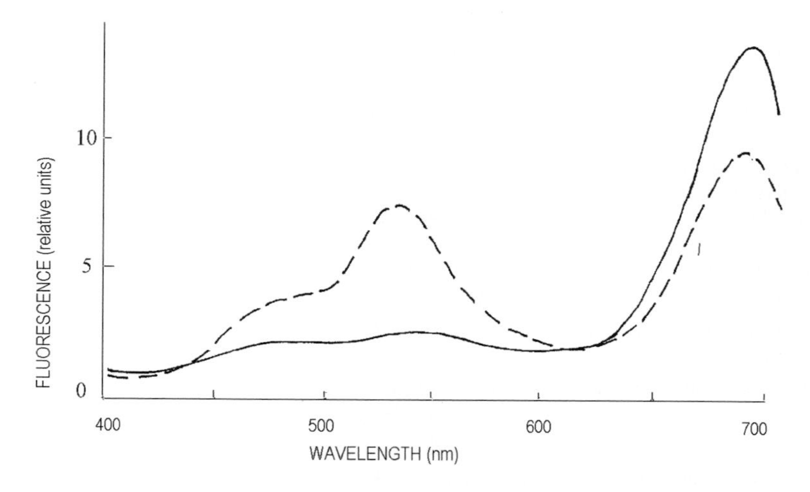

Figure 12. The fluorescence spectra of a vegetative microspore of the horsetail Equisetum arvense with (broken line) and without (unbroken line) treatment by 0.05 ppm of ozone.

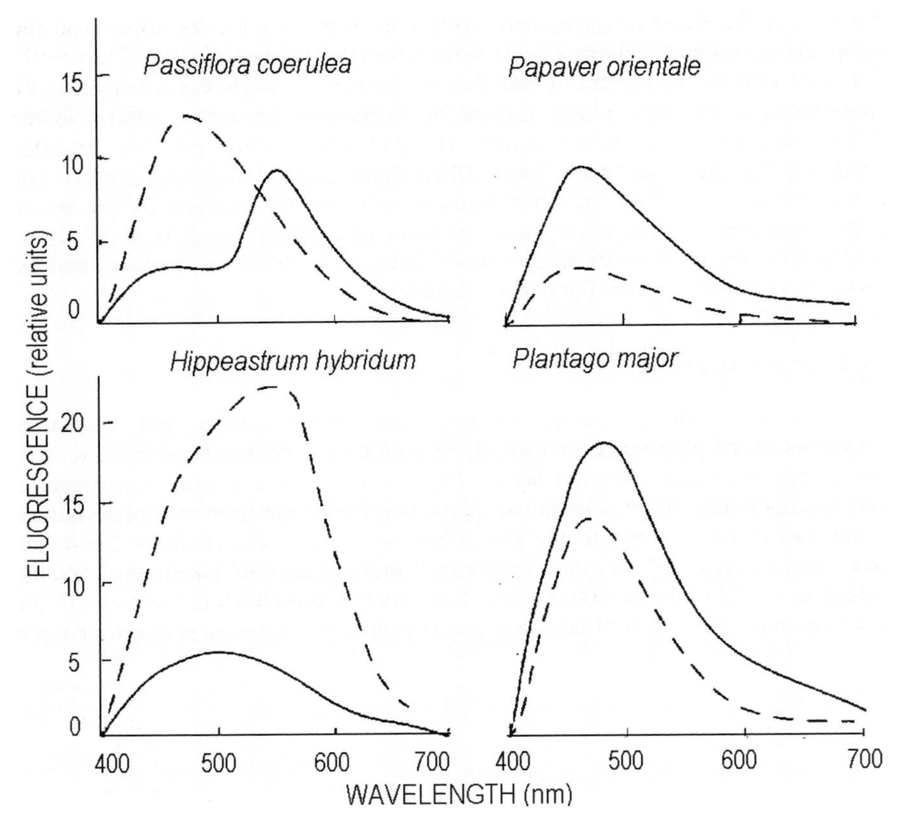

Figure 13 The fluorescence spectra of pollen under normal conditions, without any treatment (unbroken line) and after 100 hours of fumigation with ozone (0.05 ppm per h), resulting in a total dose of 5 ppm (adopted from Roshchina and Melnikova, 2001).

response to the presence of ozone in the vegetative microspores of *Equisetum arvense* (Fig.12). It appears to be produced by the fluorescing products of lipid peroxidation that occur under conditions of oxidative stress (see section 3.1). Fluorescing products will also result from ozonolysis in the generative cells, in particular, pollen (Roshchina and Melnikova, 2000; 2001). The characteristic fluorescence spectra of the pollen from various species are shown in Fig. 13. The appearance of the maxima depends on the composition of the pollen exine, which includes such fluorescent products as phenols, carotenoids, anthocyanins, etc. For instance, a shift in the fluorescent spectra of the Hippeastrum hybridum pollen occurs due to oxidation of phenols upon O_3. The fluorescing pots on leaves and microspores of plants appear to be related to the Schiff bases formed from free radical reactions in the lipid phase of membranes. Deeper changes result in the occurrence of the fluorescing pigment lipofuscin, which frequently appears in the older growing tissues of plants as a final product of lipid peroxidation (Merzlyak, 1989) (see section 3.1.2). This appearance depends on the chemical structure of the sporopollenin components. The interaction with ozone sharply changes the emission intensity and the position of the maxima in the fluorescence spectra of the pollen of various species (Fig.13). The maximum fluorescence after treatment with ozone was moved into the short-wavelength region, and the intensity of the light emissions sharply increased. Possible mechanisms of the phenomenon for individual components of sporopollenin will be considered in Chapter 3. Thus, the ozone damage of pollen is expressed by a change in the spectrum and intensity of luminescence. Both these parameters depend on plant species and ozone concentration. Such marked changes appear to be the earliest evidence of ozone damage.

2.2. TRANSPORT OF OZONE THROUGH CELLULAR MEMBRANES

As seen in previous sections, some ozone molecules that escape the antioxidants in the cell wall and extracellular space (see also chapter 4) may reach the plasmalemma. At this point, it either does or does not penetrate plasmic membrane. If the gas does not enter the cell interior, there are two possibilities: 1. ozone can be trapped by the antioxidant system of plasmalemma, or 2. O_3 interacts with some proteinous sensor-receptor or sensor-associated enzymes (SAE) of the membrane, acting outside as chemosignal (Fig.14). At chemosignaling, intracellular structures appear to receive the information from the O_3 binding with the receptor by the help

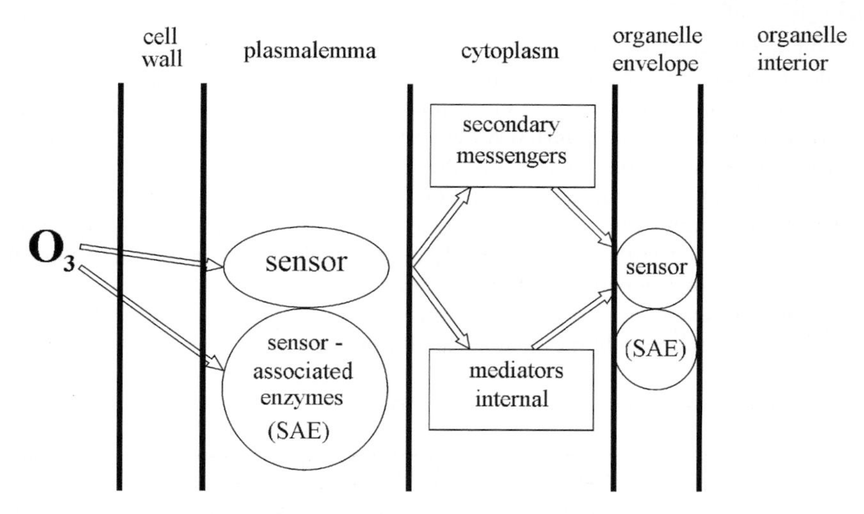

Figure 14. Possible scheme of ozone action on the plasmalemma and cellular organelles

of secondary messengers. Since ozone may form reactive oxygen species in cellular wall, extracellular spaces and plasmalemma itself (see section 3.2.), these agents also appear to act on plasmalemma by similar modes. Even some part of ozone enter cytoplasm (in the case of high O_3 concentrations), all above-mentioned mechanisms are likely, nevertheless, to occur. Fig.14 shows a possibility that the same mechanisms could be applied to the ozone penetration through the membranes of organelles.

2.2.1. *Pathways through plasmalemma to cytoplasm and cellular organelles.*

The ozone entrance into cytoplasm and then into other organelles includes the overcoming of membraneous barriers. Effects of O_3 on the membranes appear to be not only direct. The products of ozone transformation in cellular cover and air-filled cavities of a plant may also contribute, in particular active oxygen species do.

The common principles governing ozone interaction with membranes derive from their structural peculiarities. Membranes are asymmetric and represent ensembles of lipid and protein molecules linked together with the help of non-covalent bonds. The functional differences between membranes are caused by the structural variety of the protein and lipid components. The difficulties that ozone has penetrating into a cell are due to its slow diffusion through a powerful lipid layer (lipid concentration can reach 50 %), and to the responses of membranous systems, which regulate the quantities of the active forms of oxygen.

When ozone interacts with membranes, ozonolysis of the major membrane components can take place. An effector may be not only ozone but also products of ozonolysis - other related active radicals and peroxides that appears to modify membranes. The proteins and lipids in membranes are the targets of ozone action itself and its reaction-capable derivatives. Fig. 14 represents a general scheme of potential sites where ozone may act on cellular membranes. First of all, ozone reacts with proteins located on the external surface of the membranes (they can function as enzymes, sensors, receptors, etc.). Only the O_3 that escapes such reactions reaches the intramembranous proteins and lipids. For this reason, it is usually argued that the proteins of membranes are more sensitive to ozone than the membrane lipids are (Mudd, 1982), as the latter serve primarily as a structural and information barrier. Any remaining ozone penetrates the membrane into the cell or compartment of an organelle (if the event occurs in the cytoplasm). According to Mudd (1982), the elements in intact cell membranes that act as potential targets for ozone action are:

External surface **Within cell**

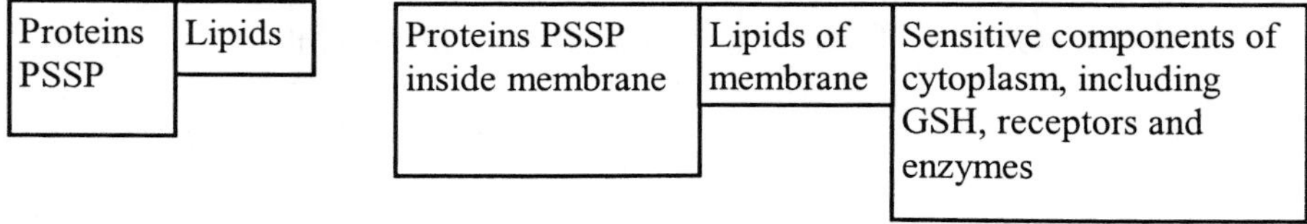

Thus, ozone first (and most frequently) interacts with proteins on the external surface of a membrane, especially PSSP proteins with disulphide bonds. Ozone breaks the SS-bond and induces the formation PSH, as it does when proteins inside the membrane are involved. Other (less frequent) targets of O_3 are the membrane lipids. As a second order reaction, ozone interacts with proteins on the internal surface of a membrane and, as part of a final cellular response, with sensitive components of the cytoplasm, including GSH (reduced glutathione), which converts to GSSG, an SS-bonded form of glutathione.
Table 2 shows the ozone targets in membranes and the main responses of plant cells on both molecular and cellular levels. All changes induced by O_3 are due to its disturbance of the main membrane constituents—proteins and lipids (Table 2). The regulatory function of membranes involves the control of cell permeability to

Table 2. Targets of ozone in plant cell membranes and O_3-induced responses

Molecule	*Targets*	*Reaction on molecular level*	*Reactions in cells*
Proteins	Amino acid residues of tryptophan, tyrosine and phenylalanine, SH-groups (Treshow, 1984)	Enzymic activity (Tingey et al., 1975)	Increase in membrane permeability
Lipids	Double bonds of unsaturated fatty acids (Hewitt and Terry, 1992).	Lipid peroxidation, formation of hydroxymethyl peroxide (Salter and Hewitt, 1992). Formation of malondialdehyde as a product of lipid peroxidation (see section 3.1)	Formation of fluorescing products (Castillo and Heath, 1990)

various substances, compartmentalization, the formation of DNA and RNA complexes with membranes, etc. Moreover, some receptors of biologically active compounds located on membranes are engaged in hormonal regulation. This function can be modified by ozone and associated compounds such as free radicals and peroxides (see chapter 3).

One of the important effects of ozone is the transformation of lipids located in the membranous systems of cells (such lipid damage impedes the functioning of large submolecular complexes). Ozone acts on lipids, and so do the radicals and peroxides that are produced in these reactions. In natural conditions, membranes are largely ozone tolerant, and only their damage accelerates lipid destruction. With long exposures to high ozone concentrations, changes in the composition of unsaturated fatty acids are observed. Ozone, as was mentioned above, destroys the double bonds in unsaturated fatty acids. Thus, ozonides are accumulated as the intermediate products of oxidation. Besides the effects of ozone itself, the associated formation of peroxides and radicals instigate, to various degrees, further activity. It is even possible that some of them react with membranes more heavily than ozone does (Hewitt and Terry, 1992). Polar, very hydrophilic molecules of ozone do not penetrate into the sites on membranes where unsaturated fatty acids are located (Heath et al., 1974). However, ozone in water solution produces hydroxyl radicals, which react with the fatty acids contained in the lipid structure. Another possibility is that the hydroxyl ion damages membranes and improves the ability of ozone to enter the lipid layer (Frederick and Heath, 1975). Damage to the lipids of membranes is due to the formation of hydroximethylperoxide, a substance produced by the reaction of ozone with unsaturated hydrocarbons, perhaps with isoprene and ethylene. In this case, lipid damage can be more intensive than that caused by ozone alone (Salter and Hewitt, 1992).

Common reactions of ozone, free radicals and peroxides with proteins and lipids in model systems will be considered in sections 3.1; 3.2; 3.3. Their interactions with these compounds in living cells, both those found in the plasmalemma and those situated in intracellular compartments, will each be given separate consideration below.

2.2.2. Plasmalemma

The membrane damage caused by ozone is readily discernible as disturbances to membrane permeability and regulatory functions. Such effects are especially noticeable in the plasmalemma. Most studies focus on the influences that O_3 has on the membrane permeability to water, ions and some organic substances (Evans and Ting, 1973; Ting et al., 1974). Exposure to ozone (0.18 ppm) for a few minutes results in the rapid loss of the turgor (Guderian, 1985). Increasing the concentration of ozone or the duration of exposure causes transparent pots to appear on leaf plates as a result of water infiltration into the intercellular space (Tingey and Taylor, 1982). Both phenomena indicate that the plasmalemma has become more permeable to water as well as to some organic compounds (Table 3).

Permeability. Ozone increases the conductivity of membranes for ions K^+, Ca^{2+} and Mg^{2+} (Table 3), and their concentration in the extracellular fraction is increased. Ozone acts, mainly, on the passive permeability of membranes. Heath (1980) used the unicellular algae *Chlorella sorokiniana* as a model to illustrate the primary action of ozone on vegetative cells. Permeability to K^+ was increased by exposure to ozone in concentrations of approximately 120-150 ppm for 2-3 min. Significant changes in the plasmic-membrane permeability of *Euglena gracilis* were found to occur with ozone exposures at concentrations of 240 μl/l, administered at a gas-flow rate of 1 μmole/min (Chevrier et al., 1990). According to Chevrier et al. (1990), the emission of K^+ does not result from membranes breaking down in reaction to ozone but is more readily caused by the oxidation of sulfhydryl groups in the acid residues of membrane (K^+ ,Mg^{2+}) ATPases. Probably, ozone only has an effect on the passive efflux of K^+ from vegetative cells, although data indicates that ozone inhibits the K^+-stimulated ATPases in the plasmic membranes of the leaf cells of string bean *Phaseolus vulgaris* (Dominy and Heath, 1985). The calcium flow into and out of a cell is also subject to ozone stress. An effect on this movement in isolated and purified vesicles of the plasmatic membranes extracted from string-bean (*Phaseolus vulgaris*) leaves has, in fact, been demonstrated (Castillo and Heath, 1990). The treatment of leaves with ozone increased the passive permeability of plasmic membranes to calcium but did not show any visible signs of damage, and even displayed restored levels of Ca^{2+} ions. The occurrence of visible effects depends on the rate at which the ozone concentration is varied. A sharp increase in ozone intake (from the atmospheric levels to 0.15 μl/l or 0.15 ppm.), followed by a period of maintenance at a stationary level resulted in a sharp decrease in the passive output of Ca^{2+} without any leaf necrosis being observed. A slow rise in the concentration of O_3 (up to 0.14 μl/l or 0.14 ppm over a period of 1.5 hours) followed by a slow decrease in the subsequent 1.5 hours caused necrosis on more than 10 % of the leaf surface and even higher output of Ca^{2+}. As the total dose of the gas in both experimental versions was identical (0.6 μl/ hour), it seems natural to conclude that a slow increase in ozone concentration has a stronger effect than a rapid one (Castillo and Heath, 1990).

Table 3. The effects of ozone on the permeability (P) of plant cells.

K^+ - efflux	*Ca^{2+} - efflux*	*Uptake of glucose and 2-deoxy- D-glucose*	*Uptake of vitamin B_{12} and acetate*
P increase in unicellular algae *Chlorella sorokiniana* K^+ under 20-150 ppm ozone for 2-3 min (Heath, 1980) and in *Euglena gracilis* under 240 ppm ozone (Chevrier et al., 1990).	Passive P increase in vesicles from plasmatic membranes of *Phaseolus vulgaris* (Castillo and Heath, 1990)	P increase in leaves of *Phaseolus vulgaris* (Castillo and Heath, 1990).	Decrease of P in leaves of *Phaseolus vulgaris* (Castillo and Heath, 1990)

The ozone modification of membranes results in an approximate twofold increase in their permeability, not only to ions but also to a number of organic compounds. In string *bean Phaseolus vulgaris*, there is an increased permeability of membranes to radioactive glucose and 2-deoxy-D-glucose. Castillo and Heath (1990) also observed a decrease in the absorption of vitamin B_{12} and acetate after a 15-minute exposure to ozone. A return to the initial levels of absorption was observed after 12 hours for vitamin B_{12} and 20 hours for acetate. Apparently, cells are capable restoring themselves after ozone treatment.

There are various points of view concerning the mechanism causing the changes in the membrane permeability for organic compounds. Dominy and Heath (1985) have suggested that ozone damages the vitamin B_{12}-binding proteins by oxidizing the SH-groups. Chevrier et al. (1990) have shown that sulfhydryl groups are very quickly oxidized by ozone. However, Vatunava's experiments (1988) with some inhibitors have shown that there is no evidence of any participation by the surface SH groups of proteins in such a process.

The highest concentrations of ozone (50-115 ppm) cause a very rapid increase in the permeability of plant membranes (Mudd et al., 1984). But appreciable increases are also seen at lower doses of ozone (less than 0.5 ppm). In this case, the activity of membrane-linked glyceralaldehyde dehydrogenases is inhibited, and lysozyme is oxidized. It is assumed that ozone acts on membrane proteins, especially those having sulfhydryl groups.

One of the possible strategies leading to an understanding of the mechanisms of ozone damage involves a study of the role played by the free radicals produce by lipid peroxidation. Current research has shown that ozonating the string bean *Phaseolus vulgaris* for 20-25 minutes generates the products of lipid peroxidation (malondialdehyde), a loss of unsaturated fatty acids and changes in membrane permeability (Alscher and Amthor, 1988). It has also been shown that, as a result of lipid peroxidation, the membrane lipid layer becomes more permeable to ions and organic molecules. It is, furthermore, necessary to note that membrane damage from ozone is similar to damage arising from water stress, pathogenous invasion and action of metabolic toxins; in other words, it has a non-specific character.

Sensitivity of the surface sensors and enzymes. The modification of proteins or lipids due to oxidation by ozone and/or its derivatives (see chapter 3) can either strengthen or depress the activity (the sensitivity) of the surface sensors and receptors for biologically active compounds, as well as the membrane-binding enzymes. In a number of cases, the deactivation of the enzymes in membranes occurs due to the products derived from the ozonolysis of amino acids' residues. Membranes contain antioxidants, which can lower the oxidizing effect of ozone on their protein and lipid constituents. By changing the rates of oxidative reactions, antioxidants also modify these membrane structures.

In contemporary views, the physiological response by a cell to a majority of hormones and other physiologically active substances requires all these compounds to interact with a series of receptors located on the membrane. These hormone-and-receptor interactions result, first of all, in the activation of adenylate cyclase and an increase in the intracellular level of cyclic AMP, which is an intermediate for hormonal activity in several metabolic processes (Alberts et al., 1994). Adenylate cyclase is a lipoproteid compound and, like the receptors, is located on the membrane. The receptors of hormones are also comprised of proteins and lipids. For this reason, the lipid and protein components of membranes play an essential role in realizing any hormone effect. The oxidizing reactions in membranes also influence the concentration of cyclic nucleotides and, consequently, the sensitivity of cells to the hormonal action (Burlakova, 1981; Burlakova and Khrapova, 1985). As plasmatic membranes contain antioxidants inhibiting oxidative reactions, they appear to be active modifiers of the hormonal sensitivity of cells, like ozone and some of the products of ozone-related transformations. Thus, any change in membrane content, oxidative state or structure influences the hormonal regulation of cellular metabolisms.

As is already well known, membranes require effective protection against ozone damage. In addition to the protection afforded by enzymes, protective functions are provided by the presence of antioxidants in the lipid bilayer, of which tocoferol is the most important. Detecting and deactivating such toxic agents as the hydroxyl ion ($\dot{O}H$), tocoferol thus causes a break in the lipid peroxidation chain. Moreover, it stabilizes the structure of membranes.

2.2.3. Cytoplasmic matrix and cellular organelles

In considering the ozone effects on cellular level, the depth of O_3 penetration into the cell constitutes an important factor. The highest concentrations of this gas can penetrate into the cytoplasm and induce effects on the organelles (Table 4).

The effects of ozone on cytoplasm have mainly been investigated in severely damaged cells. In such cells, cellular content aggregates into a dense mass (of material) on the periphery of the cell (Pell and Weissberger, 1976; Swanson et al., 1973; Thompson et al., 1974). It was demonstrated that ozone affects the membranes of the microsomes in the coleoptiles of string bean *Phaseolus vulgaris*, causing a transition in the system from a liquid crystal state to a gel-lipid phase (Pauls and Thompson, 1981). This latter substance is formed by lipids derived from the membranes upon treatment with ozone (0.2 mM for 60 min) and differs in structure from the lipids in intact membranes. Under the influence of O_3, the sterol component of a membrane is modified so that it becomes more volumetric. Consequently, the number of pores changes in the membrane, resulting in a corresponding effect on the membrane's permeability. The ozone-induced changes in membrane content and structure are similar to those that occur when leaves wither or age (Pauls and Thompson, 1981). Ozone also induces changes in the content and structure of organelles (Table 4).

If the primary site for ozone damage is the plasmalemma, all other injuries visible under the microscope, as well as all metabolic disturbances in the cell, do not directly result from ozone action. A number of researchers have suggested that ozone does not penetrate the plasmatic membrane, as the high reactionary character of O_3 causes it to react with the external membranes and other surface structures. Since the damage effects of ozone on intracellular structures are, therefore, limited, it is considered unlikely that ozone could reach the central vacuole of a completely differentiated leaf cell (Urbach et al., 1989; Lange et al., 1989a). Several accounts indicate that only a small percent of the ozone reaching the plasmalemma diffuses into the cytozole, injuring membranes by linkage with the double bonds in the compounds and inciting accompanying reactions. Tingey and Taylor (1982) have suggested that ozone diffuses into the chloroplasts if the latter have a small number of sulfhydryl groups to serve as ozone scavengers. However, sometimes changes in intracellular structures can occur without any damage of the plasmic membrane being observed. Accordingly, Swanson et al. (1973) did not notice any changes in the structure of the fatty acids in the plasmalemma of the leaves belonging to the tobacco *Nicotiana tabacum* L after they were exposed to ozone. They did indeed observe the compression of chloroplasts and the swelling of mitochodria. To explain a possible ozone effect on the structures of chloroplasts that is prior to any change to the plasmalemma, Tingey and Taylor (1982) assumed that ozone must have an indirect effect on cellular organelles by forming free radicals. However, the opinion also exists (Mudd, 1982) that ozone and its derivatives do indeed penetrate the membrane. Some experiments would seem to confirm that the polysomes in chloroplasts decreased in size under ozone treatment (Tingey et al., 1975; Chang, 1971). However, a similar phenomenon in the polysomes of the cytoplasm has not been observed. A very sensitive parameter of ozone-induced chloroplast damage is the intensification of chlorophyll fluorescence. Many experiments have provided evidence of such occurrence before any damage to the plasmic membrane is detected. Hence, it is quite possible that ozone or any of its derivatives penetrate the plasmalemma and affect the intracellular structure.

Mitochondria. Ozone interacts with the highly specialized mitochondrial membranes, which surround these organelles. Each of the membranes contains a specific set of proteins. The outside (external) membrane contains many molecules of transport proteins, forming hydrophilic channels through the double lipid layer.

The enzymes of the electron transport chain and oxidative phosphorylation are located in internal membranes where there is a coupling of electronic transport and ATP synthesis (Skulachev, 1972). The internal membrane is characterized by a prevalence of proteins over lipids, the relation of which is usually 2:1. Among the proteins, a significant share (about 1/3 of them) can be classified as protein electron carriers—cytochromes, non-heme iron proteids and flavoproteins (Skulachev, 1972). Some of the basic components of mitochondrial membranes are lipids, which constitute about 35-40 % of total membrane content. The lipid structure of the internal and outside membranes of mitochondria differ from each other. Cardiolin, being one of the factors determining the low permeability of membranes for ions, is contained mainly in the internal membranes, and inositol phosphates (inositol-1,4,5 triphosphate) in the outside ones. Mitochondrial lipids are especially characterized by a large degree of unsaturated compounds. Each atom of phospholipid phosphate contains two or three double bonds involving fatty acids. The main reactions of mitochondria to ozone are swelling, increase in membrane permeability, formation of electron dense material, stimulation of the oxygen uptake and inhibition of coupled phosphorylation (Table 4). In sensitive plant species and depending on the dose, this gas can also injure the DNA structure. Ozone treatment (12.5 μM on 5 ml or 2.3 $\times 10^{-4}$ ppm) of mitochondria extracted from the roots and callus of the tobacco *Nicotiana tabacum* causes their rapid swelling (Lee, 1968).

Table 4. Ozone effects on intracellular organelles

Character of changes	*Mitochondria*	*Chloroplasts*	*Nucleus*
Structure	Swelling, (Swanson et al., 1973; Lee, 1968), electron-dense bodies (Thomson et al., 1974).	Pressing and disintegration of thylakoids, a decrease in the amount of ribosomes, occurrence of electron-dense materials between double membranes (Swanson et al., 1973; Wingsle et al., 1992)	Aberrations and injuries of chromosomes in cells of 2 year old seedlings of *Picea abies* (L.) Karst after fumigation of ozone 0.02-0.1 ppm (Müller et al.,1996)
Biopolymers	Injuries in DNA structure (Konstantinov et al., 1993)	Decrease in polysome sizes (Tingey et al., 1975; Chang, 1971), RNA and ribosomes (Chang, 1971).	Decrease in DNA-related fluorescence in intact cells of the microspores of *Equisetum arvense* stained with Hoechst 33342 dye (Roshchina, unpublished data
Electron transport and phosphorylation	Inhibition of phosphorylation (Lee, 1968) and stimulation of the oxygen uptake (Mudd, 1973)	Decrease in the oxygen release, total electron transport and coupling (Coulson and Heath, 1974)	
Enzymic activity		Decrease in the activity of ribulose-bisphosphate carboxylase and , therefore CO_2 fixation (Nakamura and Saka, 1978).	
Permeability (P) of membrane	Increase in P (Lee, 1968),	Increase in P of the outer envelope membrane in intact plastids for glycerol (Nobel and Wang, 1973)	

This swelling of mitochondria is completely inhibited (blocked) by protocatechuic acid, and only partially by glutathione and ascorbate. Hypertonic solutions of sucrose neither acted on any O_3-related relaxation (pressing) of mitochondria nor prevented any swelling that ozone induced (Lee, 1968). The swelling of mitochondria resulting from exposure to the gas was also observed in the leaves of tobacco *Nicotiana tabacum* (Swanson et al., 1973). The research on moderately damaged cells of string bean *Phaseolus vulgaris* detected the electron-dense formations in its mitochondria (Thomson et al., 1974).

Ozone and the associated products formed by ozonolysis in cells increase the permeability of mitochondrial membranes (Lee, 1973) and inhibit phosphorylation. It has been demonstrated that, after one hour of ozonation (1 μl/L or 1 ppm.), the phosphorylation in mitochondria is inhibited by 20 %. The products of lipid peroxidation affect the genetic system of mitochondria by injuring the DNA structure, preventing its function and inhibiting the RNA synthesis (Konstantinov et al., 1993).

Chloroplast. Chloroplasts as well as mitochondria have greater autonomy than other organelles. The multimembrane type of a structure is also peculiar to them. They contain the basic components of the photosynthesizing apparatus itself, along with components of the reproductive, nucleic and protein-synthesizing systems, all of which can be the direct targets of ozone (Table 4). Photosynthesis is a reaction sensitive to ozone and, consequently, transmits a direct or indirect influence of ozonolysis to the chloroplasts. Other O_3 targets can be the various components of the organelles. For instance, along with the changes of membrane-related reactions of photosynthesis in granal and agranal lamelles the activity of key enzyme ribulose-bisphosphate carboxylase in the matrix of the chloroplast is also changed that decreases CO_2 fixation (Nakamura and Saka, 1978).

The effects of ozone on the chloroplasts of an intact cell greatly depend on its concentration. At low concentrations, O_3 can not penetrate into a cell, and the influence can only be indirect, through a yet unknown system of intermediates. In high concentrations (> 1 ppm), ozone injuries in plasmalemma occur, so that molecules of the gas can pass into the cell and reach the chloroplasts. Chloroplast membranes are believed to be the ones most sensitive to ozone. Chloroplast has three groups of diverse membranes, which differ in their ability to resist O_3. Highly permeable to ions, the outside membrane of the chloroplast envelope is tolerant of this oxidizer, as the photosynthesis of isolated intact chloroplasts is not affected by exposure to ozone over a wide range of its concentrations. Nor is the internal membrane of the plastid envelope, which is permeable to ions to a lesser degree. Accordingly, it can be concluded that ozone does not influence the rate of reduction of the iron-containing electron carrier ferricyanide in these membranes. Usually, this reaction is only observed in the third type of the chloroplast membranes—the thylakoids—found inside the organelles. Only after the osmotic destruction of intact plastid envelope is ferricyanide reduced (Nakamura and Saka, 1978). This factor indicates that ozone does not penetrate into the region of the grana through the external and internal membranes of the chloroplast envelope. In experiments on isolated intact plastids, of which the function is supported by adding bicarbonate to the artificial medium, a decrease in oxygen release is observed. This phenomenon indicates a direct interaction of ozone with the outside chloroplast membrane (Coulson and Heath, 1974). However, in high concentrations, ozone damages the outer membrane and then reaches the thylakoids. It is not clear how ozone works in this case, whether it destroys the membrane wholly or acts specifically on separate sites of the membrane in order to render it permeable. In the isolated intact chloroplasts treated with ozone, for instance, the permeability of the outside envelope membrane for glycerol is increased, but this membrane remains impenetrable for sucrose even at very high concentration of O_3 (Nobel and Wang, 1973).

Model experiments on isolated chloroplasts from *Spinacia* sp. L. in which the outer membranes were broken allowed direct ozone action on thylakoids to be investigated (Coulson and Heath, 1974). It was shown that ozone disrupts the function of thylakoid membranes by inhibiting electron transport in both photosystems (PS 1 and 2) and the associated ATP synthesis (Coulson and Heath, 1974). The basic mechanism for decreasing photophosphorylation is, consequently, not uncoupling but the inhibition of electron flow. The data dealt with the ozone effects on the photosystems are inconsistent (even contradictory). According to one source (Coulson and Heath, 1974), ozone inhibits both photosystems, but PS 1 is, nevertheless, more sensitive than PS II. Other data (Schreiber et al., 1978) suggests that increases in ozone concentration initially injure the donor side of PS 2 (system of water destruction), and only then inhibits electron transport from PS 2 to PS 1. A study by Chang and Heggestad (1974) investigated the action of O_3 on PS 2 in chloroplasts from *Spinacia oleracea*. To a certain degree, photosystem 2 was more sensitive to ozone than photosystem 1. Ozone treatment caused a reduction in retarded luminescence of leaf plates, a phenomenon related to photosystem 2 (Veselovskii and Veselova, 1990). The decrease was observed just after the malondialdehyde formation as a final product of lipid peroxidation was found (see section 3.1). It is assumed that ozone disrupts the normal flow of electrons from the exited chlorophyll by destroying specific components of the membranes rather than by causing general disintegration. Ozone damages the structure of the chlorophyll-protein complexes, in this case, both the pigment itself and the components of the pigment-protein complexes. Such complexes of pigments and proteins are readily subject to damage. It has been demonstrated that, at low ozone concentrations, the protein components of membranes are oxidized, and, in higher ones, lipid peroxidation and pigment destruction contribute to this process (Tingey and Taylor, 1982). These changes are apparently connected to the accumulation of free radicals, which arise by using ozone to disturb electron transfer. Exposed to high concentrations of ozone, chlorophyll is destroyed very quickly, and chlorophyll "*в*" is destroyed faster than chlorophyll "*a*" (Konig De and Jegier, 1968a), although some information is available suggesting that chlorophyll "*a*" is more sensitive to ozone. Experiments by Beckerson and Hofstra, (1979c) on string bean *Phaseolus vulgaris* have shown that quantities of both forms of chlorophyll are diminished by ozone activity. O_3 not only destroys already formed pigment molecules, but it prevents the formation of new molecules of chlorophyll "*a*" and chlorophyll "*в*", as illustrated by the ozone effects on seedlings of barley *Hordeum* L (Gaponenko et al., 1988).

The ribosomes of chloroplasts are also subject to a strong ozone effect. The exposure of leaves from string bean *Phaseolus vulgaris* to an ozone concentration of 0.35 ppm. for 20-35 minutes decreased the ribosomal RNA in chloroplasts (Chang, 1971). The reduction of RNA content specifically involved RNA-23S, and continued for 1.5 days after exposure. Ozone induced a dissociation of the chloroplast polysomes (Chang, 1971). After a 40-minute exposure to ozone, the content of RNA in ribosomes was reduced by 40 %. The decrease in the RNA appears to be caused by a blocking of the SH-groups. Thus, the action of ozone on SH-groups is likely the crucial factor.

Antioxidant isoenzymes function to eliminate the free radicals formed at ozonolysis (see chapter 4 for details). In *Arabidopsis thaliana* exposed to ozone, there is an accumulation of mRNAs encoding both cytosolic and chloroplastic antioxidant isoenzymes (Conklin and Last, 1995). The steady-state levels of three mRNAs encoding cytosolic antioxidant isoenzymes (ascorbate peroxidase, copper/zinc superoxide dismutase, and glutathione S-transferase) also increase. O_3 causes a decline in the levels of two chloroplastic antioxidant mRNAs (iron superoxide dismutase and glutathione reductase) and two photosynthetic protein mRNAs (chlorophyll *a*/*b*-binding protein and ribulose-1,5-bisphosphate carboxylase/oxygenase small subunit). This decline does not include all mRNAs encoding chloroplast-targeted proteins, since ozone causes a general increase in mRNA.

The disturbance of the photosynthetic function of chloroplasts is accompanied by structural changes. Thus, the long-term influence of low ozone concentrations can indeed cause more extensive damage than high concentrations of O_3 over a short period (Schreiber et al., 1978; Frage et al., 1991). Seedlings of the pine *Pinus sylvestris* were exposed to low (70-80 ppb. or 0.07-0.08 ppm for eight hours per day for ten days) and high (300 ppb or 0.3 ppm. for eight hours per day over five days) concentrations of ozone (Wingsle et al.,1992). The amount of chlorophyll "*a*" decreased during exposure by respectively 8 and 11 %. At high concentrations of ozone, the amount of soluble proteins is increased over the last three days of the ozone fumigation without any change in the total protein content or in the level of the antioxidant enzymes superoxide dismutase and glutathione reductase. Thus, there were no significant general changes in the quantity of the membranous lipids or acyl groups in them. Appreciable shifts in the carotenoid structure were also not observed, although by the fifth day of the fumigation with ozone the level of violoxantine has raised to 19 %, and the level of zeaxanthine was lowered to 42 %.
Exposure to 0.15-ppm ozone for 8 hours causes the changes in the shape of chloroplasts, a disintegration of the thylakoids and a reduction of the ribosomes. Appreciable changes in the electron-dense material between the two membranes of the envelope are also observed. In the stroma of the chloroplast, a set of fibrils will be formed. Plastid formation of crystal structures in, for instance, string bean *Phaseolus vulgaris* can also be observed (Thompson et al., 1974). The ultrastructural changes caused by moderate short-term exposure to ozone are, however, reversible. In this case, the invagination of chloroplasts, granulation and increased electronic

density in the stroma are observed no longer than 7 hours after the ozone treatment. At higher concentrations of ozone, the plastid membranes are destroyed, and the process of photoassimilation therefore stops.

Nucleus. Reactions of nucleus with ozone depend on its concentration. Small doses (< 0.05 ppm) can stimulate growth reactions, such as the germination of pollen (Roshchina and Melnikova, 2000). Without penetrating the plasmalemma, O_3 appears to trigger nuclear division via sensor-receptory mechanisms and secondary messengers, as seen in Fig.14. At higher ozone concentrations, the organelles are disturbed (Table 4).

The first stage in any damage to the nucleus occurs by direct deactivation of the nucleus biopolymers, along with a growth in free radicals, peroxides and hydroperoxides, which can react with such components of nuclear membranes as enzymes. Perhaps, at various levels of the structural organization, the above-mentioned active products modify DNA and protein molecules. Ozone is also found to cause even more strong biological damage seen as a chromosome ruptures than some doses of radiation which do not disrupt normal life but, nevertheless, also rupture chromosomes (Fetner, 1962; Pryor, 1979). Moreover, O_3 ,for instance, may accelerate the loss of nuclear-encoded mRNAs in senescing leaves of potato *Solanum tuberosum* (Glick et al., 1995). The greatest damage to nuclei is rendered by the reactions of ozonolysis, which cause membranous lipids to be modified. Ozone reacts with the double bonds in lipids based on glycerophosphocholine to form ozonides. The maximal quantity of ozonides will arise if esterified phospholipids have polyunsaturated ester-acyl groups (Harrian and Murphy, 1996). These are then decomposed to aldehyde and derivatives of carbonic acid. In such a manner, lipid peroxidation and free radical processes, resulting from fumigation with 0.02-0.1 ppm ozone (Muller et al.,1996), have been seen to instigate aberrations and other damage to the chromosomes in the roots of two-year-old fir tree seedlings of the species *Picea abies* (L.) Karst. A consequence of such toxic effects is the stacking of chromosomes and the formation of bridges between chromatids. Upon the ozone (0.2 ppm) treatment the decrease in DNA-related fluorescence emitted by microspores of horsetail *Equisetum arvense* stained with Hoechst 33342 dye was observed as an earliest symptom of the chromatin changes (Table 4). For pollen grains similar experiments with the same fluorescent probe on DNA were not adequate.

The state of the nucleus is strongly affected by O_3 pollution. Its analysis is especially important at ozone induced oxidative stress (Pell et al., 1997). Effects of acute or long-time chronic exposures may lead to such profound cellular changes as hypersensitive response, senescence and even cell death. The causes of ozone-induced cell death include nuclear DNA fragmentation and chromatin condensation both of which are observed prior to lesion formation and are characteristic symptoms of the hypersensitive response leading to cell death (Rao and Davis, 2001).

CONCLUSION

Movement of O_3 from the surface into the interior of the plant cell and the subsequent cell reactions depend on the ozone dose and on the state of constant protectory systems. The first barrier to this gas is comprised of excretions, cellular covers and extracellular spaces filled with secretions. All these systems contain antioxidants. If any O_3 reaches the plasmic membrane, it interacts with the sensory and antioxidant components of the plasmalemma. In such cases, small concentrations of ozone may act on intracellular structures without penetration inside the cell, perhaps, via the external reception and participation of intracellular secondary messengers. At high concentrations, the gas passes through the plasmalemma and has direct contact with the cytoplasm and cellular organelles. Cellular responses, observed before the formation of adaptive reactions (see chapter 5), include changes to the following: (1) the fluorescence of intact cells; (2) the permeability of membranes to ions; (3) the sensitivity and activity of the surface sensors-receptors and enzymes, and (4) structure.

CHAPTER 3

MOLECULAR MECHANISMS OF THE OZONE INTERACTION WITH PLANT CELLULAR COMPONENTS

In this chapter, we shall consider O_3- induced changes in the basic classes of organic substances important for understanding the molecular mechanisms of ozone's activity in living cells.

Ozone is one of the most reactive oxygen species (Kanfer and Turro, 1981), but other reactive oxygen species are also formed at the ozonolysis of cellular components (section 3.1.), such as free radicals and peroxides. They also contribute to the overall effects of ozone, especially at higher concentrations of O_3 .

Consequently, ozone activity can occur in two ways, either directly or indirectly through the formation of reactive oxygen species. The reactive oxygen species that can be formed during ozone exposure include free radicals (superoxide anionradical, hydroxyl radical, peroxyradical, etc), peroxides (hydrogen peroxide and organic peroxides) and singlet oxygen. Below in sections 3.2. and 3.3., each step in such reactions will be considered in a detail.

3.1. OZONOLYSIS OF BIOLOGICALLY IMPORTANT CELLULAR COMPOUNDS

The most important ozone reaction involves the breaking of double bonds (unsaturated links) and the ozonides' formation to the site of the break (Fig.15). This reaction proceeds very quickly, whereas other types of chemical groups in organic substances react with O_3 more slowly.

The interaction of ozone with the double bonds of organic molecules is known as the Criegee reaction (Fig.15). It results in the destruction of such bonds and a splitting of the O_3 molecule itself. Initially, ozone attaches to a double-bond site to form a primary ozonide, which then is split by rupturing the -O-O- and -C-C- bonds. The process can then proceed in various ways. The reaction of a short-living diradical Criegee formation with carbonyl is the best known one. Upon recombination, these products are transformed into secondary ozonides, which then convert into alcohols and carbonylic acids.

In a water medium like in a living cell, the reaction takes a different direction. In this case, hydroxyhydroperoxide is formed from a primary ozonide or diradical Criegee, and it, in turn, decomposes to produce carbonyl and hydrogen peroxide (Fig.16). A further step involves the disintegration of hydrogen peroxide (Fig.16) resulting in the formation of two particles—hydroxyl anion (OH) and hydroxyl

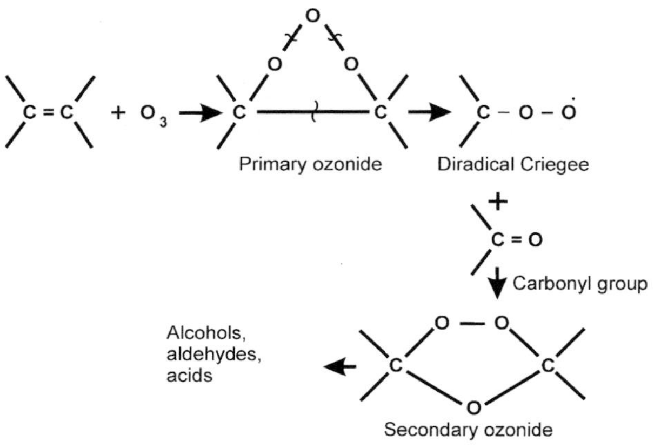

Figure.15. Criegee reaction (Criegee 1959; 1975).

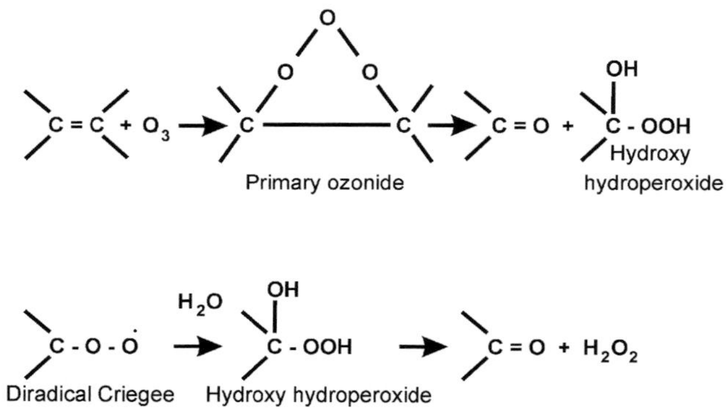

Figure.16. Criegee reaction in a water medium (Criegee, 1975).

radical ($H\dot{O}$). Hydroxyl radical is a highly active chemical and can participate in reactions of lipid peroxidation in biological membranes.

Ozonolysis is a reaction of great importance for the modification of membrane lipids. Ozone reacts with the double bonds of lipids based on glycerophosphocholine, to produce ozonides. The maximum quantity of ozonides is produced if the esterified phospholipids have polyunsaturated acyl groups (Harrison and Murphy, 1996). Ozonides are then broken down into aldehydes and the products of carbonic acid.. A characteristic product of such lipid disintegration is malondialdehyde, the concentration of which could by used to determine the extent of lipid decomposition.

3.1.1. Interaction of ozone with compounds in the cell cover.

The interaction of ozone with plant cells occurs initially in the cell cover and then in the membranes of plasmalemma and organelles. In the first case, ozonolysis of components in cell wall, apoplast, excretions and secretions, leads to break of double bonds of lipophilic and phenolic compounds. Among these compounds, as exemplified by pollen exine (see section 2.3), there may be phenols or carotenoids that have numerous double bonds.

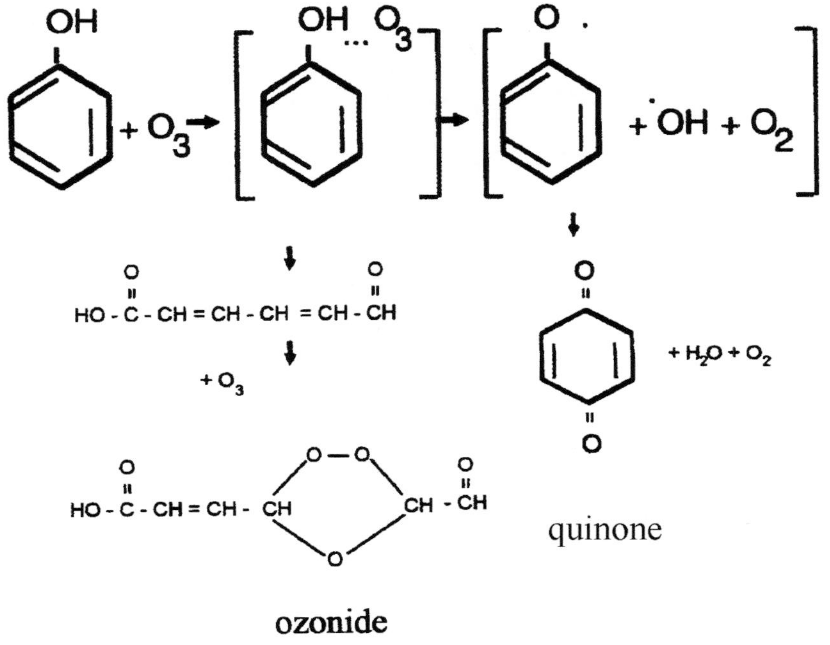

quinone

ozonide

Figure 17. Ozonides and quinones formed by the interaction of ozone with phenols.

Phenols (rutin, esculetin, etc) react with ozone via a radical mechanism involving the missing of hydrogen from the oxygroup and the breaking of the aromatic cycle. Intermediate products of these reactions are ozonides and quinones (Fig.17). Carotenoids, in particular β-carotene, link with ozone causing double bonds to rupture, mainly on the molecule's unsaturated chain. The free radicals of R, forming as a primary product of the ozonolysis (Figs.14, 15), then initiate the chain reactions peculiar to lipid peroxidation (see section 3.1.2) and resulting in the formation of malondialdehyde, the final product of the process (Fig.18).

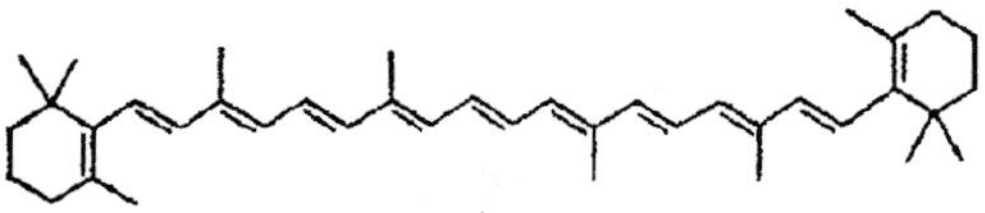

β-carotene

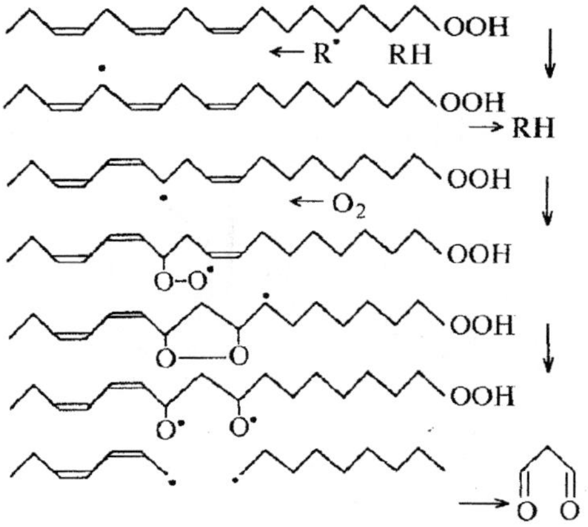

Malondialdehyde

Figure18. Possible scheme for the peroxidation of β-carotene resulting in malondialdehyde

In addition, the ozonolysis of other lipophilic compounds in exine (outer layer of cover) of microspores may occur. Blue-pigmented azulenes are also found in the pollen of various plant species (Roshchina et al., 1995) and in the vegetative microspores of the horsetail *Equisetum arvense* (Roshchina et al., 2002a). Perhaps, undergoing ozonolysis of the double bonds in sesquiterpene lactone, they serve as antioxidants (Chapter 4). In the exine of the microspores, there can be found secretions of a lipophilic nature, which are excreted out the cell (Roshchina et al., 1998). For instance, they can be seen in the pollen of *Papaver orientale* (Fig.19). These components also undergo ozonation to form the free radicals that initiate lipid peroxidation (see section 3.1.2).

Ozonolysis of the individual crystal carotenoid β-carotene and the phenols flavonoid rutin and coumarin esculetin leads to changes in their fluorescence spectra (Fig.20). The maximum in the orange region of β-carotene and rutin disappears, and the maximum of esculetin is shifts to the short wavelength region (Fig.20).

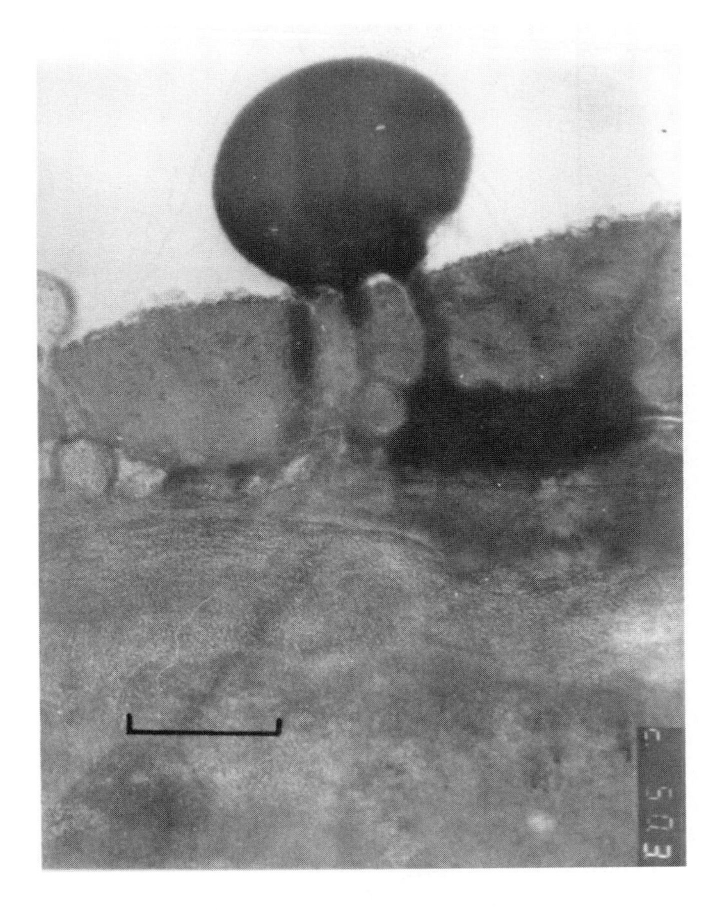

Figure 19. Electron microscopy of the ultrathin slice of the pollen cover from Papaver orientale (adopted from Roshchina et al., 1998a) Bar. 0.5 μm. Microchannels of exine are filled with lipophilic secretion. A droplet of the secretion is excreted onto the surface of the pollen.

Ozonation also influences the biological activity of the compounds, so that *in vitro* pollen germination is, for instance, affected (Roshchina, 2001). Crystalic esculetin and β-carotene (1 mg) were ozonated for 3 hours (total dose 0.15 ppm) and then in concentration 10^{-5} M were tested on the pollen germination of *Hippeastrum hybridum*. The ozonated esculetin inhibited germination to a lesser degree (~2 fold) than the corresponding non-ozonated compound, while the reverse was true for the β-carotene (perhaps due to the contribution of the products formed as a result of lipid peroxidation).

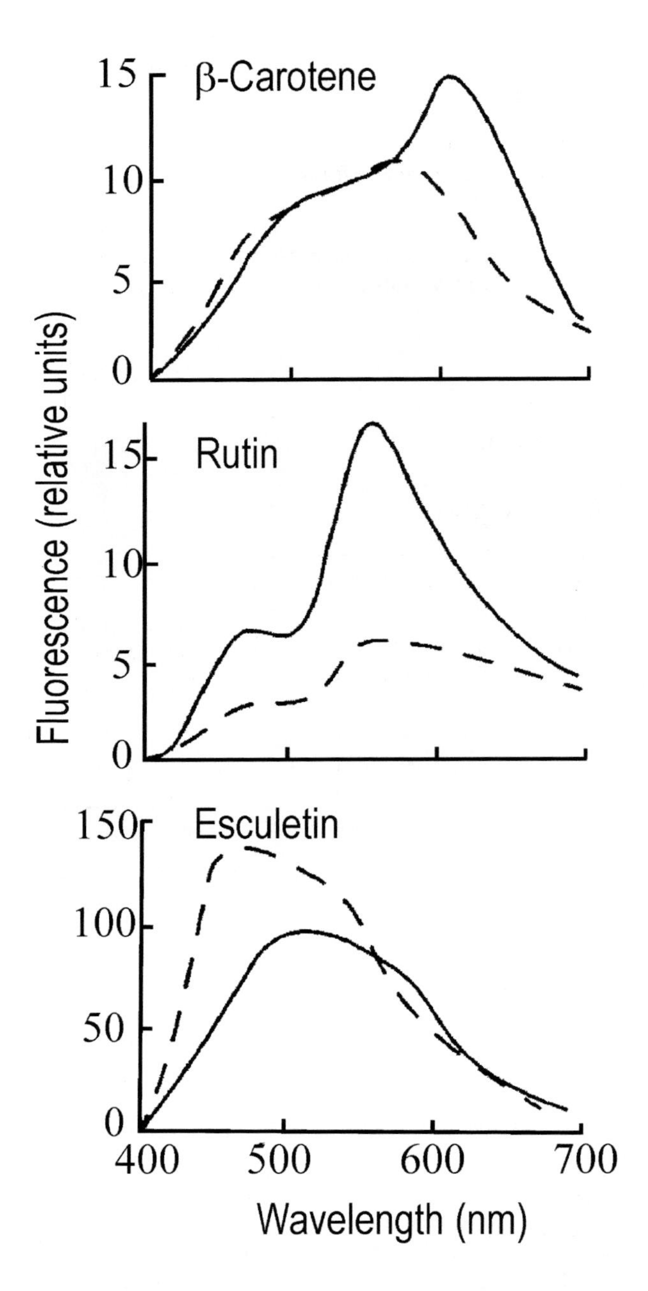

Figure 20. The fluorescence spectra of the individual components of the cell cover of pollen with (broken line) and without (unbroken line) treatment by a total dose of 5 ppm of ozone).

Thus, the components of the cellular cover, such as secretions and excretions, may interact with ozone. Some of them, such as phenols and carotenoids, can themselves also serve as antioxidants (see chapter 4).

3.1.2. Interaction of ozone with membrane lipids

Ozone reacts with the unsaturated chains of membranous lipids to form ozonides and, subsequently, free radicals and, finally, stable products, as shown in Figs. 15 and 16. Moreover free radicals may additionally initiate lipid peroxidation, as seen in Fig.21.

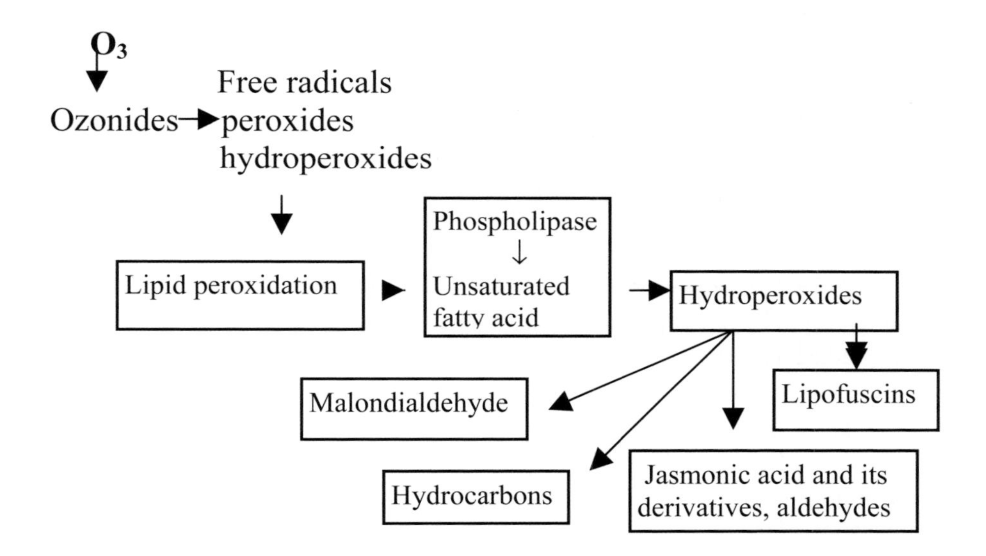

Figure 21. Scheme of the ozone interaction with lipids.

The mechanism of lipid peroxidation differs from ozonolysis. In both lipid peroxidation and ozonolysis, a long reaction chain instigated by high concentrations of ozone leads to the production of malondialdehyde. However, the products of the reactions are, on the whole, quite different. In general, there is no uniform opinion in the literature dealing with the mechanism in which particular fatty acids disintegrate when exposed to ozone. For instance, Mudd et al. (1971) and Mudd in a review (1997) suggest that ozone can cause any fatty acid to decompose by means of ozonolysis. However, it was discovered that, while oleic acid only undergoes ozonolysis, linoleic and linolenic acids are subject both to ozonolysis and peroxidation, with a different product resulting from either of these reactions.

Lipid peroxidation occurs only in the event that there are free radicals in the system (Vladimirov and Archakov, 1972). A peculiarity of the reaction consists in the formation of a free radical chain. Ozone can induce lipid peroxidation because it generates radicals. In the end, the chain reaction of the lipid oxidation leads to a loss of semi-oxidized fatty acids, consumption of oxygen and accumulation of hydroperoxides (Elnster, 1982; Rabinovitch and Fridovich, 1983; Merzlyak and Pogosyan 1988). The scheme of the basic reactions involved in lipid peroxidation is well known and has been considered in a number of papers (Vladimirov and Archakov, 1972; Pryor, 1979; Vladimirov et al., 1980; Vladimirov, 1986; Salter and Hewitt, 1992). The most whidely acknowledged scheme for ozone is described below.

There are five elementary processes involved in lipid peroxidation: (1) initiation, (2) origin of a free radical chain (oxidation of a radical), (3) continuation of the free radical chain, (4) branching of the chain and (5) breakage of the oxidation chain. In the initiation stage, a hydroxyl radical ($\dot{O}H$) reacts with an organic substance (R_1H to form an organic radical ($\dot{R}_1$).

$$\dot{O}H + R_1H \rightarrow H_2O + \dot{R}_1. \tag{1}$$

The radical ($\dot{R}_1$) interacts with a molecule of organic substance (RH) to produce a new radical ($\dot{R}$) and a new molecule of substance R_1 H (one of the originators of the free radical chain).

$$\dot{R}_1 + RH \rightarrow \dot{R} + R_1H \tag{2}$$

In the following stage, $\dot{R}$ quickly joins with O_2, forming a peroxy radical $R\dot{O}2$ (continuation of the lipid peroxidation chain).

$$\dot{R} + O_2 \rightarrow R\dot{O}_2 \tag{3}$$

$R\dot{O}_2$ attacks the unsaturated lipids (RH). As a result of this reaction, both an organic peroxide (ROOH) and a new radical ($\dot{R}$) will be formed, the former of which promotes the continuation and branching of the oxidizing chain.

$$R\dot{O}_2 + RH \rightarrow ROOH + \dot{R} \tag{4}$$

Organic peroxides, resulting from the oxidation, can join together in a process that leads to the generation of free radicals.

Not all $\dot{R}$ and $R\dot{O}_2$ radicals continue to participate in chain reactions; some of them, as a result of their interaction, turn into stable molecular products (thus breaking the free radical chain)

$$\dot{R} + \dot{R} \rightarrow RR$$

$$R\dot{O}_2 + \dot{R} \rightarrow ROOR \qquad (5)$$

$$R\dot{O}_2 + R\dot{O}_2 \rightarrow ROOR + O_2$$

Recombination reactions involving free radicals result in the disruption of the free radical chain of oxidation, since two free radicals are, at once, lost in each act of recombination. In the lipid peroxidation chain, the primary molecular products are peroxides (reaction 4), which are all represented as hydroperoxides (ROOH). Disruptions of the chain reactions involved in lipid peroxidation are also possible when free radicals react with antioxidants (see chapter 4). Based on the above, it is possible to conclude that lipid peroxidation represents a multi-step process in oxidative degradation, during which highly active intermediates will be formed but which, in the end, produces volatile low-molecular compounds.

The final products of lipid peroxidation are worthy of special consideration. Secondary products of lipid peroxidation may combine with alcohols, epoxicompounds, aldehydes, ketones, acids, ketoesters and other compounds (reaction 5), as lipid peroxides are highly active substances. Added exogenously in high concentrations to living organisms, they have a clearly negative effect on the organism (see section 3.3). Among the final products is toxic malondialdehyde, which is an indicator of lipid peroxidation. Its aldehyde groups interact with any NH_2 groups in amino acids and proteins, forming Schiff bases as fluorescent products (see section 3.1.3).

Jasmonic acid and its derivatives are found among the products of the peroxidation of fatty acids that occurs under ozone treatment (Koch et al., 2000). This compound can participate in the defense of gene expression (Rao and Davis, 2001). A significant portion of the final products of lipid peroxidation is comprised of volatile hydrocarbons. Their composition depends on the nature of the compounds that have undergone oxidation. In this regard, experiments on the isolated thylakoids from various cyanobacteria are of special interest (Sandmann and Böger, 1982). These structures from *Anacystis nidulans*, *Anaebena variabilis*, *Spirulina platensis* will form a variety of short-chain volatile hydrocarbons, which arise as a result of peroxidation of the photosynthesizing membranes. *Anaebena* releases ethane, ethylene, pentane and pentene, whereas *Spirulina* only emits pentene and pentane. Unlike these cyanobacteria, *Anacystis* does not contain polyunsaturated fatty acids and therefore does not excrete any of the above-mentioned hydrocarbons. But in model experiments in which some exogenous unsaturated fatty acids were added, the release of short-chain hydrocarbons was observed. Qualitative structure and the ratio of formed low-molecular hydrocarbons depend on the fatty acid that was added to the medium and that underwent peroxidation. Upon peroxidation of oleic acid, only 0.003 nmoles of ethane is formed, whereas linoleic acid predominantly produced (in nmoles) pentaine (8.14) and small amounts of such hydrocarbons as ethane (0.32), ethylene (0.19), propane (0.09) and pentene (1,19). Ethane (6.31) and ethylene (3.71) were produced by linolenic acid; while the isomer of linolenic acid, formed ethane (0.29), ethylene (0.25), propane (0.21), pentane (17.42), and pentene (4.30). As for arachidonic acid, the amounts of ethane, ethylene, propane, pentane and pentene released were respectively 0.46, 0.28, 0.11, 28.64 and 2.62 nmoles. Thus, the dominant products of the peroxidation of the above-mentioned fatty acids were: ethane from oleic and linolenic acids, pentane from linoleic, the isomer of linolenic and arachidonic acids.

A suspension of the diatomic algae *Phaeodactylum tricornutum* was found to contain excretions of hexanal, ethane and ethylene (Shobert and Elstner, 1980). In these conditions chlorophyll and carotenoids are partially bleached. The length of chains of volatile hydrocarbons released in this manner depends on the location of double bonds in a fatty acid, in particular how far is such bond from carboxylic group. Malondialdehyde will be formed only when a fatty acid contains at least three (and no less) double bonds. About one percent of the carbon skeleton of a vegetative cell can be excreted as volatile hydrocarbons: hexenal (from linolenic acid), nonal (from oleic acid), etc. These compounds appear to be the products of lipid peroxidation (Frankel, 1983).

Apparently related to the products of lipid peroxidation are the so-called pigments of aging—the lipofuscins. Similar pigments with characteristic fluorescence in the visible part of the spectrum may accumulate in animal cells under under extremal factors. According to widely held views (e.g. Vladimirov and Archakov, 1972; Tappel, 1975), the occurrence of liposoluble fluorescing substances is connected with the interaction of lipid peroxidation products (in particular malondialdehyde) with compounds containing, free amino acids. Karnaukhov (1988; 1990) suggested that this yellow pigment of aging is formed by the transformation of carotenoids. Similar products are also formed in plants as a result of ozone activity. For instance, upon O_3 fumigation pigments of aging that fluoresce in the region 420-440 nm are found in the lipid fractions of soya *Glycine soja* and corn *Zea mays* (Brooks and Csallany, 1978). Ozone stimulated synthesis of lipofuscin by 2.7 times. The pigment is also found in ripening fruits (Maguire and Haard, 1976) and aging leaves (Wilhelm and Wilhelmova, 1981). Furthermore, the formation of lipifuscin-like pigments have been found in the pollen of some plants (Roshchina and Karnaukhov, 1999). Fig. 22 shows that chronic ozonation (0.05 per hour for 100 hours at 3 hours per a day so that total dose is 5 ppm) altered the fluorescence spectrum of the carotenoid-enriched pollen from the mock-orange *Philadelphus grandiflorus*. The maximum fluorescence in the wavelengths of 540-550 nm, characteristic for carotenoids, disappeared, and a new maximum in the 440-480 nm

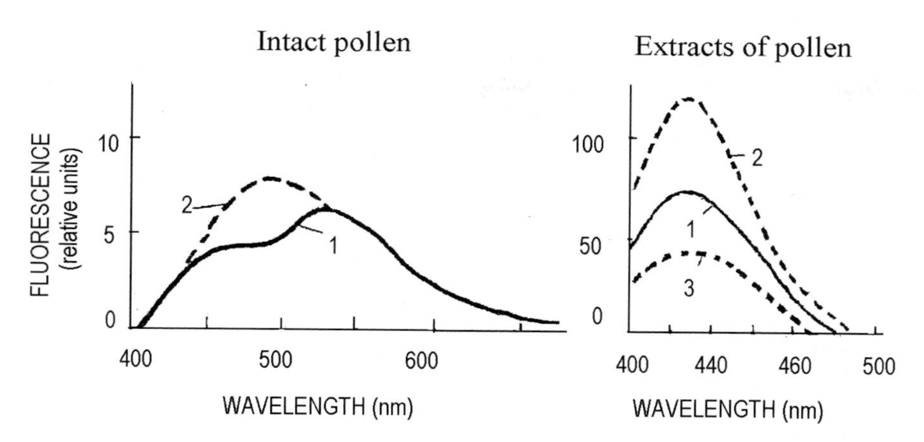

Figure 22. Formation of new fluorescing products in the intact pollen of the mock-orange Philadelphus grandiflorus as a result of ozone treatment (Roshchina and Karnaukhov, 1999). At the left - the fluorescence spectra of intact pollen in the control (1) and after chronic ozonation (2) for 100 hours at 3 hours per day (Total dose of ozone - 5 ppm). At the right -the fluorescence spectra of chloroform/ethanolic extracts (1:1 in volume) from imtreated pollen, (1) and after treatment with ozone without purification (2) or after purification of the fluorescing component (Rf 0.2) by thin-layer chromatography on Silicagel (3).

range appeared. An analysis of chloroform/ethanol extracts from this pollen detected the presence of a pigment with the fluorescence maximum of 430 nm, which is similar to that of lipofuscins (Fig. 22). In objects, in which the concentration of carotenoids is small, and was shielded by other pigments, or in which any of the pigments were absent (as is the case with the plantain *Plantago major*), lipofuscin was not formed. Arising in plants, the pigments of aging have a diverse nature, and the mechanism of their formation cannot be described within the framework of any uniform mechanism (Merzlyak, 1989).

3.1.3. Reaction of ozone with amino acid and proteins.

When exposed to ozone, practically all amino acids undergo oxidizing transformations to various degrees (Fig.23), especially tryptophan, tyrosine, histidine and cysteine.

Among the amino acids, the most responsive to O_3 are cysteine and cystine, as they contain SH-groups. More moderately reactive are tyrosine, phenylalanine and tryptophan. The highly reactive nature of the sulfur-containing amino acids is due to the rather easy polarization of SH and SC groups, which resultantly provide a

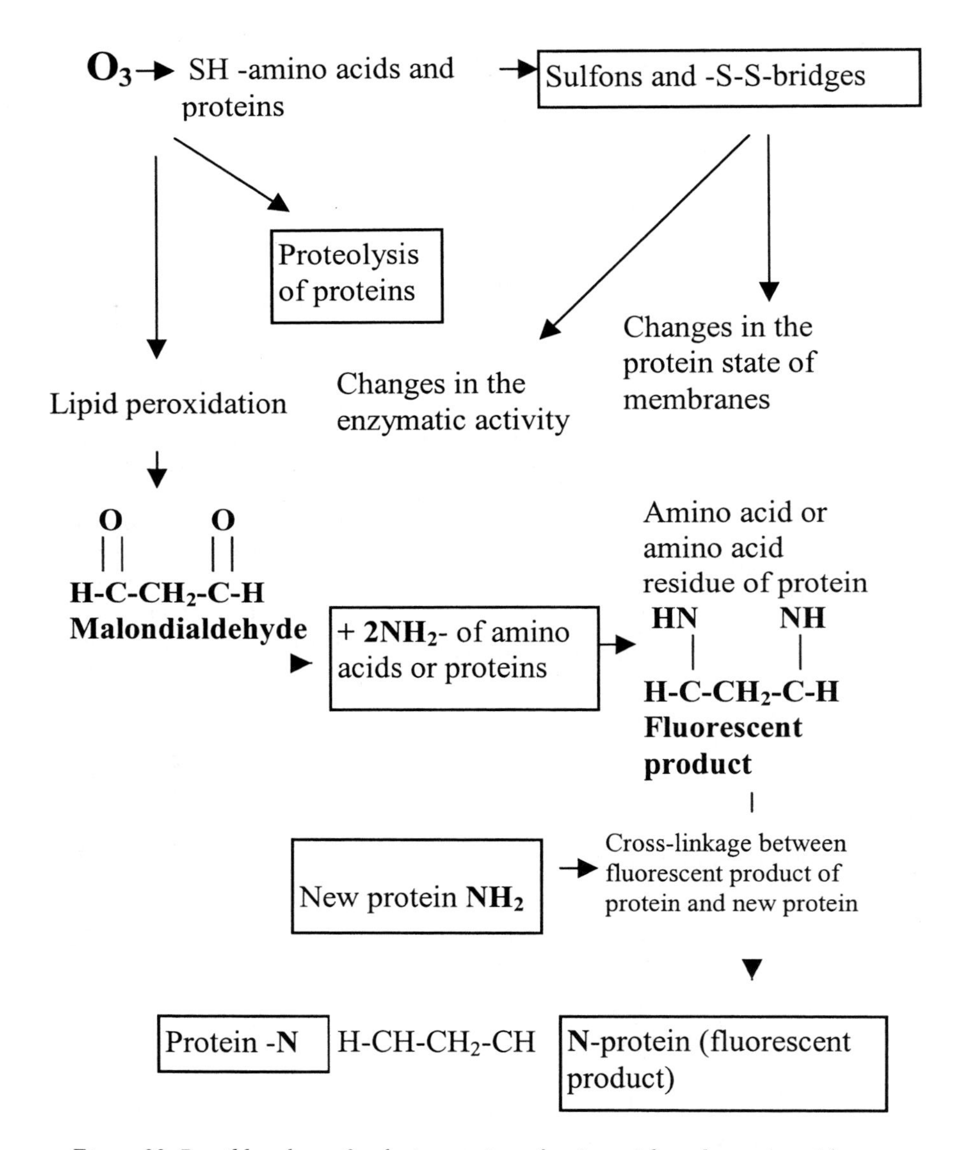

Figure 23. Possible scheme for the interaction of amino acids and proteins with ozone

place for the attack with ozone (Kayushin et al., 1976). Under the influence of ozone, various products of ozonolysis are possibly formed:

$$\text{cysteine (RSH)} \rightarrow \text{cystine (RSSR)},$$
$$\text{cysteine (RSH)} \rightarrow \text{cysteinic acid } (RSO_3\,H)\ \text{RSSR}),$$
$$\text{methionine } (CH_3\,SR) \rightarrow \text{methioninesulfoxide } (CH_3\,SOR),$$
$$\text{tryptophan}(C_{11}H_{12}N_2O_2) \rightarrow \text{N -formylkinugenine}$$

First of all, ozone reacts with tryptophan, methionine, tyrosine, and phenylalanine residues of amino acids, which are oxidized at physiological values of pH. Ozone acts on SH-groups of these amino acids (Fig.23). These gtoups are oxidized to form sulfonic compounds containing an SO_2 group that can be linked to two organic radicals. An example is provided by dimethylsulfon $(CH_3)_2SO_2$. Thiol agents do not reduce sulfons, i.e. the reaction is irreversible. Thus, sulfhydryl groups of proteins in membranes are directly modified by ozone, that changes the membraneous permeability. O_3 can attack the protein components of plasmalemma, enabling further its penetration into the cell and even into the cellular organelles. Ozone also oxidizes the protein thiols, so that disulfur bridges between the protein chains are formed. Due to an ability to react with the thiol groups of cysteine, a greatest sensitivity to ozone is displayed, for instance, by the protein papaine from the latex of the *Papaya* tree. Lysozyme and ovidine are sensitive to O_3 due to the oxidation of the tryptophan residues. Under the influence of the free radicals generated in ozonolysis, peroxides of both amino acids and proteins can also be formed (Gebicki and Gebicki, 1993).

Ozone, in a dependence on dose, also causes the destruction of membranous proteins, and they, as seen in Fig.23, become more sensitive to endogenous proteolysis (Treshow, 1984; Cross and Haliwell, 1994). The specificity of their oxidizing modifications depends on the composition of the amino acids and the structural organization of the protein molecules. It is assumed that the residues of the amino acids of proteins react with ozone (Mudd, 1982).

Enzyme activity also changes under ozone influence (Fig.23). O_3 lowers the activity of the enzymes-markers of the internal part of the cytoplasmic membrane such as acid phosphotase and glyceralaldehyde-3-phosphate dehydrogenase (Tingey et al., 1975). The decrease in their activity occurs due to local disturbances in the active center of the enzyme, where thiols are located (Menzel, 1971). When cysteine is allowed to act on the inhibited enzyme, its activity is only partially restored because, it is believed, the thiol group in the enzymic active center is also transformed to such oxidized compounds as sulfonic acid. Ozone also inactivates acetylcholinesterase (Mudd, 1982), which is a marker of the outside part of plasmic membrane. The active center of the enzyme contains serine and histidine. When the amino acid residues are oxidized, it appears to cause the loss of enzyme activity. Enzymes, however, can be protected against ozonolysis by a substrate, which covers the amino acid residues and, thus, interferes with (prevents) their oxidation (Mudd, 1982).

Malondialdehyde, a product of lipid peroxidation, also may have an effect on amino acids and proteins. Its aldehyde groups interact with any NH_2 groups in amino acids and proteins to form Schiff bases as fluorescent products, a reaction represented in Fig.23. Their fluorescence (excited by ultraviolet light 360-380 nm) is in the visible area of the spectrum (440-560nm). Antioxidants usually retard the formation of the fluorescent products based on amino acids, as well as their residues in proteins. The fluorescent proteinous substances, in turn, will form cross-links with new protein molecules (Fig.23). The toxicity of malondialdehyde that inhibits the activity and biosynthesis of enzymes, can be explained by the formation of cross-links in the membranous proteins (Kikugawa and Beppu, 1987; Leshem et al., 1992). Some visible changes in plant cell autofluorescence (section 2.2, Fig.9-13) are, perhaps, due to the malondialdehyde interaction with amino acid and proteins.

Thus, ozone reacts with amino acids and proteins via ozonolysis, proteolysis or products of preliminary lipid peroxidation.

3.1.4. Reactions of ozone with amines

The reactions of ozone with amines proceed rather easily. The mechanism of the reaction depends on the structure of the initial amine (Razumovskii, 1979).

According to Bailey et al. (1968), the amine R_3N reacts with ozone as follows:

$$R_3N + O_3 \rightarrow R_3N^+{-}O{-}O{-}O^{-\cdot}$$

As a result, ozonide is formed, which is decomposed by missing of molecular oxygen :

$$R_3N^+ {-}O{-}O{-}O^- \rightarrow R_3N^+ {-}O^- + O_2$$

Frequently found in plants are quaternary amines and their derivatives, such as the neurotransmitters acetylcholine, dopamine, nor adrenaline, adrenaline, serotonin, and histamine (Roshchina, 1991, 2001), as well as polyamines including spermine, spermidine and others (Goodwin and Mercer, 1978). Thus, the possibility is not excluded that their ozonation takes place in accordance with the previously described schemes. The products of the lipid peroxidation induced by O_3, such as malondialdehyde, also may react with amines, forming fluorescing Schiff bases (see section 3.2.1).

3.1.5. Interaction of ozone with nucleic acids

If O_3 penetrates inside a cell (in particularly high concentrations), it may directly interact with nucleic acids, but if not, then an indirect mechanism is activated involving, in particular, the formation of free radicals (see section 3.2). Malondialdehyde, a product of lipid peroxidation (see section 3.1.2), also can inhibit DNA replication and synthesis in plants (Dhindsa, 1982). As a whole, the direct interaction of ozone with nucleic acid has not been well studied, but it is generally belived that ozone can act as a mutagen in animals, plants and microorganisms on account of this interaction (Victorin, 1992). Even low concentrations of ozone, 0.1-1 ppm in short exposures, can induce mutations. As the concentration of O_3 and the duration of the treatment increase, the number of mutations also becomes higher. The mutagenic features of ozone concern its ability to induce defects in the structure of nucleic acids (Hamelin and Chung, 1976). There are two main types of possible effects: a degradation of some nitrogen bases, perhaps, by ozone acting on double bonds to produce ozonides (Ishizaki et al., 1981), and the instigation of ruptures in single or double spirals (Hamelin, 1985; Sawadaishi et al., 1985). However, it is more likely that the effects are induced by the active oxygen species resulting from ozone interaction with water (see section 1.5). This damage causes a decrease in the normal genetic function of DNA. Moreover, ozone prevents the inclusion nitrogen bases in the DNA structure at the time of its synthesis. Doses of ozone (≤ 5 ppm) have been shown to decrease the inclusion of ^{3}(H)-uridine in the DNA of the bacteriophage T7 (Mura and Chung, 1990). Such effects were not observed in experiments with oxygen alone.

3.1.6. Reaction of ozone with NADH and NADPH

Biologically important reducers, such as nicotineamideadeninenucleotide (NAD) and nicotineamideadeninenucleotide phosphate (NADP) in its reduced form as NAD(P)H, are readily oxidized by ozone (Menzel, 1971), as indicated in Fig. 24.

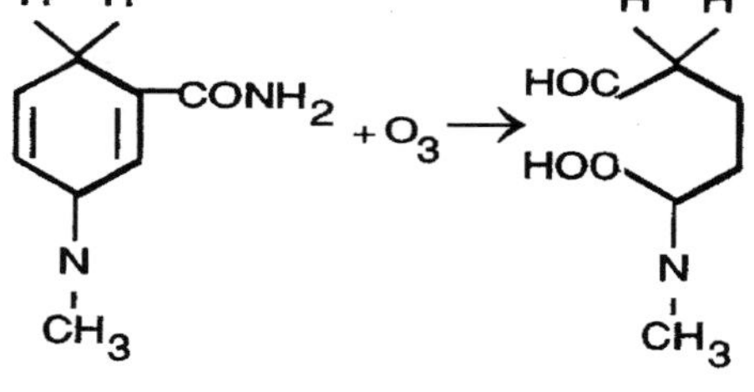

Figure 24. Reaction of ozone with NADH

Ozone acts to break the aromatic ring, so that aldehyde groups are formed at the site of the break (Mudd et al., 1974). At first, ozonolysis involves 5-6 double bonds and, subsequently, 2-3 bonds. The reaction depends on pH and proceeds with at a high rate in neutral and alkaline media. NAD and NADP are more tolerant of ozone than their reduced forms are. The products of the reaction of NADH and NADPH with ozone have no biological activity.

3.1.7. Reaction of ozone with plant regulators of growth and development

In biological researh on ozone reactions with the regulators of growth and development, phytohormones are especially important, as these substances are active in very small quantities (10^{-5}-10^{-11}M), and can be easily deactivated when treated with O_3, active species of oxygen or free radicals (Adepipe and Ormroid, 1972). Five groups of phytohormones are known: auxins, gibberellins, cytokinins, abscisic acid and ethylene. Their significance consists in the chemical initiation of the physiological programs in plants (Derfling, 1985). Table 5 indicates the main ozone effects and the possible mechanisms by which its interaction with phytohormones occurs.

The main representative of auxins, a group of natural compounds stimulating cellular division (mitosis), root formation, respiration and protein synthesis, is indoleacetic acid (Fig. 25). Upon ozone, the double bond in the indole ring of indoleacetic acid is degraded so that ozonides are formed (Bailey, 1978). Major gibberellin-gibberellic acid, as well as other representatives of this class, consists in 4-th isoprene residues, forming 4 rings, which undergo ozonation (Table 5).

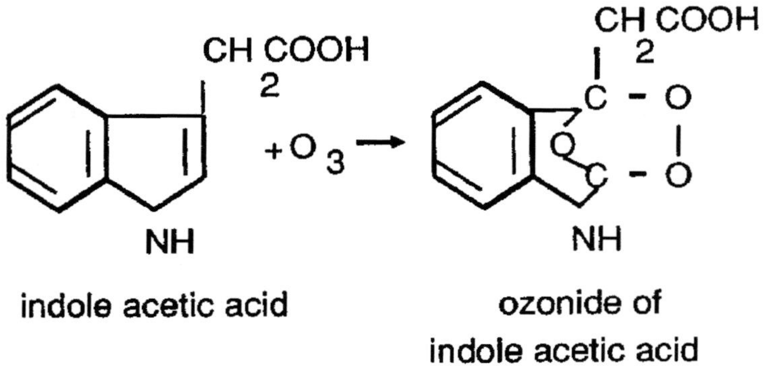

indole acetic acid $+O_3 \rightarrow$ ozonide of indole acetic acid

Figure 25. Reaction of ozone with indole acetic acid.

Table 5. The interaction of ozone with the growth regulators

Formulae of regulator	*Biological effect*	*Target for ozone*
Auxins 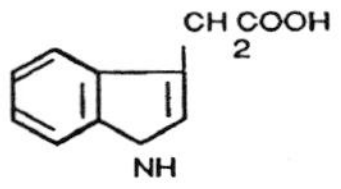 CH_2COOH, NH	Stimulation of cellular division (mitosis), root formation, respiration and protein synthesis	Double bond in indole ring of indoleacetic acid is broken and ozonide will be formed (Bailey, 1978). All products of the oxidation of indoleacetic acid are capable of inhibiting plant growth.
Gibberellins 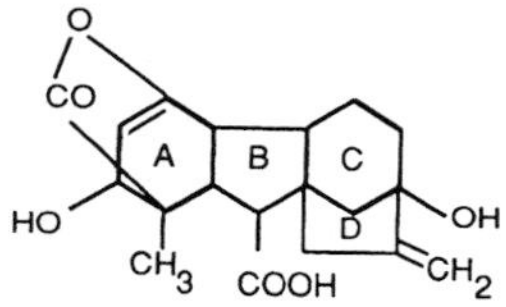 O, CO, A, B, C, D, HO, OH, CH_3, COOH, CH_2	Stimulation of stem growth, of fruit sizes, seed germination, changes in the flower's form and size, of the membranes' properties, set RNA and functioning enzymes.	Double bonds of rings A and B are first of all broken that leads gibberellin to be inactivated
Cytokinins 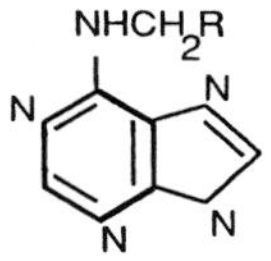 $NHCH_2R$, N, N, N, N	Stimulation of cellular division (cytokinesis) and retarding of the aging processes.	Structure of the purine ring that results in a decrease and even a complete loss of phytohormone activity
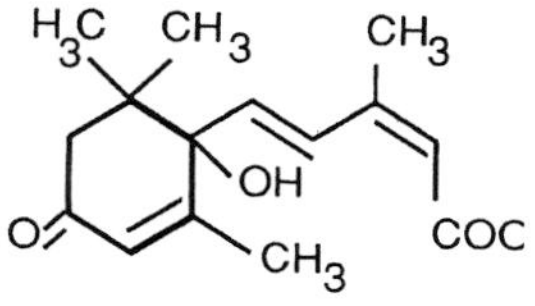 Abscisic acid H_3C, CH_3, CH_3, OH, O, CH_3, COC	Regulation of stomata closing	Double bonds of vulnerable part of a molecule. The loss of activity due to oxidation of one of the methyl groups.
Phenolic inhibitors	Retardation or stimulation of the effects that main hormones have on plants.	Double bonds break; ozonides and quinones are formed via the radical mechanism, causing hydrogen be removed from oxygroup and the aromatic cycle to be broken.

Cytokinins, derivatives of adenine with a lateral chain of a various structure, also have double bonds to which ozone can be linked. Abscisic acid is the most widespread inhibitor of plant growth and the antagonist of auxins, gibberellins and cytokinins. On the chemical nature, abscisic acid is comprised of a group of sesquiterpenes (C_{15}), consisting of three isoprene units with single carboxylic, hydroxylic and oxygroups, all of which can be oxidized. The plant resistance to O_3 after the preliminary treatment by abscisic acid is increased due to a stomata closing that prevents ozone penetration into the leaf (Rosen et al., 1978). Phenols and their derivatives, when added exogenously, frequently retard or stimulate the effects of the main hormones of plants. Razumovskii and Zaikov (1974) argue that they react with ozone via a radical mechanism in which ozonides and quinones are formed (see Fig. 17). Ethylene has also all properties of phytohormones. It causes a number of physiological effects, of which the major one is a maturing of juicy fruits and stimulation of leaves' fall. The basic mechanism by which the reaction with ozone occurs is the Criegee-reaction, in which double bonds are splitting (see section 1.3.2 and 2.1), and as a result free radicals and peroxides are formed. As has been shown (Finlayson-Pitts and Pitts, 1986), ethylene is a precursor to the 4-th kinds of Criegee radicals (Fig.26). These free radicals are highly reactive species that start a chain of non-controlled redox reactions, at first in lipids of membranes and then in cytoplasm of cells. The stability of the products decreases from dioxirane to peroxymethylene.

dioxirane	methylene-bis(oxy)	planar peroxymethylene	perpendicular peroximethylene

Figure 26. Products of the reaction (Criegee radicals) resulting from the interaction of ethylene and ozone

In solution, Criegee intermediates react with aldehydes and ketones; as a result, secondary ozonides will be formed, which are stable peroxides.

```
                              R O - O R
                               \/    \/
RCOO + RC = O →                C      C
 |       |                     /\    /\
                                  O
```

Reactions of the ethylene oxidation are possible when this phytohormone react with either hydroperoxide radical or atomic oxygen (Kurchii, 1990). Epoxide will be produced.

$$H_2C = CH_2 + H\dot{O}_2 \rightarrow H_2C - CH_2 + H\dot{O}$$

$$\backslash /$$

$$O$$

$$H_2C = CH_2 + (O) \rightarrow H_2C - CH_2$$

$$\backslash /$$

$$O$$

Last years data received, leads us to assume that ozone and the products of free-radical oxidation participate in the synthesis and catabolism of some phytohormones (Leshem et al., 1986).

The reader can also find more information on other groups of organic compounds in the fundamental study by Bailey (1978) and the review by Cueto et al. (1994).

3.2. FREE RADICALS AS INTERMEDIATES IN THE PROCESS OF OZONOLYSIS

The mechanism of ozone's biological activity includes, at least partially, some involvement of free radicals—atoms or molecules that have one or more unpaired electrons in their external orbits (orbital). The main sources of free radicals are: (1) ozonolysis of organic and inorganic compounds (section 3.1.and Bailey, 1978); (2) ozonolysis of water (section 1.4); (3) peroxides formed in various reactions initiated by O_3, such as hydrolysis (section 1.3) or lipid peroxidation (section 3.1). Quinones can also participate in the generation of free radicals (Weiner, 1994). Highly reactive free radicals usually exist in very low concentrations of about 10^{-11} to 10^{-9} mol/L, and have short lifetimes (Saran et al., 1988; Vladimirov et al., 1991). They can easily interact with biologically important compounds to modify them and, consequently, to change physiological processes. Free radicals are found in plant, animal and microbial cells. The reactions with ozone that form free radicals occur on the surfaces of both vegetative cells (Dodd and Ebert, 1971; Priestly et al., 1985; Melhorn et al., 1990; 1993) and generative cells (Priestly et al., 1985). The reactions, when the radicals are accumulated, intensify with aging.

3.2.1. Free radicals in biological systems.

In biological systems, many biochemical reactions proceed by forming free radicals as the intermediate products. By using method of EPR (electron paramagnetic resonance) these intermediates can be detected in the actively metabolizing cells of plants and animals (Fridovich, 1979; Priestly et al., 1985; Merzlyak, 1989). The concentration of free radicals can be as much as 10^{-6}–10^{-8} moles per 1 g of fresh tissue mass. The free radicals arise in reactions with a large number of inorganic and organic substances, switching on branched alkenes, alcohols, and enols (Pryor and Church, 1991). The formation of such active oxygen species as superoxide anion radical ($\dot{O}_2^-$), trioxide radical ($\dot{O}_3^-$), hydroxyl radical ($\dot{O}H$) is also observed. Among sources of radicals in a cell, the main ones are peroxides, hydroperoxides and hydrogen peroxide, which are all formed during lipid peroxidation, as well as in the ozonolysis of both air components and water under normal as well as various stresses. Peroxides can be reduced in a Haber-Weiss reaction (Haber and Weiss, 1932) by electron donation from iron (Simic et al., 1989). The result will be the formation of a hydroxyl radical ($\dot{O}H$) and free carbon radicals ($R\dot{O}$)

$$H_2O_2 + Fe^{2+} \rightarrow \dot{O}H + OH^- + Fe^{3+}$$

$$ROOH + Fe^{2+} \rightarrow R\dot{O} + OH^- + Fe^{3+}$$

These reactions occur rather fast and do not require high concentrations of double-valent iron (Simic et al., 1989). Lipid peroxides can generate radicals that initiate a chain formation of other radicals (Larson, 1978). Ozone interaction with a water phase also results in the formation of oxygen radicals $\dot{O}_2^-$, $\dot{O}H$, $H\dot{O}_2$ (see section 1.3.5).

Insofar as biological processes are concerned, these radicals of oxygen are the most important. Their characteristics greatly vary (Table 6). The most short living

Table 6. Lifetimes and diffusion radii of active oxygen species in biological systems (adopted from Zenkov and Men'shchikova, 1993).

Active oxygen species	*Lifetime (s)*	*Radius of diffusion, μm*
$\dot{O}H$ hydroxyl radical	10^{-9}	< 0,01
$\dot{O}_2^-$ superoxide anion radical	10^{-6}	0,3
O½ singlet oxygen	10^{-6}	0,3
$R\dot{O}$ alkoxy radical	10^{-6}	Depends on R
$H\dot{O}_2$ perhydroxy radical	10^{-3}	10
$R\dot{O}_2$ peroxy radical	10^{1}	Depends on R
H_2O_2 hydrogen peroxide	Depends on activity of catalase and peroxidase	

and, accordingly, most reactive of them are hydroxyl radicals, which have a very small radius of action. Peroxy radical is less reactive, but has, consequently, a greater radius of action. Hydrogen peroxide and superoxide anion radical are rather stable active oxygen species and can diffuse significant distances from the place of their formation and can even cross cellular and intracellular membranes.

A high reactivity and short lifetime in biological systems are the common properties of the oxygen radicals. Owing to the brief time of life, the radicals of oxygen are limited to small radii of action. Resultantly, the action of hydroxyl radical is limited to the size of an average organic molecule, for instance, pepsin (Zenkov and Men'shchikova, 1993). Superoxide anion radical and singlet oxygen have radii of action commensurable with the size of a cell. Organic radicals and hydroperoxides have larger radii of action, a characteristic that is also discernible when studying both tissues and organisms as wholes. On the basis of K, the rate constant of the reaction ($m^{-1}s^{1}$), the radicals are distributed in the following ranges:

$$\dot{O}H > H\dot{O}_2 > R\dot{O} > \dot{O}_2^- > RO\dot{O}.$$

Of all the known oxidizers, hydroxyl radical is, thus, the strongest. In organisms, the possible transformation of less reactive forms of oxygen ($\dot{O}_2^-$ and $RO\dot{O}$) into more reactive ones—e.g. singlet oxygen ($^1O_2$)—is possible if there is a spontaneous dismutation of the superoxide radical ($\dot{O}_2^-$) or the $\dot{O}H$ radical in the Fenton and Haber-Weiss reactions. It is necessary to note that these processes are multiple and variable, and that they result in the generation of free radicals.

The main method for the assay of the free radicals of oxygen and other active oxygen species involves a measuring of their activity in the presence of enzymes, for instance superoxide dismutase if superoxide anion radical is determined. This radical is found on animal and plant surfaces (Fridovich,1984; Labas et al., 1999) and thought to be formed in norm by the redox-system in plasmalemma of plant cells (Aver'yanov, 1991; Mehdy, 1994; Murphy and Auh,1996; Minibayeva et al., 1998; Gordon et al., 1999), and especially in root cells (Minibayeva et al., 1998;Gordon et al., 1999), in cell walls (Aver'yanov, 1991) and on the surface of pollen (Roshchina and Melnikova, 1998). It is found in etiolated seedlings during their development (Shorning et al., 2000). In leaves, formation of the radical is supposed because superoxide dismutase presents in the apoplast (see Chapter 4). Difficulties in the assaying of superoxide radical are due to its rapid conversion into hydrogen peroxide in the presence of the superoxide dismutase of water-enriched tissues.

One of the most direct methods of radicals' assaying involves the measurement of chemiluminescence by using chemiluminescent probes (Goto and Niki, 1994). The chemiluminescence of lucigenin (Fig27), a special fluorescent probe on the superoxide anion radical added to the surface of the pollen from *Philadelphus grandiflorus* and *Hippeastrum hybridum* or to the vegetative microspores of *Equisetum arvense*, is blocked by 0.013 mg/ml of endogenous superoxide dismutase but not by 0.05 mg/ml of peroxidase (Roshchina et al., 2002b; 2003). There also was no chemiluminescent reaction with luminol, a specific luminescent probe for hydroperoxides. The chemiluminescence of lucigenin in microspores occurs both

LUCIGENIN

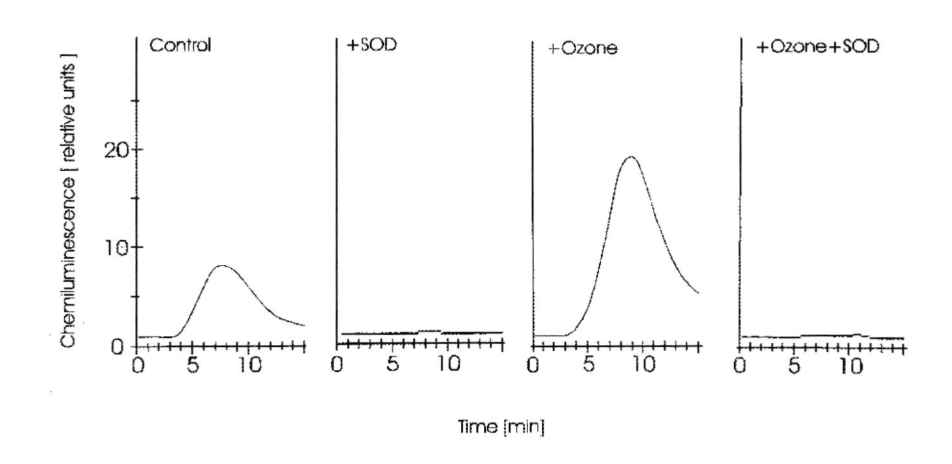

Figure 27. The chemiluminescence of 0.2 μM lucigenin added to intact cells of microspores (10 mg) of Equisetum arvense. Control – without any treatment; (+SOD) - addition of 0.013 mg/ml superoxide dismutas; Ozone - preliminary treatment with ozone (0.06 ppm); Ozone+SOD – preliminary treatment with ozone (0.06 ppm) and then addition of 0.013 mg/ml superoxide dismutase

with or without the addition of water, but, in the latter case, the emission is much less and seen only after initiation by 2.5-9 μmoles NADH. The activity of the endogenous enzyme, measured according to some method (Bors et al., 1978), was not detected in the samples. The example of the chemiluminescence kinetics for

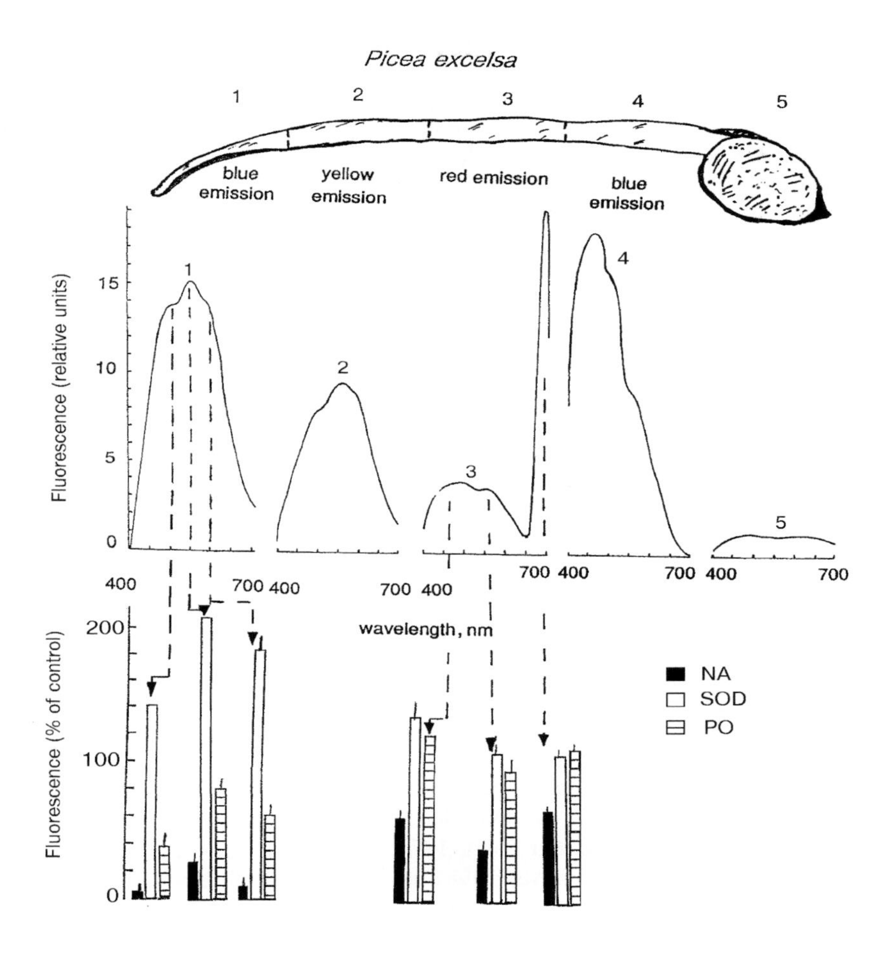

Figure 28.The fluorescence spectra of a Picea excelsa seedling (upper side) and the influence of oxidant noradrenaline (NA) 10-5 M, antioxidant enzymes 0.6 mg/ml-superoxide dismulase (SOD) or peroxidase (PE) on the fluorescence intensity determined in different maxima (lower side).

microspores of *Equisetum arvense* is shown on Fig. 27. Ozone strengthened the chemiluminescence intensity, and exogenous superoxide dismutase completely blocked the emission. The exogenous application of superoxide dismutase also changed the autofluorescence and germination of pollen *Hippestrum hybridum* (Roshchina and Melnikova, 1998a) as well as the same processes of vegetative microspores *Equisetum arvense* (Roshchina et al., 2003). The presence of the superoxide anion radical on the surface of microspores appears to have a protective function against pests and other unfavorable factors.

The root tips of the plants excrete forms of active oxygen (Streller and Wingle, 1994). In the authors' own experiments (Roshchina et al., 2000), the addition of the antioxidant enzymes affects the autofluorescence of developing 3-day-old seedlings of *Picea excelsa* and *Larix decidua* induced by ultraviolet light.). Superoxide dismutase and peroxidase diminish the superoxide anion radical and hydrogen peroxide, respectively. As shown on Fig.28, the difference is discernible in the chlorophyll-less part of the developing root tip (root meristem - part 1). The leaf meristem, with its chlorophyll fluorescence (mainly part 3 of seedling), is less sensitive to the enzymes. Superoxide dismutase stimulates the autofluorescence of root meristem, whereas the generator of superoxide anion radical noradrenaline decreased. A result of the reaction with the enzyme may be a dismutation of the radical and hydrogen peroxide, which stimulates the autofluorescence.

The occurrence of free radicals in small concentrations may require normal plant development, as is the case for the superoxide anion radical (Gordon et al., 1999; Shorning et al., 2000). But a so-called oxidative burst (enhanced formation of active oxygen species) is a peculiar resistance response to stress factors (Lamb and Dixon, 1997; Wojtaszek, 1997). For instance, the formed superoxide radical induces stomatal closure (Mori et al., 2001).

A significant amount of oxygen free radicals may be generated in ozonolysis and contribute to the visible effects of O_3.

3.2.2. *Formation of free radicals during ozonation and their biological effects*

The effects of ozone on living cells involve the formation of free radicals. Both direct and indirect methods can be used to detect their presence, under normal conditions, in the vegetative tissues of plants (Merzlyak, 1989), the cellular surfaces of the generative organs, such as the pollen (microspores) of *Osmunda regalis* and *Sycopodium clavatum* (Dodd and Ebert, 1971; Priestley et al., 1985), the stigma of pistils (Roshchina and Mel'nikova, 1998), and the seeds of various plant species (Priestley et al., 1985). As a stress-inducing factor, ozone stimulates the formation of free radicals in vegetative cells and pollens of several species (Bennett et al., 1984; Melhorn et al., 1993). One of the processes that frequently results in the formation of free radicals *in vivo* is the reaction of ozone with compounds containing double bonds. The radicals in this reaction are Criegee-radicals of olefins, superoxide radicals and hydroxyl radicals. O_3 initiates lipid peroxidation causing peroxides, hydroperoxides and hydrogen peroxide to be generated in a cell, all of which can then be reduced in Haber-Weiss reactions (see section 3.2.1). These processes are multiple and variable, so that they result in the generation of free radicals.

Using the EPR method, it has been demonstrated (Melhorn et al., 1990) that the formation of free radicals occurs in plants exposed to ozone. Free radicals are found in leaves of pea *Pisum sativum* and string bean *Phaseolus vulgaris*, after they have been fumigated with ozone in concentrations of 70-300 nl/l (0.07-0.03 ppm) for 21 days (4 - 7 hours per day). During the treatment, ethylene release increased from 2,5 nmole/g of dry weight per hour (control) up to 6.2-16.5 nmole/g of dry weight per hour (in experimental conditions). Radicals were produced before any appreciable damage to the leaf was observed. The authors assumed that the cause of the toxic action of ozone is a reaction between and ethylene, which forms in plants under ozone stress. The use of the inhibitor of ethylene biosynthesis—aminoethoxivinylglycine—prevented ozone damage in most plant parts (Melhorn and Wellburn, 1987). The experiments indicate that, when such damage does occur, it stimulates the formation and release of alkenes, making plants more sensitive to ozone.

It is now necessary to consider the cellular reactions of the main oxygen radicals formed during ozonation and indicate their significant effects.

Superoxide anion radical. Unlike other oxygen radicals and singlet oxygen, the superoxide anion radical has a rather long lifetime in water solutions—about several milliseconds (Table 6), and, in especially clean water, can even be a few orders longer. In this connection, the assumption is that, in living cells, this radical is capable of moving significant distances from its place of formation. The superoxide anion radical is found experimentally in the nuclear, plasmic and mitochondrial membranes, as well as in chloroplasts and chromatophores. It can diffuse through the cellular membrane but, as an anion, has a negative charge. Therefore, it migrates through a membrane with much more difficulty than hydroxydioxide $H\dot{O}_2$ does (Takahàshi and Asada, 1983).

The superoxide anion radical participates in reactions of oxidation and reduction. It oxidizes sulfur-containing substances, such as cytochromes, ascorbate, catecholamines and NADPH. In water solutions, $\dot{O}_2^-$ easily reduces quinones to semiquinones (Afanas'ev, 1979; Afanas'ev and Kupriyanova, 1993). Probably, the interaction of $\dot{O}_2^-$ with hydrogen peroxide (the Haber-Weiss reaction, see section 3.2.1) causes hydroxyl radicals to form in living cells.

$$\dot{O}_2^- + H_2O \rightarrow O_2 + HO^- + H\dot{O}$$

The mechanisms of superoxide radical action may be largely similar to those involving ozone itself (see section 3.1.) and consist in the interaction with the double bonds of various compounds, such as unsaturated fatty acids in membrane lipids, phenols, terpenoids and other substances. The radical can oxidize the CH-groups in antioxidants, a process illustrated in fig.29 by the interaction of cytokinin with $\dot{O}_2^-$, one that leads to the formation of an appropriate radical, then to hydroperoxide and, finally, to ketone (Leshem, 1984; Leshem et al., 1986). In high concentrations, the superoxide anion radical can, however, break both the hydrogenic and covalent bonds in proteins and nucleic acids.

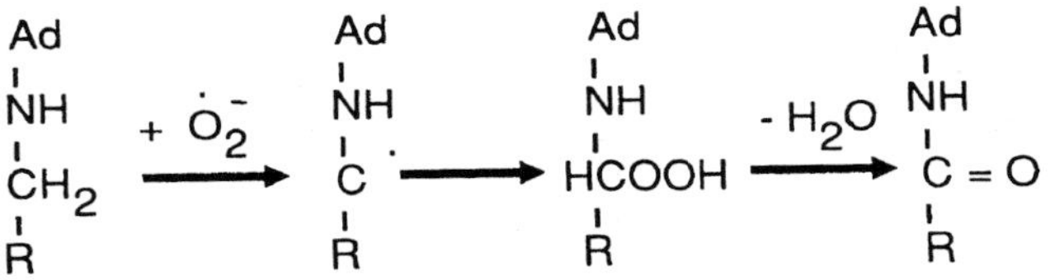

$$\begin{array}{c} Ad \\ | \\ NH \\ | \\ CH_2 \\ | \\ R \end{array} \xrightarrow{+\ \dot{O}_2^-} \begin{array}{c} Ad \\ | \\ NH \\ | \\ C^{\cdot} \\ | \\ R \end{array} \longrightarrow \begin{array}{c} Ad \\ | \\ NH \\ | \\ HCOOH \\ | \\ R \end{array} \xrightarrow{-\ H_2O} \begin{array}{c} Ad \\ | \\ NH \\ | \\ C = O \\ | \\ R \end{array}$$

(Ad - adenosine, R- radical)

Figure 29. Interaction of cytokinin with a superoxide anion radical

In living systems, superoxide anion radical is, depending on concentration, either a necessary cellular agent (Minibayeva et al., 1997; Gordon et al., 1999; Shorning et al., 2000) or a damaging factor (Fridovich, 1981, 1983). The main effects are shown in table 7. Low concentrations of the radical may have messenger and regulatory functions. Recently data have become available that low concentrations of free radicals are possibly involved in signaling activity. This is especially the case for the superoxide radical, which causes hyperpolarization of the plasmic membrane (Gamalei et al., 1998). There are also indications that this radical likely functions as an informational agent (Saran and Bors, 1990) or a secondary intermediate (messenger) implicated in the activation and proliferation of animal cells (Hancock, 1997). The basis for such a supposition is the fact that many animal cells produce very small amounts of superoxide radical (Heinecke et al., 1987; Steinbrecher, 1988). Like NO, it is supposed to play, in vivo, a messenger role (Gabbita et al., 2000). The radical interacts with soluble guanylate cyclase (either directly or indirectly, via the range of reactions with a participation of NO) to regulate the formation of the secondary messenger cyclic guanosyl monophosphate, which is itself a regulator of many intracellular processes (Saran and Bors, 1990).

Large amounts of the superoxide anion radical may cause enzyme inactivation, as it certainly does for catalase (Giabrielli, 1983), glutathione peroxidase (Fridovich, 1986) and acetylcholinesterase (Goldstein et al., 1972; Shinar et al.,1983). $\dot{O}_2^-$ participates in such reactions as lipid peroxidation, instigates membrane damage and causes DNA ruptures (Halliwell and Gutteridge, 1985).

In tissue culture, superoxide anion radical destroys plant and animal cells, and kills bacteria (Salin, 1987). Severe injuries may result that even lead to plant cell death (Overmyer et al.,2000; Rao and Davis, 2001). The accumulation of $\dot{O}_2^-$ is ethylene-dependent, and occurs before ozone induces any visual symptoms of leaf damage (Overmyer et al., 2000).

Table 7. The biological effects of the superoxide anion radical

Biological effects	*References*
Signaling and regulation of many intracellular processes	Saran and Bors, 1990; Zenkov and Men'shchikova, 1993; Hancock, 1997
Regulation of plant cell elongation	Shorning et al., 2000
Hyperpolarization of plasmic membrane	Gamalei et al., 1998
Detoxication of xenobiotics in cells	Minibayeva et al., 1998; Gordon et al., 1999
Induction of lipid peroxidation,	Halliwell and Gutteridge, 1985, Zenkov and Men'shchikova, 1993
Formation of $H\dot{O}_2$; H_2O_2; $\dot{O}H$ и O_2^1	Zenkov and Men'shchikova, 1993
Reduction of cytochrome C.	Zenkov and Men'shchikova, 1993
Stimulation of the chloroplasts' ATP synthesis in darkness	Tyszkiewisz and Roux, 1989
Membrane damage	Halliwell and Gutteridge, 1985,
Inactivation of many enzymes, in particular catalase and glutathione peroxidase, acetylcholinesterase	Giabrielli, 1983; Fridovich, 1986 Goldstein et al., 1972; Shinar et al.,1983
DNA damage	Halliwell and Gutteridge, 1985
Cell death	Overmyer et al.,2000

The superoxide radical additionally serves the apparently significant purpose of detoxifying xenobiotics in root cells (Minibayeva et al., 1998; Gordon et al., 1999). The contribution of electron transport chains located in plasmic membranes to xenobiotic formations on the roots of *Triticum* wheat seedlings is well established. Thus, the addition of substances non-penetrated plasmalemma, such as natural (NADH or NADPH) and artificial (ferricyanide) electron donors, to the medium in which the roots are grown enhances superoxide radical formation during 1 hour incubation. Moreover, the inhibitors of flavin enzymes located in plasmalemma, chlorpromasine and quinine, markedly decrease this effect, although they have no influence on the control sample to which electron donors were not added. The flavin component of the plasmalemma redox system is supposed to participate in the generation of superoxide radical.

In comparing the effects of ozone (Chapter 2) and superoxide anion radical, it is necessary to note the similarity of their activity, especially in high concentrations and, in particular, involving the initiation of lipid peroxidation as well as damages of proteins and nucleic acids. In not so high concentrations, the radical may be necessary for seedling development over the first 2-4 days that in wheat coleoptiles correlates with the period when DNA and protein content is increasing (Shorning et al., 2000). Only after 6-7 days, when superoxide anion radicals have sufficiently accumulated, arising of apoptotic DNA and protein desintegration have been observed (Shorning et al., 2000).

The possibility of there being some similarity between effects of O_3 and $\dot{O}_2^-$ at low concentrations should not be excluded. The effects of small ozone doses on plant cell development are considered in chapter 6.

Hydroxyl radical. The precursors of hydroxyl radical are superoxide anion radical and hydrogen peroxide. Hydrogen peroxide participates in Fenton reaction (see section 3.3.), which is the main source of hydroxyl radical. The reaction occurs according to the following scheme (Halliwell and Gutteridge, 1985; 1986):

$$H_2O_2 + Fe^{2+} \rightarrow Fe^{3+} + \dot{O}H + OH^-$$

The hydroxyl radical will be produced and initiate Haber- Weiss reaction (1932), so that hydroxylanion (OH^-) and hydroxyl radical ($H\dot{O}$) are then formed :

$$H_2O_2 + \dot{O}_2^- \Leftrightarrow O_2 + H\dot{O} + OH^-$$

The lifetime of hydroxyl radical in a cell is about 100 μs, and the distance over which it can pass from its place of formation to a target molecule cannot exceed 100 nm (Slater, 1976). Consequently, the locations of formation and reaction of the $\dot{O}H$ radical must be in direct affinity with each other. Hydroxyl radical reacts with all compounds present in biological systems. It is the strongest of all known oxidizers and can rupture C-H and C-C bonds. Due to this ability, hydroxyl radical has a significant effect on living cells (Table 8). It is, for instance, the main radical that

Table 8. Biological effects of hydroxyl radicals

Effect	*Reference*
Induction of lipid peroxidation	Zavigelskii and Paribok, 1971; Afanas'ev, 1984; Simic et al., 1989; Schreck et al., 1991
Rupture of any CH- bonds	Simic et al., 1989; Schreck et al., 1991
Damage of proteins and nucleic acids	Davies, 1987;Simic et al., 1989; Schreck et al., 1991
Mutagenic	Zenkov and Men'shchikova, 1993
Carcinogenic	Zenkov and Men'shchikova, 1993
Cytostatic	Zenkov and Men'shchikova, 1993
Inhibition of the chloroplasts' ATP synthesis	Tyszkiewisz and Roux, 1989

initiates lipid peroxidation, and represents one of the strongest cytostatics, mutagens and carcinogens. $\dot{O}H$ radicals can participate in reactions of three basic types (Larson, 1978):

1. Detachment of hydrogen atoms

$$RH_2 + \dot{O}H \rightarrow (\dot{R}H) + H_2O$$

2. Linkage to double bonds

$$H\dot{O} + R \rightarrow (\dot{R}OH)$$

3. Electron transfer

$$H\dot{O} + R \rightarrow \dot{R} + OH^-$$

The interaction of hydroxyl radicals with unsaturated fatty acids involves the first type of reaction, which is considered as one of the basic reactions for initiating lipid peroxidation in biological membranes. It is also a reaction implicated in the interaction of $\dot{O}H$ radical with ribose and desoxyribose in nucleic acids. Consequently, this interaction underlies the mutagenic action of hydroxyl radicals (Oberley et al., 1986). The second type of reaction is illustrated by the interaction of the radicals with the purine and pyrimidine bases of nucleic acids. The binding of an $\dot{O}H$ radical to the timing molecule can disrupt the complimentary of the bases in a DNA chain and, ultimately, lead to mutation or cell death (Youngman, 1984).

It is, however, also necessary to note that the biological role of the reactive species of oxygen and, in particular, hydroxyl radical is not just a negative one. In animals and plants, these radicals help to protect organisms against pathogenic microorganisms (Aver'yanov and Lapikova, 1988a, b).

Other oxygen radicals. The roles played by other radicals of oxygen (table 9) in the

Table 9. Biological effects of some oxygen radicals (adopted from Zenkov and Men'shchikova, 1993).

Radical	*Biological effects*
$H\dot{O}_2$	Induction of lipid peroxidation. Formation H_2O_2 and $\dot{O}_2^-$ in interactions with organic molecules. Cytotoxic effects
$R\dot{O}_2$	Interaction with unsaturated lipids, formation of $\dot{R}$ и ROOH. Recombination reaction is accompanied by biochemiluminescence.
$R\dot{O}$	Induction of lipid peroxidation. Cytostatic and carcinogen.

ozone effects on plants have not yet been sufficiently studied. Table 9 provides some information concerning living cells. It seems that these radicals induce lipid peroxidation and have cytotoxic features, just like the above-mentioned superoxide and hydroxyl radicals.

3.2.3. Mechanisms by which free radicals act on cellular components.

We shall consider some of significant possible mechanisms involved in free-radical activity. Among them are those that effect proteins, nucleic acids and lipids, and that result in cellular damage.

Oxidative modification of proteins. Radicals are able to induce the oxidative modification of proteins, nucleic acids, glucose amines, lipids and pigments. The first of these changes is the earliest indication of cell damage. The disulfide-, sulfhydryl-, carbonyl- and amino- groups in proteins provide sites for oxidation because their ability to capture electrons make them good acceptors of free radicals (Kayushin et al., 1976). The reactions proceed at very high rates, of which the K (the rate constants) are $\sim 10^{-9}$-10^{-10} $M^{-1}s^{-1}$. The value of such constants depends on the amino acid composition of the proteins. Any reactive oxygen species acts on proteins to effect a change in the structure of the protein molecule, along with its corresponding physico-chemical and biological properties (Dubinina and Shugalei, 1993). The character of the protein modification depends on the type of reactive oxygen species. Davies et al. (1987) studied 17 proteins (bovine serum albumin, catalase, hemoglobin, globins, casein, transferase and others) in order to determine the extent to which free radicals of oxygen caused aggregation, fragmentation and modification of their amino acid residues, as well as their sensibility to proteolysis induced by these radicals. In the experiments, hydroxyl radical ($\dot{O}H$) formed covalently-bonded protein aggregates, but there were only a few if any products of fragmentation found in the medium. Although tryptophan was removed from the modified proteins. Most of them underwent proteolysis 50 times faster than untreated proteins. Unlike hydroxyl radical, superoxide anion radical did not, in the study, induce either aggregation or fragmentation of proteins, nor did it result in the removal of tryptophan. With the simultaneous treatment of both hydroxyl radical and superoxide anion radical, protein fragmentation occurred, the molecule charge was changed, tryptophan was removed and bityrosine was formed. Some of the effects, although less expressed, were similar to the modifications occurring when $\dot{O}H$ alone was administered. Thus, proteins are very sensitive to the free radicals of oxygen. In some cases, the oxidative modification of proteins is accompanied by changes in their enzymic activity. The superoxide anion radical, hydroxyl radical and hydrogen peroxide can induce the inactivation of enzymes as a result of the modification of certain amino acid residues necessary to support the native properties of the enzymes (Archakov and Mokhosoev, 1989). For instance, superoxide inhibits, *in vitro*, the activity of catalase (Giabrielli, 1983), acetylcholinesterase (Goldstein et al., 1972; Shinar et al., 1983), glutatione peroxidase and other enzymes. These changes in their activity primarily are explainable by local disturbances in active centers, in particular by the oxidation of such amino acid residues as that of histidine, as well as changes in the valence of the metals included in the enzymes (Davies, 1987). There are, however, still other reasons. Exposure of the proteins hydroxyl radical, either alone or in combination with superoxide radical, induced changes in the primary protein structure and increased protein sensitivity to proteolysis (Davies et al., 1987). Oxidative modification also involved changes in the secondary and tertiary structure of the protein. Results of experiments with various chemical scavengers (traps) showed that all observed changes were attributable to the $\dot{O}H$- radical.

The toxic effects of the radicals were studied, mainly in connection with mechanisms for ionizing radiation in living organisms and involving processes that give rise to reactive oxygen species (Zavigelskii and Paribok, 1971; Afanas'ev, 1984; Simic et al., 1989; Schreck et al., 1991). At present, evidence has been accumulated which testifies to the fact that free radicals have a signaling function, and this is especially true of superoxide anion radical, which induces the hyperpolarization of the plasmic membrane (Gamaley et al., 1998).

Destruction of nucleic acids. The main free radical processes have mutagenic and lethal effects due to the destruction of nucleic acids. As in other cellular reactions, $\dot{O}H$ interacts with pyrimidine and purine fragments as well as with desoxyribose (Simic et al., 1989). The interaction of hydroxyl radical with a pyrimidine base leads to the rupture of the double C_5-C_6 bond in its ring, and yields three fragmented pyrimidine radicals at ratios of 56, 35 and 9%, the most important of which is thymine (Fig.30). The $\dot{O}H$ radical either removes the hydrogen from the C-H bond or includes itself in the double bonds. The detachment of hydrogen from methyl groups also occurs in thymine and 5-methylcytosine.

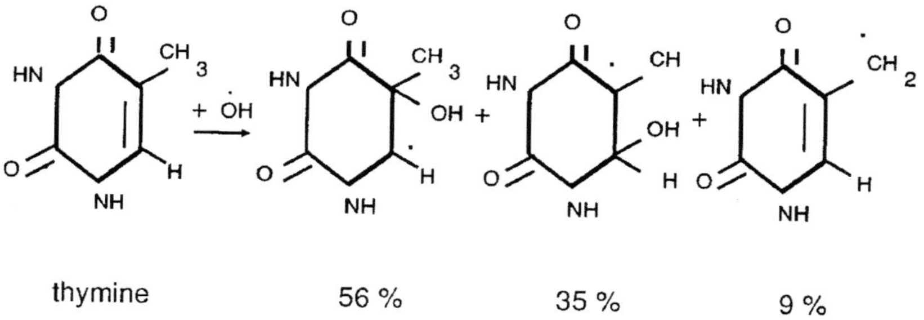

Figure 30. The reaction of the hydroxyl radical with thymine (Simic et al., 1989).

Similar reactions involving hydroxyl radical are observed in purine bases. For instance, guanine generates 4 isomeric products (Simic et al., 1989). Reactions of the $\dot{O}H$ radical with desoxyribose (dR) implicate the C-H bond. In this case 5 types of the desoxyribose radicals are formed •dR(-H):

$$dR + \dot{O}H \rightarrow \bullet dR(-H) + H_2O$$

These radicals are not equivalent, each having different properties (Simic et al., 1989). In some of them, the breaking of the bonds occurs either with or without participation of oxygen. In the first case, •dR(-H) radical reacts with oxygen, forming the peroxy radical $dR(-H)O\dot{O}$

$$\bullet dR(-H) + O_2 \rightarrow dR(-H)O\dot{O}$$

Subsequent reactions, involving peroxy radicals, lead to further breaks in the DNA-molecules. Superoxide anion radical can also attack nucleic acids, especially DNA. Thus, under the influence of ozone, formation of the above-mentioned free oxygen radicals and the subsequent destruction of chromosome are thought to occur.

Free radical oxidation of lipids. The most significant effect of free radicals is the initiation of the chain oxidation of lipids. This process involves a series of reactions, as a result of which unsaturated lipids are oxidized, forming short-chain hydrocarbon fragments. Among the initiators of lipid peroxidation are such reactive oxygen species as $\dot{O}_2^-$; $H\dot{O}_2$; $\dot{O}H$, as well as hydrogen peroxide, hydroperoxides ROOH, organic peroxides and epoxides. The main reactions of lipid peroxidation are considered in section 3.1. Here we present some details of this process involving the effects of free radicals. The most possible candidate for undertaking this initiation is hydroxyl radical ($H\dot{O}$), which, interacting with a molecule of unsaturated fatty acid (LH), forms the radical of this acid ($\dot{L}$)

$$LH + H\dot{O} \rightarrow H_2O + \dot{L}$$

Radical $\dot{L}$ forms lipodioxyl ($LO\dot{O}$) in interaction with oxygen.

$$\dot{L} + O_2 \rightarrow LO\dot{O}$$

Lipodioxyl, in its turn, reacts with a molecule of unsaturated fatty acid (LH) to form hydrodioxide (LOOH):

$$LO\dot{O} + LH \rightarrow \dot{L} + LOOH$$

Branching of the chain occurs in the hydrodioxide (LOOH) reactions. In the presence of Fe^{2+} ions or ultraviolet irradiation, lipoxyl ($L\dot{O}$) is produced, which creates a new oxidation chain by reacting with a new molecule of unsaturated fatty acid:

$$LOOH + Fe^{2+} \rightarrow Fe^{3+} + HO^{-} + L\dot{O}$$

$$L\dot{O} + LH \rightarrow LOH + \dot{L}$$

The precipice of the chain occurs as a result of a reaction between radicals, which leads to the simultaneous disruption of both free radical chains. This step involves either reactions between two $\dot{L}$ or two $LO\dot{O}$ radicals or between $\dot{L}$ and $LO\dot{O}$

$$LO\dot{O} + LO\dot{O} \rightarrow LOH + \text{alcohol (ketone)} + L = O + O_2 + \text{photon}$$

In estimating the biological effects of free radicals, it should be understood that they might be transformed into each other. Therefore, any view suggesting that only one participates in any effect or specific response is not realistic. Perhaps, all of the species contribute in the observed effects. Ozone can act, inducing oxidation or peroxidation of biomolecules directly or indirectly, through the formation of reactive oxygen species in free radical processes. The sequence of the events may include lipid peroxidation, loss of the functional activity of enzymes, changes in membrane permeability, cellular damage or even the cell death (Mustafa, 1990). Free radicals demonstrate a wide spectrum of the effects on living cells (Tables 7-9). Their toxicity could be neutralized by many antioxidants (see Chapter 4).

3.3. PEROXIDES AND HYDROPEROXIDES AS INTERMEDIATES IN THE PROCESS OF OZONOLYSIS

Ozone effects are assumed to involve the formation of hydrogen peroxide and organic peroxides. Peroxides and hydroperoxides are formed in reactions between ozone and unsaturated hydrocarbons, which may occur both in atmospheric air and in the apoplasts of plant vegetative cells, or on the surface of microspores (pollen and vegetative microspores). Other sources of peroxides can be lipid peroxidation, initiated by the free radicals formed from ozone interaction in living cells.

Peroxides have $-\overset{|}{\underset{|}{C}}-O-O-C-$ groups, whereas hydroperoxides have $-\overset{|}{\underset{|}{C}}-O-O-H-$ hydrogen links with peroxy group. Upon ozonolysis, hydrogen peroxide and organic peroxides will be formed (see section 3.3).

3.3.1. Hydrogen peroxide

Hydrogen peroxide, the most stable among the reactive oxygen species, is constantly present in biological systems, as it is formed in many processes, especially the ozonolysis of water (see section 1.4.). It is not a radical, i.e. does not have any unpaired electrons. The formation of hydrogen peroxide and the reactions involving it are closely associated with the free radical oxygen chain. The stability and absence of any electrical charge allow H_2O_2 molecules to penetrate biological membranes, a factor that distinguishes it from hydroxyl radical $H\dot{O}$, hydroperoxy radical $H\dot{O}_2$, and superoxide anion radical $\dot{O}_2^-$. For similar reasons, hydrogen peroxide can travel significant distances from the place of its formation and diffuse into a cell.

Hydrogen peroxide will be formed in any biological system in which the superoxide radical appears, in particular due to mitochondrial and chloroplast metabolic activity (Gabbita et al., 2000). For instance, water deficiency increases the endogenous content of hydrogen peroxide in *Vigna* seedlings (Mukherjee and Choudhuri, 1983). As a result of the dismutation of $\dot{O}_2^-$, hydrogen peroxide and singlet oxygen are formed.

$$\dot{O}_2^- + \dot{O}_2^- + 2H^+ \rightarrow H_2O_2 + O_2$$

H_2O_2 can be formed in reactions of ozone with unsaturated hydrocarbons, according to the following scheme (Stachelin and Hoigne, 1982):

$$\text{Unsaturated hydrocarbons} + O_3 + H_2O \rightarrow \text{oxidized product} + H_2O_2$$

species contribute to the observed effects to different degrees. Ozone can directly act to induce the oxidation or peroxidation of biomolecules, or do so indirectly by forming reactive oxygen species in free radical processes. The sequence of these events may include lipid peroxidation, loss of the functional activity of enzymes, changes in membrane permeability, cellular damage or even cell death (Mustafa, 1990). Free radicals demonstrate a wide spectrum of effects on living cells (Tables 7-9), but their toxicity can be neutralized by many antioxidants (see chapter 4).

In a molecule of hydrogen peroxide, the -O-O- bond is fragile; therefore the molecule easily turns into hydroxyl radical (see section 1.4). Reactions between superoxide radical and hydrogen peroxide will also from hydroxyl radicals:

$$\dot{O}_2^- + H_2O_2 \rightarrow \dot{O}H + OH^- + O_2$$

The toxicity of H_2O_2 is only connected to its ability to form hydroxyl radicals, which induce extreme chemical activity.

Like superoxide radical, hydrogen peroxide can work both as an oxidizer and a reducer. It is capable of oxidizing such compounds as thiols, in which reduced glutathione (GSH) is transformed into an oxidized form (GSSG):

$$2GSH + H_2O_2 \rightarrow GSSG + 2H_2O$$

The removal of both superoxide radicals and hydrogen peroxide from biological systems prevents the spontaneous formation of hydroxyl radical, which cannot, itself, be removed by any specific enzymic activity.

As a source of $\dot{O}H$ radicals, the biological toxicity of H_2O_2 has been well known for a long time, along with the main mode of its action—the oxidation of SH-groups (Jocelin, 1972). In the presence of metals, this toxic capacity increases, also possibly due to the hydroxyl radical formation (section 3.2). Hydrogen peroxide also causes protein to form aggregates and, resultantly, cross-bridges between molecules. However, toxicity is observed only under the non-physiological conditions involving high concentrations of H_2O_2 (Fridovich, 1976). In living cells, the concentration of hydrogen peroxide is around 10^{-5} M (Mueller et al., 1997). Moreover, it is possible that, instead of water, this compound may be a donor of electrons in oxygenic photosynthesis, as doses of $5x10^{-7}$ M and $5x10^{-5}$ M stimulate the development of maize seedlings 3 and 5 times, respectively (Apasheva and Komissarov, 1996). Although, as shown in some experiments (Samuilov, 1999), autotrophic growth of cyanobacteria was inhibited by the addition of exogenous H_2O_2 in a wide range of the concentrations. Role of hydrogen peroxide in living systems , perhaps, is a perspective field for the future research.

The presence of hydrogen peroxide as a generator of hydroxyl radicals in a medium can be determined by using a spectrophometric method in the ultra violet region of the spectrum or through various colorimetric reactions, controllable in the presence of catalase or peroxidase (Hildebrandt et al., 1978). The chemical method of H_2O_2 detection could be based also on its ability to oxidize epinephrine (adrenaline) into adrenochrome, a process that can be measured by the fluorimetric method (Fridovich, 1979). Chemiluminescence with luminol is also used to assay of H_2O_2 and other organic peroxides are present in biological tissues (Haugland, 1996), but this reaction was not found to be applicable to the vegetative microspores of horsetail and the generative male (pollen) microspores of other plants (Roshchina et al., 2003).

3.3.2. Organic peroxides and hydroperoxides in atmospheric air

Rainwater and atmospheric air, as were shown by means of high-performance gas chromatography (Hellpointner and Gäb, 1989), contain hydrogen peroxide (H_2O_2) and other organic peroxides, such as hydroxymethyl peroxide ($HOCH_2OOH$), hydroxyethyl peroxide ($CH_3CH(OH)OOH$) and methyl peroxide (CH_3OOH). Rainwater contains stable amounts of the compounds, respectively 5.3-63; 0.2-0.8;

0.1-0.3 and 0.1-0.5 μmoles/L, whereas the volumes in the aerozole are -453-914; 4-15; 0 and 40-179 parts per 10^{12}. Higher concentrations in the air were characteristic only of H_2O_2 and CH_3OOH, smaller concentrations were detected for $HOCH_2OOH$, and hydroxyethyl peroxide was completely absent. According to Melhorn et al. (1990), the amount of organic peroxides in rainwater were found to be 0.01-50 μmoles/L or 0.24 - 1200 ppm.

Organic peroxides will be frequently formed in reactions involving ozonized hydrocarbons and organic radicals (Larson, 1978). They will also be formed when O_3 reacts with unsaturated alkyl compounds. For instance, the formation of methyl hydroperoxide from methane, an elementary hydrocarbon, is represented by Hanst and Gau (1983) as follows. In the beginning, a reaction with the hydroxyl radical causes the hydrocarbon to lose hydrogen and to form an alkyl radical $\dot{C}H_3$:

$$CH_4 + OH \rightarrow CH_3 + H_2O$$

Then, the alkyl radical, interacting with oxygen, forms an alkylperoxide radical:

$$\dot{C}H_3 + O_2 \rightarrow \dot{C}H_3O_2.$$

The alkyl peroxide radical reacts with a hydroperoxy radical to produce methyl hydroperoxide:

$$\dot{C}H_3O_2 + H\dot{O}_2 \rightarrow CH_3OOH + O_2$$

Hydroethyl peroxide is formed from ethyl peroxide in a similar way, as the ethyl peroxide radical $\dot{C}_2H_5O_2$ reacts with a hydroperoxide radical

$$\dot{C}_2H_5O_2. + H\dot{O}_2. \rightarrow C_2H_5OOH + O_2$$

As a result, hydroethyl peroxide and oxygen are formed.

3.3.3. Formation of peroxides and hydroperoxides in reactions between ozone and the gas excretions of plants

Living plant organisms release volatile compounds of various structures—from very simple to extremely complex. Went (1960) has calculated that the biosphere liberates approximately $2x10^8$ tons of terpene-like or weakly oxidized hydrocarbons per year. In addition, similar plant excretions are found to contain various hydrocarbons, such as ethylene, propylene, isobutylene, isoprene and others. The reactions of these (mainly unsaturated) hydrocarbons with ozone result in the formation of peroxy radicals, which convert, in subsequent reactions, to such oxygen-containing compounds as peroxides, alcohols, ketones, acids, etc.

Experiments with nine alkenes (ollefins)—ethylene, propylene, 1-butene, isoprene, *cis*-2-butene, α- and β-pinenes, 2-carene, and limonene—were performed in both laboratory and field conditions (Hewitt and Kok, 1991). In special chambers of the laboratory, they reacted with ozone in the presence of surplus water. The quantities of the organic peroxides in gas and aerosol factions were determined by means of high-performance liquid chromatography. In the reaction medium, which included ethylene, propylene, 1-butene, and isoprene, a main product of the ozonation was hydroxymethyl hydroperoxide. Hydrogen peroxide was not formed in these conditions. Conversely, in the cases, involving α- and β-pinenes, 2-carene, and limonene, hydrogen peroxide and methyl hydroperoxide were the prevailing products of the reactions.

In 1989, parallelly with the laboratory experiments, the organic peroxides were found among the gas components of air by Colorado state researchers conducted field tests (Hewitt and Kok, 1991). In the natural atmosphere, the concentration of H_2O_2 was approximately 0.5 parts per billion (or 0.0005 ppm). The maximum concentration of hydroxymethyl hydroperoxide was about the same, but the amount of methyl hydroperoxide was much lower. In additoin, 3 non-identified organic hydroperoxides were also detected. Other authors (Gäb et al.,1985), reporting on experiments with various alkenes and ozone, detected the presence of bis(hydroxymethyl)peroxide. The most significant quantity (0.64 - 0.66 mmoles of this peroxide) was observed to result from the ozonolysis of isoprene and ethylene and, to a smaller degree (0.06-0.2 mmoles), from that of α- and β-pinenes, limonene and 2-carene. Except for this main product of the ozonolysis, only small amounts of hydroxymethyl peroxide were found in the reaction medium. Moreover, other products that were detected included formic acid, formaldehyde, carbonyl oxide, the ozonide of ethene (ethylene) and hydroxymethyl hydroperoxide .

According to Gäb et al. (1985), the formation of hydroxymethyl peroxide and bis(hydroxymethyl) peroxide is possible in the following reactions:

1. In the ground-level layer of the air, a source of hydroxymethyl peroxide is likely the reaction of carbonyl oxide with water:

$$CH_2OO + H_2O \rightarrow HOCH_2OOH \quad (1)$$

2. Hydroxymethyl peroxide, reacting with formaldehyde, forms bis(hydroxymethyl) peroxide

$$HOCH_2OOH + CH_2O \rightarrow HOCH_2OOCH_2OH \quad (2)$$

3. Hydroxymethyl hydroperoxide will also be formed via a reaction of formaldehyde with hydrogen peroxide

$$CH_2O + H_2O_2 \rightarrow HOCH_2OOH \quad (3)$$

The carbonyl oxide (CH_2OO) in reaction 1 and the formaldehyde in reactions 2 and 3 can be formed in reactions of alkenes with ozone

$$CH_2{=}CR_2 + O_3 \rightarrow CH_2OO + R_2CO$$

$$CH_2{=}CR_2 + O_3 \rightarrow R_2COO + CH_2O$$

3.3.4. Organic peroxides in plant leaves

Organic peroxides, produced in reactions between ozone and hydrocarbons, are found in the leaves of plants, and their subcellular localization is also shown (Pellinen et al., 1999). By using high-performance liquid chromatography, it was possible to separate and to identify 9 organic peroxides and hydrogen peroxide in foliage of plants exposed to ozone (Hewitt et al., 1990a; Salter and Hewitt, 1992).

Experiments with *Eschscholtzia californica* (Hewitt et al., 1990a), have shown that plants in a pure or open air environment do not contain any pracitical levels of peroxides. But plants exposed to ozone (0.120 ppm.) possess quantities of hydroxymethyl hydroperoxide ($HOCH_2OOH$), hydroxyethyl hydroperoxide ($CH_3CHOHOOH$), methyl hydroperoxide (CH_3OOH) and two other hydroperoxides still to be identified. The same authors received similar results when the velvet bean *Mucuna* sp was used as the experimental subject.

Peroxides will be formed in plants that mainly release ethylene, isoprene and monoterpenes. Possible mechanisms for forming peroxides from the interactions of ozone with isoprene and ethylene have been considered by Salter and Hewitt (1992):

Hydroxymethyl hydroperoxide will result from ozonolysis in which carbonyl oxide is formed and, subsequently, reacts with water:

$$-\overset{|}{C}=\overset{|}{C} + O_3 \rightarrow -\overset{|}{C}=O + -\dot{C}H_2O\dot{O} \xrightarrow[H_2O]{} \overset{OH}{\overset{|}{C}}H_2OOH$$

Hydroxymethyl hydroperoxide

or

$$CH_2=\overset{CH_3}{\overset{|}{C}}\text{-}CH=CH_2+O_3 \rightarrow O=\overset{CH_3}{\overset{|}{C}}\text{-}CH=CH_2+\dot{C}H_2O\dot{O} \xrightarrow[H_2O]{} \overset{OH}{\overset{|}{C}}H_2OOH$$

Hydroxymethyl hydroperoxide

In order to know, whether hydroperoxides will be formed inside or outside of a leaf of velvet beans *Mucuna* sp. exposed in ozone, Hewitt et al. (1990a) have done the original experiments. All plants of this species release isoprene. From old leaves (sixth circle), this gas is excreted in larger amounts, whereas younger leaves (fourth circle) have a rate of release that is 50-100 times less. The rate of isoprene emission has correlated with formation of peroxides in amounts of 0.042 to 620 ng/g of dry weight. In young leaves they are absent, but in old ones they are always found.

3.3.5. Biological effects of peroxides

Plant cell response to ozone may also include effects of peroxides themselves. In addition to being formed by ozone, peroxides are usual products of cellular metabolism. They are universal raw materials, which may be used by the organism itself for the most diverse purposes (Baraboi et al., 1992). Lipid peroxidation is necessary for normal physiological processes, and peroxides are also the common products of the normal metabolism of cells. They are intermediates in the synthesis of such biologically active substances as prostaglandins and steroid hormones. Under normal conditions, the concentration of the main peroxide found in the living organism, hydrogen peroxide, is no more than 10^{-5}M. Experiments with peroxides have demonstrated its various effects. The main cellular processes sensitive to them are listed in table 10. Depending on their concentrations, peroxides nay serve as secondary messengers, inductors of various syntheses, as agents that cause damage, especially to chromosomes, so that mutations result. Although only the accumulated amounts of peroxide can induce any visible negative effects, the responses to low doses of peroxides have now also been studied.

Ozone appears to have phytotoxic effects due to the formation of hydrogen peroxide and organic peroxides (Hewitt et al., 1990a,b). In regions where loss of acidic deposits is observed, such effects are especially significant because there is the increased stability of hydroperoxides in an acidic medium. In these conditions, woody species, releasing such reactive alkenes as isoprene and monoterpenes, are mainly damaged. For example, the red fir tree *Picea rubra* releases isoprene (Rasmussen, 1978; Zimmerman et al., 1978), and the ordinary fir tree *Picea excelsa* excretes both these substances (Corkill, 1988).

It should, however, be noted that peroxides, toxic for organisms, are not only the collateral and undesirable products of oxidation (Table 10). The formation of peroxides of many natural plant compounds has been observed in vivo, for instance, reactions, involving sesquiterpene lactones, produce hydroendoperoxides (Adekenov and Kagarlitskii, 1990). Due to the process of lipid peroxidation, hydroperoxides will be formed in plants growing under conditions in which concentrations of ozone are increasing (Hewitt et al., 1990 a). It has been suggested that free radicals will initially be formed, which are then transformed into peroxides (Melhorn et al., 1990). Hydrogen peroxide has been implicated as a second messenger of hormone-stimulated metabolic changes in some animal cells (Gabbita et al., 2000). Its possible role in transduction of the defense signal in plants was also examined (Apostol et al., 1989).

Table 10. Effects of peroxides on living organisms

Process	*Low concentrations* $< 10^{-4}M$	*High concentrations* $>10^{-4}M$
Development of sunflower seedlings	Stimulation (Apasheva and Komissarov, 1996)	
Elongation of maize coleoptile cells		Decrease (Schopfer, 1996)
Autofluorescence of cell walls		Accumulation of fluorescent insoluble material (Schopfer, 1996)
Red autofluorescence of vegetative horsetail microspores	Stimulation (Roshchina et al., 2003)	
Pollen germination	Stimulation (Roshchina and Melnikova, 1998; 2001)	Depression (Roshchina and Melnikova, 2001)
Pollen and pistil fluorescence	Changes in the spectra of fluorescence	Decrease (Roshchina and Melnikova, 1998a,b;2001)
Signal transduction estimated on the fluorescence of dye pyranine	Changes in the fluorescence (Apostol et al., 1989)	
Potassium ion channel opening		Activation (Kuo et al.,1993)
Calcium ion channel opening		Change (Krippeit Drews et al., 1995
Membrane potential		Change (Krippeit Drews et al., 1995
Enzyme's activity (plant peroxidases, dehydrogenases, etc)		Inhibition (Keutsch, 1971; Marklund, 1973)
Transcription of gene information	Activation (Schreck et al, 1991,1994;Gamaley and Klyubin, 1998)	
Mutations		Induction (Mallant et al., 1986; 1987)
Chromosome state		Damage
Synthesis of DNA		Decrease (Keutsch, 1971)
Synthesis of protective agents against pathogens	Induction	Induction (Chiron et al., 200a,b)
Phytoalexin (glyceollin) production	Stimulation (Apostol et al., 1989)	

In normal living cells, average concentrations of peroxides in lipids are very small—about 10^{-9}M/mg of lipids, but these amounts of peroxide substances are also biologically active. The possibility cannot be excluded that, under the influence of ozone, the formation of low concentrations of peroxides have a positive effect. Some examples of the biological activity of these compounds relevant to their concentration will be considered below. At low (10^{-9}–10^{-5} M) concentrations of hydrogen peroxide, stimulating effects are frequently observed. Micromolar concentrations of hydroperoxides can stimulate respiration by 3 fold (Sukacheva et al., 1999), a finding that illustrates their possible role in the regulation of energetic processes. In small doses, hydrogen peroxide may act as a signaling substance (Gamaley and Klyubin, 1996), an initiator of transcription activation (Schreck et al, 1991,1994), a secondary intracellular messenger (Gamaley and Klyubin, 1996), or a trigger molecule for the products of free radical reactions (Gutteridge., 1995).

The development of sunflower seedlings has be stimulated by treatment with 10^{-6} -10^{-5} M hydrogen peroxide (Apasheva and Komissarov, 1996). 100-200 μM/L concentrations of hydrogen peroxide increases the rate of the somatic embryogenesis of *Lycium barbarum*, whereas increasing the concentration to 300 μM/L has the inverse effect (Kaizong et al., 1999). The stimulative effect of low concentration of this peroxide and *tert*-butylhydroperoxide has been established in experiments by Roshchina and Melnikova (2001) on the germination of pollen in an artificial nutrient medium (Fig. 31). In concentrations $>10^{-4}$ M, suppression of the process is observed. The participation of peroxides in initiating pollen-tube generation can be shown by adding two enzymes, peroxidase and catalase, that decompose the peroxides in the medium (Table 11). In millimolar concentrations, reactive oxygen

Table 11. The effects of catalase and peroxidase on the fluorescence and germination of Hippeastrum hybridum pollen after 2 hours exposure (Roshchina and Melnikova, 1998a)

Enzyme	*Autofluorescence (% of control)*	*The germination index after 2 hours (% of control)*
Catalase		
0.1mg/ml	100	100
0.3 mg/ml	90	67
0.6 mg/ml	81	38
Peroxidase		
0.1 mg/ml	105	88
0.3 mg/ml	102	57
0.6 mg/ml	51	9.6

species, including hydrogen peroxide, activate intracellular calcium production (Krippeit Drews et al., 1995). The chloroplast fructose bisphosphatase of *Spinacia oleracea*, which is related to the thiols, is inhibited by millimolar concentrations of hydrogen peroxide (Charles and Haliwell, 1980). Hydrogen peroxide accumulation in *Medicago truncatula* roots induced a formation of colonii by the arbuscular mycorhiza-forming fungus *Glomus intrara* (Salzer et al., 1999). It is furthermore

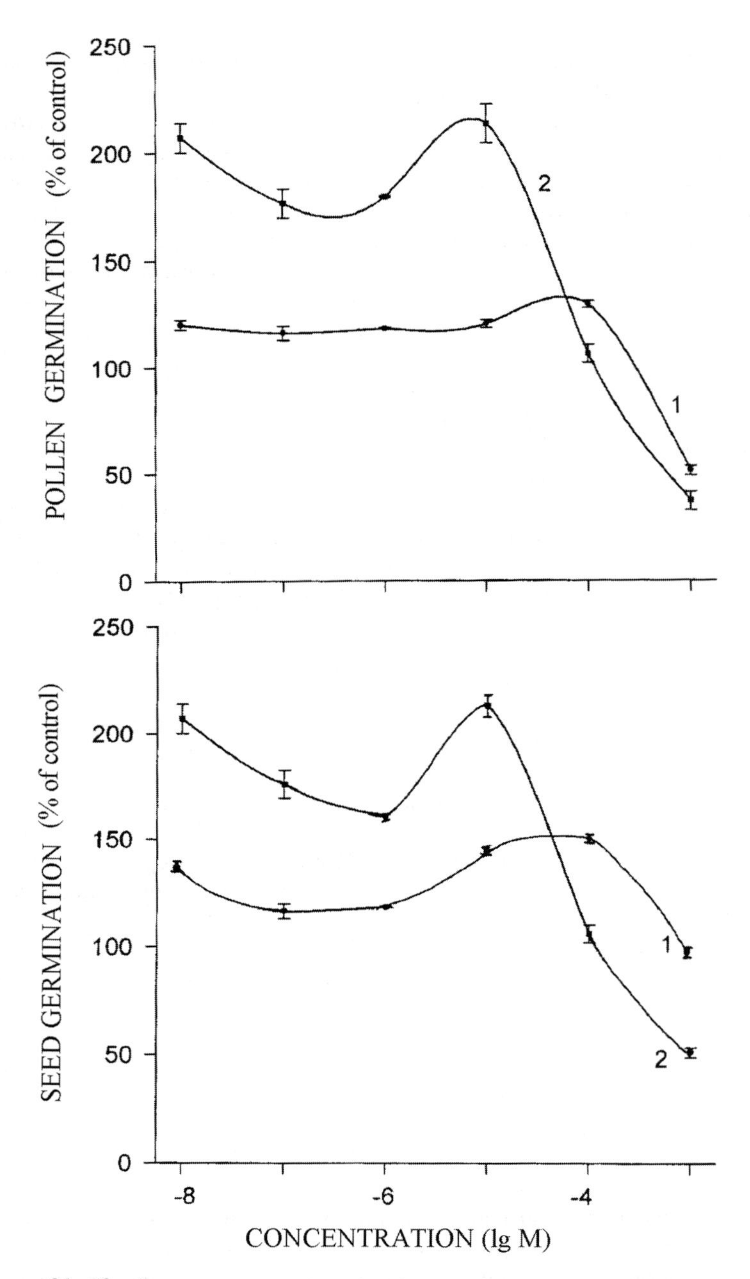

Figure 31. The dose-response curves for the germination of the pollen from Hippeastrum hybridum treated with hydrogen peroxide (1) or tert- butylhydroperoxide (2). (Roshchina and Melnikova, 2001).

known that inhibition of catalase, which destroys hydrogen peroxide, causes interruption of seed rest (Hendricks and Taylorson, 1975).

In our experiments on the treatment of the pistil stigma of *Hippeastrum hybridum* by peroxidase 1-6 mg/ml, seed yield was diminished (Roshchina and Melnikova, 1998a,b). The enzymes decreased the fluorescence of pollen, along with the total rate of pollen germination after two hours. Therefore, the role of peroxides as triggers of pollen tube germination would seemed to be confirmed.

The germination of vegetative microspores of horsetail *Equisetum arvense* was also sensitive to peroxides (Roshchina et al.,2003). 10^{-7}-10^{-5}M hydrogen peroxide or *tert*-butylperoxide stimulated this process, and peroxidase (1-2 mg/ml) depressed it.

There are, additionally, important results from research into toxically high (10^{-3}-10^{-2} M) concentrations of peroxides. For instance, peroxides (10^{-3}M) induced the accumulation of specific agents and protected plants against pathogens. Accordingly, treatment of the leaves and hypocotyls of the soya *Glycine max* with hydroperoxides (*tert*-butylhydroperoxide, hydroperoxide of linoleic acid and hydrogen peroxide), all of which will result from activated lipid peroxidation in membranes under stress, produced phytoalexin glyceollin, a compound belonging to the group of isoflavonoids (Montillet and Dedousee, 1991; Degousee, et al., 1994). Mutagenic effects of millimolar concentrations of hydrogen peroxide on plants are also well known (Mallant et al., 1986; 1987). Hydroxymethyl hydroperoxide (0.86-2.0 mM) is a mutagen and inhibits many enzymes (Keutsch 1971; Marklund, 1973). For example, irreversible inactivation of plant peroxidases occurs in vitro upon treatment with H_2O_2 (Marklund, 1973).

Any change in the peroxide quantities depends on many factors. For instance, amounts of H_2O_2 formed under the influence of ozone treatment can be approximately estimated. The concentration of this peroxide may be assessed in terms of the Criegee reaction (1975), in which one molecule of ozone produced one molecule of peroxide by interacting with one double bond of organic substance. Thus, a dose of ozone of 2.4 ppm or 9.84 x 10^{-8}M, causing a complete blockade of the pollen germination in the investigated species, should cause a formation of 9.84 x 10^{-8}M of a peroxide (Roshchina and Melnikova, 2001). However, at this concentration, peroxides act as stimulants of pollen germination (Fig. 31). The highest dose of ozone that induced distinct changes in the fluorescence of the pollen surface (5 ppm: see fig.13), should produce only 2.05 x 10^{-7}M/l of hydrogen peroxide. These results indicate that there is no direct correlation between ozone-induced effects and peroxide production.

The accumulation of high and therefore toxic concentrations of peroxides depends on certain conditions. The formation of high peroxide concentrations under the influence of ozone on and within living cells is probably connected to (1) high doses of ozone created in controllable or natural conditions, (2) the activity of enzymes maintaining peroxide balance, (3) lipid peroxidation. In the second case, peroxide accumulation depends on the content of the decomposing enzymes, peroxidase and catalase, that destroy any peroxide and only hydrogen peroxide, relatively, as well as superoxide dismutase, which catalyzes the production of H_2O_2 in response to superoxide radical dismutation (see section 3.4). For instance, it has been demonstrated that the addition of peroxidase to the pollen of various species caused significant changes in the autofluorescence of its surface and the rate of pollen tube formation (Table 11). These data correlate well with information about the quantities of peroxidases maintaining the balance of hydrogen peroxide in pollen cells (Stanley and Linskens, 1974). The third source of peroxides is represented by the reactions of lipid peroxidation (Hewitt et al., 1990; Gutteridge, 1995).

3.3.6. Mechanisms of the peroxide action

The mechanisms of peroxide action are, apparently, diverse. The activation of cells by low concentrations of peroxides is considered as a signaling reaction, associated with the redox properties of these substances (Gamaley and Klubin, 1996). It is assumed that, in this case, a molecule of peroxide interacts with a sensor (receptor) located either on the surface of a cell (if the peroxide is secreted from one cell and is perceived by another) or on the surfaces of intracellular organelles (if the peroxide is formed inside a cell and enters into interactions with other cellular components). It is unknown how peroxides link with such sensors, perhaps via the oxidation of their sulfhydryl groups. The transfer of the information received from a signaling molecule, such as peroxide, into a cell seem to occur by means of secondary messengers, which are, primarily, calcium ions. However, any non-receptive mechanism of signaling by peroxides, for instance intracellular signal transmission via activation of the redox molecules in the cytoplasm and organelles, is also possible. In this case hydrogen peroxide is considered as a secondary messenger (Gamaley and Kyubin, 1996).

The mechanism of peroxide action in high concentrations is also connected with the redox-characteristics of these compounds. Probably, they initially participate in redox reactions. One of the effects of 10^{-6}–10^{-4} M hydrogen peroxide and *tert*-butylhydroperoxide on the redox-components of pollen grains involves changes in the fluorescence spectra (Fig. 32). The maximum, peculiar to carotenoids, at 545-560 nm, seems to dissappear (Roshchina and Melnikova, 2001). In the fluorescence spectrum of the pollen from *Passiflora coerulea*, the intensity of the emission in the blue region (maximum 450 nm) is reduced, and the maximum at 550-560 nm disappears but, after 24 hours, is again restored. The disappearance of a maximum at 550-560 nm was also observed for the pollen of mock-orange *Philadelphus grandiflorus* and the yellow-day lily *Hemerocallis fulva,* enriched in carotenoids. Redox reactions with peroxides can be reversible. They are associated with reactions of the known components of sporopollenin. Fig.33 shows how hydrogen peroxide and *tert*-butylhydroperoxide act on the fluorescence of the individual components in the pollen cover. Enriched in carotenoids, the pollen of *Passiflora* and *Philadelphus* manifest changes in the orange spectral region at 590-600 nm, where β-carotene fluoresces. A novel peak at 460 nm was observed after hydrogen peroxide treatment. In the pollen of *Hippeastrum hybridum*, lacking carotenoids but containing phenols, a maximum at 560-600 nm appeared due to redox reactions involving the peroxides of such phenolic compounds as esculetin,

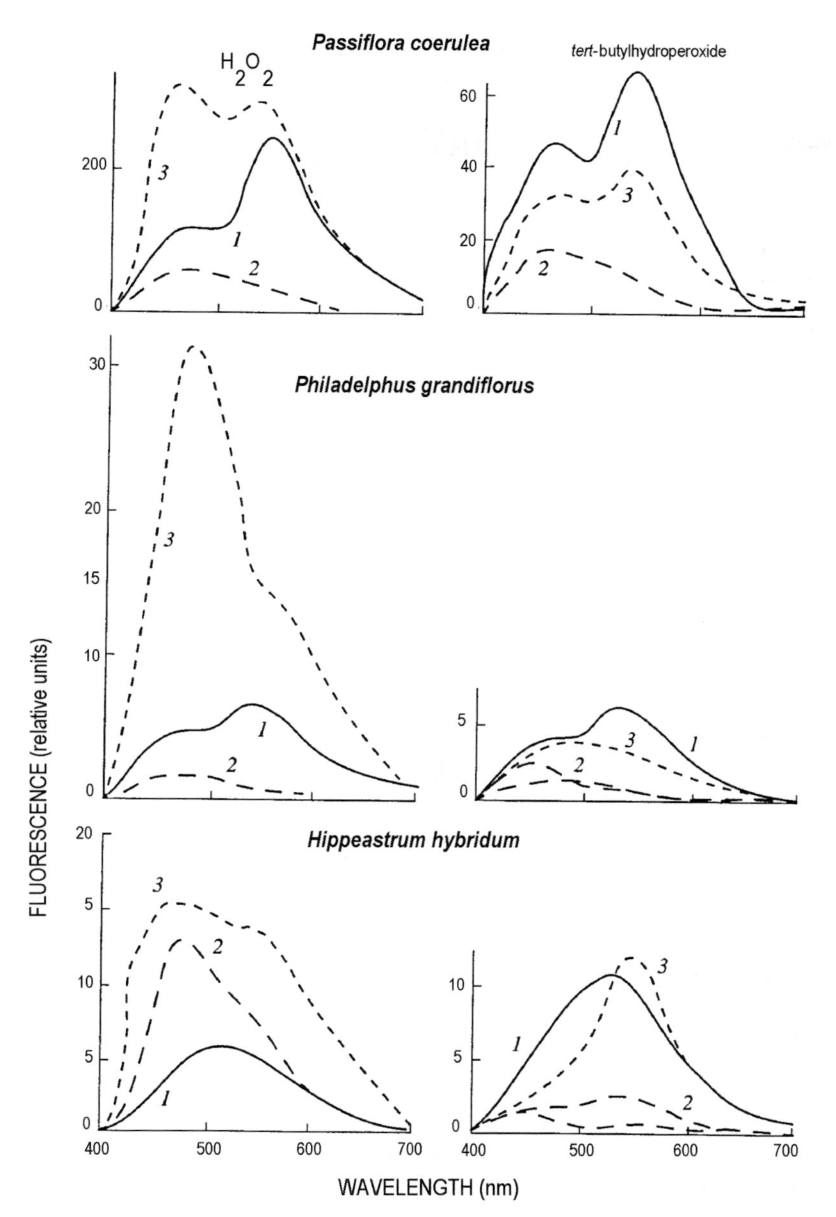

Figure 32. The fluorescence spectra of pollen without treatment (1) and after treatment with10^{-4} M hydrogen peroxide or tert - butylhydroperoxide just after 2-5 min (2) and after 24 hours (3) (Roshchina and Melnikova ,2001).

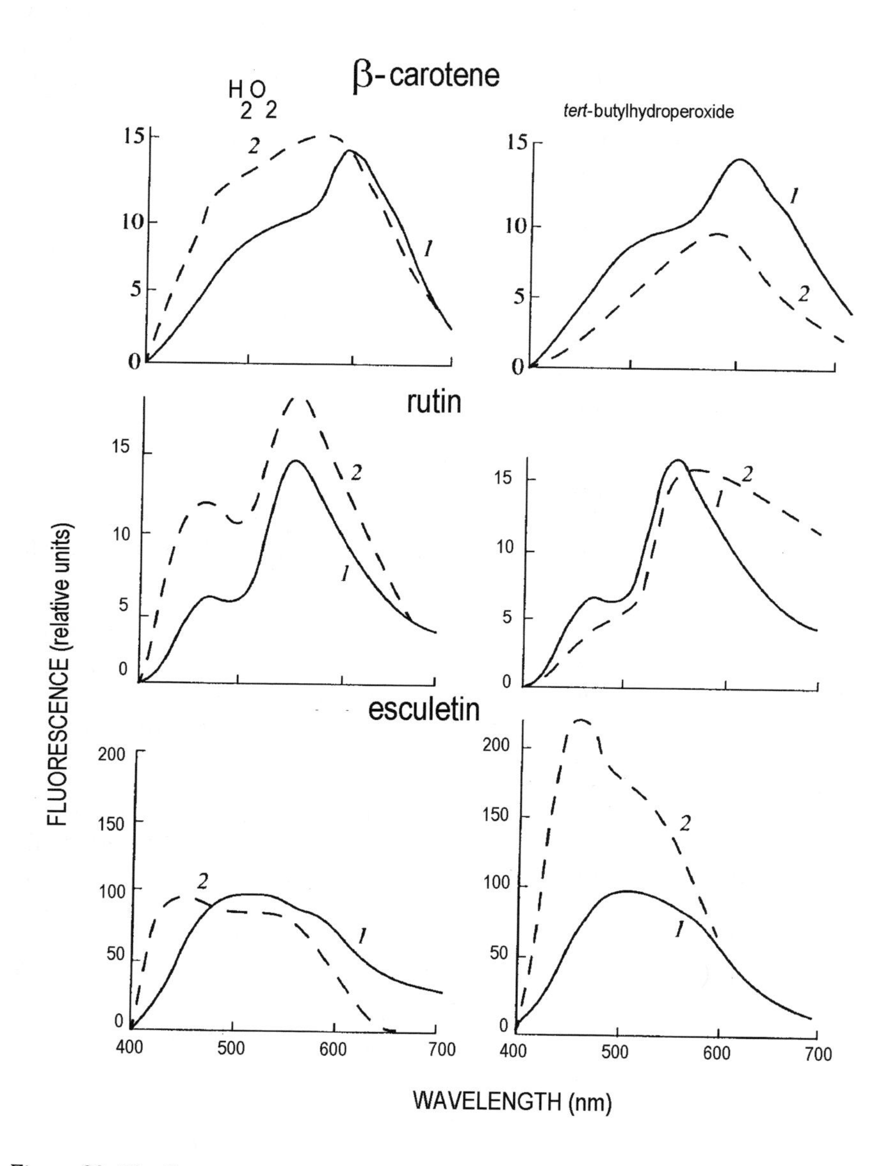

Figure 33. The fluorescence spectra of crystalline pollen constituents treated with peroxides without treatment (1) and after treatment with 10^{-4} M hydrogen peroxide or tert - butylhydroperoxide just after 2-5 min (2) and after 24 hours (3) (Roshchina and Melnikova ,2001).

(Roshchina and Melnikova, 2001), which are also found in sporopollenin (Stanley and Linskens, rutin and quercetin (Roshchina, 1974). A response common to both the peroxides used was noted in the case of esculetin: a novel peak at 440-460 nm was especially pronounced after treatment with organic peroxide. In other cases, hydrogen peroxide and *tert*-butylhydroperoxide exerted different effects. For example, rutin treated with *tert*-butylhydroperoxide displayed an additional fluorescence in the red region. The fluorescence spectra of other tested compounds, such as azulene, menthol, and riboflavin, were not affected by the peroxides (Roshchina and Melnikova, 2001).

In plasmalemma, redox components, such as the b-types of cytochromes, flavins, quinones, NAD(P)H-dependent oxidase, and the salts of iron, participate in the formation of hydrogen peroxide and superoxide anion radical from NAD(P)H and O_2 (Dahse et al., 1989; Rubinstein and Luster, 1993). Thus, at high concentrations of hydrogen peroxide exogenously formed from ozone, formation of endogenous H_2O_2 is inhibited and changes in the state of the redox chain of the plasmalemma also occur. Perhaps, they influence both the chemosensory reactions of the plasmalemma, as well as the subsequent intracellular responses.

Other targets of peroxide action on plant substances can be proteins and nucleic acids. For instance, the thiol-linked fructose bisphosphatase is inhibited by millimolar concentrations of hydrogen peroxide (Charles and Halliwell , 1980). In this case, H_2O_2 may act directly on the enzyme.

The signaling function of small doses of peroxides is connected with their indirect effect: reception on the cell surface or on the surface of the organelles inside the cell (see above). In high concentrations, their action is probably direct. It is thought that the effect of millimolar concentrations of organic peroxides is, first of all, displayed as enzyme inactivation. For instance, hydroxymethyl hydroperoxide causes irreversible deactivation of isolated plant peroxidases in vitro (Marklund, 1973), but the same thing also takes place in the intact organism. This peroxide attacks the heme group of enzymes, so that this ring is opened. Bis(hydroxymethyl) peroxide inhibits some enzymes that have sulfhydryl groups SH, especially peroxidase, from being irreversibly inactivated. In experiments on the cells of the root system of *Vicia faba*, bis(hydroxymethyl) peroxide was found to damage chromosomes (Keunsch, 1971). This peroxide acted as a cytostatic, interrupting mitosis approximately in the G_2-phase. The rate of DNA synthesis decreased during the S-phase in comparison with other periods of the mitotic cycle. Alkyl hydroperoxides are believed to oxidize the nitrogenous bases of DNA (Ball et al.,1992). As follows from the above-mentioned information, the realization of this or any other mechanism of peroxide action depends on its concentration within and without the cell.

The ozone effects on plant cells can be either direct or mediated by peroxides. When the latter's concentrations are low, stimulation of some processes is observed (Table 10). In high concentrations, there is no direct correlation between the toxic influence of O_3 and the peroxides which may be accumulated (see section 3.3.5). High concentrations of peroxides $>10^{-4}$M usually suppress plant reactions only if catalase and peroxidase are blocked (see chapter 4).

CONCLUSION

The mechanisms of ozone action on plant cells are both direct and indirect, and depend on the concentration of O_3. Direct interaction of ozone with cellular components at small concentrations is possible for the surface ingradients of the cell wall, excretions and plasmalemma (lipids, proteins, phenols, terpenoids, etc). High concentrations of this gas permit it to enter the cell by damaging plasmic membrane, as well as to affect intracellular ingradients (even nucleic acids). Indirect action of ozone consists in the ozonolysis of cellular components and/or the water in the cell so that such active products as ozonides result. The formation of free radicals and peroxides, including such reactive oxygen species as superoxide anion radical, hydroxyl radical and hydrogen peroxide, also contributes to the ozone effects. In this case, small concentrations of the products of ozonolysis and reactive oxygen species may play a role as chemosignals outside of plasmalemma, inducing a cascade of secondary messenger reactions within cells. At high concentrations, these O_3 derivatives also damage cellular components. The main targets of both ozone itself and the products of ozonolysis are the double bonds of lipids and any secondary metabolites, as well as the sulfhydryl groups of proteins. As a result of their interaction with lipids, lipid peroxidation occurs, and forms products that, in their turn, influence other cellular components. Interactions with proteins and amine-containing groups in amino acids lead to changes in the activities of enzymes, including their deactivation and the formation of fluorescing products.

CHAPTER 4

PROTECTIVE CELLULAR REACTIONS AGAINST OZONE AND STEADY-STATE PROTECTIVE SYSTEMS.

Each cell requires protection against the oxidizing stress caused by ozone, as well as by the oxygen radicals $\dot{O}_2^-$, $\dot{O}H$ and peroxides formed by the ozonation of living cells. Cellular excretions and the surface of the cell wall constitute the first line in the cell's defense against ozone itself. In response to oxidation by ozone or any of its derivatives, a cell will mobilize its protective antioxidant system, to which two groups of substances belong—high-molecular antioxidant enzymes and low-molecular antioxidants (Kangasjarvi et al., 1994; Larson, 1997; Langebartels et al., 2000). They may be located in plant excretions as well as on the cellular surface, in the extracellular space or inside of the cell itself. These antioxidant enzymes are: superoxide dismutase, catalase and peroxidase, including glutathione peroxidase along with its coupled enzyme glutathione reductase. Low-molecular antioxidants such as ascorbic acid, tocoferol, thiol (SH) compounds and other substances, interact with radicals and peroxides to form weak reactive products. The active oxygen species that may be scavenged or of which the formation can be inhibited by the main high and low-molecular antioxidants are listed in table 12. Both groups of antioxidants (high and low-molecular) will be considered in this chapter. Special attention will be given to plant excretions containing compounds with antioxidant features, due to both their enzyme activity (high-molecular antioxidants) and their double bonds (high and low-molecular antioxidants). These substances are named antiozonants because they interact with ozone and the intermideates of ozonolysis, completely entrapping them at low concentrations of O_3.

4.1. ANTIOXIDANT ENZYMES.

Antioxidant enzyme systems include superoxide dismutase, peroxidase and catalase, all of which are catalyzers of the reactions that eliminate reactive oxygen species. Glutathione reductase can also participate in the process.

4.1.1. Superoxide dismutase (SOD) in plants

Superoxide dismutases are considered to be the mechanism that protects plants against superoxide anion radical (Bowler et al., 1992; van Camp et al., 1994a). In plants, three types of superoxide dismutases are found, all of which can be distinguished by the structure of their active centers: superoxide dismutase contains

Table 12. Reactive oxygen species, their scavengers or the inhibitors of the processes of their formation

Reactive oxygen species	*Scavengers or inhibitors*
Ozone (O_3)	Most substances with double bonds (phenols, terpenoids, etc), located on the plant surface may scavenge this gas.
Superoxide anion radical ($\dot{O}_2^-$)	←Mn-SOD, Cu, Zn-SOD ←ubiquinone ←ascorbic acid
Perhydroxyl radical ($H\dot{O}_2$)	←ascorbic acid ←uric acid ←ubiquinone ←α-tocoferol
Hydrogen peroxide (H_2O_2)	←catalase ←peroxidase
Hydroxyl radical ($\dot{O}H$)	←uric acid ←ascorbic acid ←α-tocoferol ←uracyl
Singlet oxygen (O_2^1)	←ascorbic acid ←histidine ←carotenoids ←uric acid ←α-tocoferol
Peroxy radical ($R\dot{O}_2$)	←ascorbic acid ←uric acid ←ubiquinone ←α-tocoferol
Alkoxy radical ($R\dot{O}$)	←α-tocoferol ←uric acid ←ubiquinone ←ascorbic acid

SOD – superoxide dismutase

copper and zinc (CuZn-SOD), manganese (Mn- SOD) and iron (Fe- SOD) in its active center. The basic distinctive feature of the different types of superoxide dismutase—the sensitivity to cyanides—is used to identify the presence of these enzymes in crude extracts.

CuZn-SOD, which is sensitive to cyanides, can be found in nuclei, the cytoplasmic matrix, peroxisomes, and the intercellular space (Asada et al., 1980). Extracellularly active Cu-Zn-superoxide dismutase is also contained in the needles of the pine *Pinus sylvestris* (Streller and Wingsle, 1994). Mn-SOD, which is resistant to cyanides, in the mitochondria of eukaryotes, and in bacteria. The cyanide-resistant forms of Fe- SOD are present in bacteria, protozoa and plants belonging to the Ginkgoaceae, Cruciferae and Nymphaeaceae families (Asada et al.1980). In chloroplasts, there are two classes of SOD: the Cu-Zn and Mn forms (Hausladen and Alscher, 1993).

The synthesis and activity of superoxide dismutase depends on many external factors. For example, ozone fumigation of leaves of the tobacco *Nicotiana plumbagifolia* induces the synthesis of SOD and causes an increase in its activity (Camp et al., 1994 a, b). In iron deficient leaves of the grapes *Vitis* and *Lupinus luteus*, the activity of SOD increases by 20-30 % (Ostrovskaya et al., 1990). The NADH-dependent superoxide dismutase located on the plasmic membrane is activated immediately after a pathogen or elicitor enters the plants of the Solanaceae, Fabaceae and Graminae families (Doke et al., 1994). Many other factors have similar effects.

All superoxide dismutases catalyze the reaction of dismutation at approximately identical rates. As a result, hydrogen peroxide and oxygen will be formed:

$$\dot{O}_2^- + \dot{O}_2^- + 2H^+ \xrightarrow{SOD} H_2O_2 + O_2$$

The reaction consists in the transportation of an electron from one superoxide radical to another and proceeds in two stages. In cases in which an CuZn- SOD copper atom is included in the active center of the enzyme, the dismutase serves as an intermediate acceptor of the electron. Schematically, the process can be represented as follows (Vladimirov et al., 1991):

$$SOD\text{-}Cu^{2+} + \dot{O}_2^- \rightarrow SOD\text{-}Cu^{+} + O_2$$

$$SOD\text{-}Cu^{+} + \dot{O}_2^- + 2H^+ \rightarrow SOD\text{-}Cu^{2+} + H_2O_2$$

Zn is contained in the active center of the enzyme but does not participate in the catalytic cycle. H_2O_2 , formed in this reaction, is a strong oxidant itself. Its quantities are reduced by the activities of specific antioxidant enzymes—such as catalase—which catalyzes the reaction transforming hydrogen peroxide into water and oxygen.

$$2H_2O_2 \xrightarrow{catalase} 2H_2O + O_2$$

or non-specific enzymes, such as peroxidase (See section 3.4).

Superoxide dismutase, which scavenges superoxide, inhibits the reactions in which this radical participates. The protective function of the enzyme is connected not only with the elimination of the superoxide radical itself. A superoxide anion radical can, in the presence of metals (Me^{n+1}) and especially the 3-valent iron (Fe^{3+}), interact with $2H_2O_2$ to form a hydroxyl radical. In this case, the superoxide dismutase prevents the Haber-Weiss cycle from forming the extremely toxic hydroxyl radical:

$$H_2O_2 + \dot{O}_2^- \xrightarrow[(Me^{n+1})]{(Fe^{3+}).} \dot{O}H + OH^- + O_2$$

Hydroxyl radical ($\dot{O}H$) is very reactive and serves as a generator of free radicals. Although specific scavengers of this radical are not known, it is clear that superoxide dismutase and catalase do indeed suppress $\dot{O}H$ formation or, at least, decrease its concentration (Fridovich, 1981; Rabinovich and Fridovich, 1983).

The role played by superoxide dismutase and other related enzymes, such as peroxidase and catalase, in protecting plants against oxidizing stress and, in particular, against the toxic $\dot{O}_2^-$ radical has been widely discussed in the literature (for a survey, see Bennett et al., 1984). The effects of exogenous superoxide dismutase on various processes (section 3.2, Figs.27 and 28) serve as a marker for the activity of the superoxide anion radical. The accumulation of endogenous enzyme depends on many factors, for instance, deficiency of microelements. Although a manganese deficiency does not seem to reduce the contents and activity of Mn-SOD upon O_3. However, there is a possibility that SOD accumulation is not related directly to ozone-induced stress.

Levels of intracellular superoxide dismutases are under genetic control. Increases in the concentrations of hydrogen peroxide or superoxide anion radical are accompanied by the activation of the transcription that initiates the synthesis of superoxide dismutase and catalase. Ozone stimulated the activity of the enzymes-scavengers of free radicals (Fukunagu et al., 1992).

Recently, transgenic plants were used to study the processes of ozone resistance involving accelerated formation of superoxide dismutase. In an experiment with such plants of the tobacco *Nicotiana tabacum*, the increase in enzyme activity had a small effect on plant tolerance of ozone, whereas its superproduction in chloroplasts correlated with 3-4- times higher than normal plant stability (Camp et al., 1994b). Thus, after 4-7 days of exposure to ozone (6-8 ppm), no leaf damage was observed. However, earlier work (Pitcher et al., 1991) indicated that above-normal levels of superoxide dismutase did not significantly raise plant tolerance to ozone. In these experiments on the superproduction of the enzyme, a surplus of hydrogen peroxide appears to have been formed. According to the Haber-Weiss reaction (1932; see above), this extra H_2O_2 also yields surplus hydroxyl radical, which can, as is well known, react with DNA, proteins and lipids. Hence, one consequence of enhanced levels of superoxide dismutase in transgenic plants is the incomplete removal of H_2O_2 and a subsequent accumulation of hydroxyl radicals. Increased quantities of superoxide dismutase in plants would, by itself, be insufficient to lower ozone toxicity.

Kurtikara and Talbot (1990) assumed that organisms respond to the stress caused by free radicals by using signaling mechanisms to induce the production of enzymes providing appropriate protection. Exposure to ozone causes reactive oxygen species to be formed, and these, in their turn, induce the formation of antioxidants via trigger mechanisms like those in the leaf tissues of the tomato *Lycopersicon esculentum*. After 4 exposures to ozone, the amounts of such antioxidant enzymes as superoxide dismutase and glutathione-SH-reductase increased up to 200 %.

4.1.2. Catalase.

Hydrogen peroxide is toxic for a living cell and, consequently, needs to be eliminated. The catalase enzyme, which is found in all plant and animal tissues, performs the task by splitting this peroxide (Saunders, 1978; Scandalios, 1994). Catalase protects against the accumulation of H_2O_2, which can damage various cellular components. The enzyme catalyzes the reaction:

$$H_2O_2 + H_2O_2 \rightarrow 2\ H_2O + O_2$$

It can also react non-specifically with lipid hydroperoxides. Catalase is structurally and functionally similar to peroxidase. It also belongs to the heme-protein group of enzymes.

In plants, catalase is found in the peroxisomes-organelles, which are enclosed by an elementary bilayer membrane, and in glyoxisomes (Miroshnichenko, 1992; Men'shchikova and Zenkov, 1993). Glyoxysomes represent a form of peroxisomes that contain enzymes involved in the lipid metabolism. In an overwhelming number of plant species, multiple forms of catalase have been detected, although for some species, for instance the string bean *Phaseolus vulgaris* and the lentil *Lens Adans* only a single form has been identified. The generation of hydrogen peroxide mainly occurs in the plant cell wall, the nuclear and plasmic membranes, as well as in the membranes of chloroplasts, microsomes and peroxisomes. Only an insignificant amount of the compound is formed in the cytozole. If the peroxide is formed outside of the peroxisomes, it diffuses into a peroxisome, where it is decomposed by catalase.

4.1.3. Peroxidase

Many types of peroxides can be decomposed by peroxidase, which is widespread in natural organisms, and has been studied in such detail and from such a variety of perspectives as is the case for only a few other enzymes. All plant tissues contain both acidic (pH 3.5 –5.0) and alkaline (pH 7.5 –9.5) isoforms of peroxidase. Acidic isoforms have higher specific activity. The quantitative content of peroxidases depends on the plant species, the nature of the tissue, and also the influence of various external factors. All peroxidases contain protoheme in the prostetic group and appear to have almost identical catalytic activity. However, their physico-chemical properties differ somewhat from each other. Many reactions, catalyzing peroxidase, also involve the formation of heavily colored compounds used to identify the enzyme activity (Saunders, 1978).

In cells, ascorbate-dependent peroxidases are found in isolated chloroplasts and soluble supernatant (Gillham and Dodge, 1986), but additionally in the cell wall, cytoplasm, vacuole, and, mainly, tonoplast (Gazaryan, 1992; Andreeva, 1988).

Depending on the functions that they serve, plant peroxidases can be divided into two groups:

1. Peroxidases (especially ascorbate peroxidase) that decompose peroxides. They are formed upon ozone stress (Creissen et al., 1994) from a substrate (AH_2) that is oxidized at the expense of the atomic oxygen in hydrogen peroxide

$$AH_2 + H_2O_2 \leftrightarrow A + 2\ H_2O$$

The substrate (AH_2) of peroxidase can be comprised of aromatic amines, cytochrome C, ascorbate, indoleamines, phenols (which have strong oxidizing action when oxidized into quinones), and other compounds

2. Peroxidases that form free radicals in a process of lignification. Two of their main functions involve assembly of the cell wall and its restoration after wound damage to the cell. For this task, they are transported from the cytoplasm, where they were synthesized, to the free space of the cell (Heath and Castillo, 1988).

The structure of the cell wall includes lignin, which is a three-spaced polymer of various chemical compounds. The transformations in some of them occur with the participation of peroxidase (Halliwell and Gutteridge, 1985), as shown in fig.34.

The precursors of lignin are phenylalanine and tyrosine. These aromatic amino acids are deaminated and, then, transform, via a number of intermediates, into aromatic alcohols, which serve as a substrate for the peroxidases located in cell wall. Peroxidases of the cell wall and hydrogen peroxide, forming at the cell surface, participate in the oxidation of oxycinnamic alcohols that to produce appropriate phenoxy radicals. The latter participate in the process of polymerization involved in lignin formation (Westermark.,1982).

Deactivation and destruction of organic peroxides may also be performed by the glutathione peroxidase enzyme, which catalyzes the reaction in which glutathione (GSH) is oxidized by hydrogen peroxide

$$H_2O_2 + 2\ GSH \rightarrow 2\ H_2O + G\text{-}S\text{-}S\text{-}G \text{ or}$$

$$2ROOH + 2\ GSH \rightarrow ROH + H_2O + G\text{-}S\text{-}S\text{-}G$$

Oxidizable glutathione is again restored as a result of the related activity of another enzyme: glutathione reductase. After exposing the tomato *Lycopersicon esculentum* to ozone, the level of glutathione-SH-reductase increased up to 200 % and the level of glutathione up to 400% (Kurtikara and Talbot, 1990)

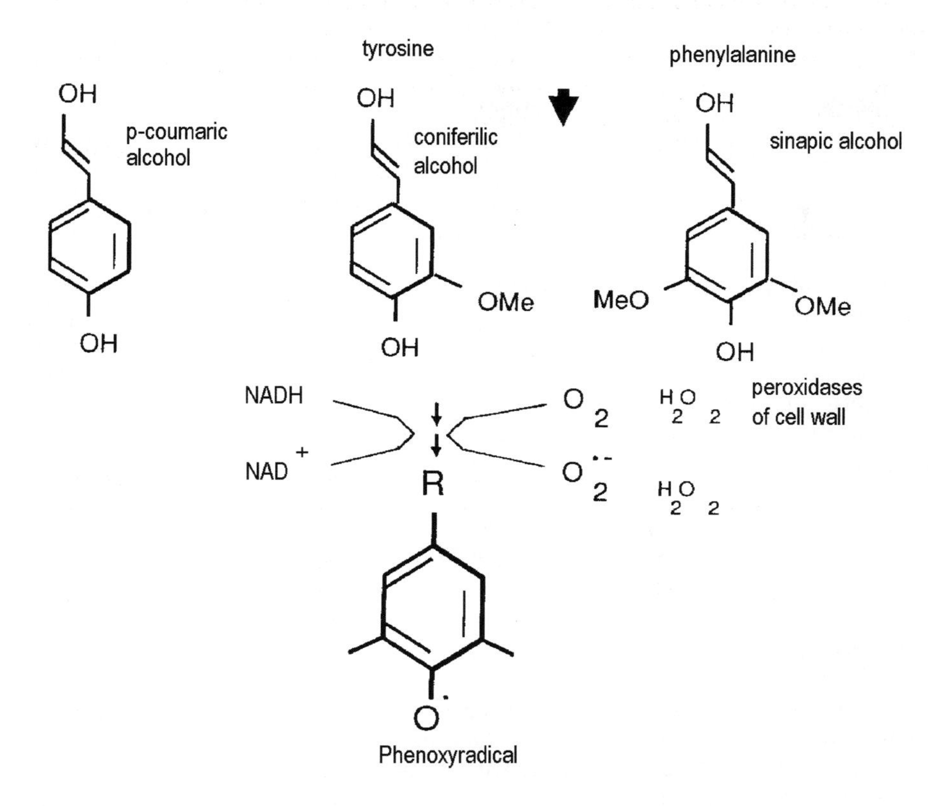

Figure 34. Scheme for the formation of the lignin precursor— phenoxy radical

Unlike the majority of peroxidases this enzyme is not a heme-protein, but contains an atom of selenium, connected with a peptide of molecular weight of approximately 22 000 Da. In many organisms, glutathione peroxidase is a primary means of preventing the accumulation of H_2O_2 and organic peroxides in a cell

Besides the above-mentioned high-molecular enzyme-antioxidants, the proteins containing Fe and Cu are also capable of protecting cells against oxidative stress. They can also act as catalysts of free radical processes outside cell because the enzymes poorly penetrate through membrane and tissue barriers. Chasov et al. (2002) have thought that the activity of the cell surface peroxidase itself correlates with the generation of superoxide anion in wounded wheat root cells.

The effects of exogenous peroxidase on various processes serve to indicate the contribution that peroxides make in plant cells (see Fig.28 in section 3.2. and section 3.3).

4.2 LOW-MOLECULAR ANTIOXIDANTS AND INHIBITORS OF REACTIVE OXYGEN SPECIES.

Low molecular substances, having small molecular mass and constant rates of interaction with free radicals, can additionally serve as protectors against the damaging action of reactive oxygen species. Plants synthesize a huge quantity of various compounds, including secondary metabolites. Many of them have antioxidizing ability. Only a few will be mentioned here, and their antioxidizing properties discussed in greater detail. These compounds include multiatomic alcohols (glucose, ribose and others), phenols (α-tocoferol, ubiquinones, flavonoids, phenolcarbonic acids, etc), ascorbic acid, some amino acids, polyamines, terpenoids (mono-, di-, triterpenes, sesquiterpene lactones), as well as a majority of plant pigments of various chemical natures, such as carotenoids, flavonoids, phenolcarbonic acids, etc (Table 13). They are all easily oxidized in the presence of hydroxyl radicals (Vladimirov et al., 1991). The reaction proceeds by forming the less toxic radicals of alcohols, which can attach to a molecule O_2 with the subsequent formation of aldehydes and other products of oxidation. The substances, serving as antioxidants, also carry out antioxidant functions in organism alongside the other functions specific to them.

4.2.1. Phenols

Insofar as antioxidant functions are concerned, phenols are the most active because they have an aromatic ring in their structure; these include mono- and polyphenols, naphthols, oxyderivatives of other aromatic compounds, and others. On the whole, phenolic antioxidants effectively inhibit the production of superoxide anion radicals, hydroxyl radicals, peroxy radicals, and the processes of lipid peroxidation (POL) initiated by the active oxygen species. The main active component of the antioxidant substances, giving them the capacity to retard the oxidation processes involving radicals, is a hydroxyl group, bound to the aromatic nucleus (Zhou and Zheng, 1991). The rather easy estrangement of the hydrogen atom from the hydroxyl leads to the formation of various isomers of pheno radicals. Grouped among the phenols are a lot of substances: simple phenols, phenolic acids, flavonoids, quinones, naphthoquinones, stilbenes, tannins, coumarins and others. Among the phenols, such flavonols as quercetin often participate in the detoxification of superoxide radicals, but their contribution in this process is not too great. Redox reactions involved in the interaction of ozone with aromatic compounds like flavonoid rutin and coumarin esculetin are detectable by their fluorescence—one of the basic physical parameters of a living cell (Roshchina and Melnikova, 2001). The appearance of new maxima, the displacement of maxima in the fluorescence spectra or the quenching of light emissions can serve as indicators of the fluorescing products generated by ozonolysis (Bernanose and Rene, 1959).

Antioxidant activity is also a feature of many hydroquinones in general. Their activity in plants treated with ozone was confirmed by measuring the protoplasts of the rice *Oryza* (Yoshida et al., 1994 b). Hydroquinones decreased the rate of lipid peroxidation but did not influence ATPase activity.

Table 13. Low-molecular antioxidants and mechanisms of their action

Antioxidant	*Oxidant of which the effect is inhibited*	*Defense Mechanism*
Phenols (flavonoids, naphthols, phenolcarbonic acids, polyphenols, etc)	$R\dot{O}_2$, $\dot{O}H$ and $\dot{O}_2^-$.	Formation of phenoxy radicals, less toxic
Tocoferol	$R\dot{O}_2$, $\dot{O}H$, $\dot{O}_2^-$ and $R\dot{O}$	Formation of complexes with the double bonds in the fatty acid residues of lipids
Ubiquinones	$R\dot{O}_2$, $\dot{O}H$, $\dot{O}_2^-$ and $R\dot{O}$	Linkage with $\dot{O}H$ and $\dot{O}_2^-$ to prevent cleavage of double bonds during lipid peroxidation
Ascorbic acid	$H\dot{O}_2$; $\dot{O}_2^-$, $\dot{O}H$, $R\dot{O}$, H_2O_2; O_2^1	Donation of electron to $\dot{O}_2^-$, $\dot{O}H$ and $H\dot{O}_2$, forming less toxic products
Thiol (SH) compounds, in particular glutathione	$R\dot{O}_2$, $\dot{O}H$ and $\dot{O}_2^-$.	Donation of proton, leads to the oxidation of $\dot{O}H$ and $H\dot{O}_2$, forming less toxic products and $2GSH + H_2O_2 \rightarrow$ $G\text{-}S\text{-}S\text{-}G + 2H_2O$
Uric acid	$\dot{O}H$ and $R\dot{O}$	Linkage with $\dot{O}H$, forming less toxic products, to prevent lipid peroxidation
Multiatomic alcohols (glucose, ribose, mannitol)	$\dot{O}H$	Oxidation, formation of less toxic alcohols
Terpenoids Carotenoids	$\dot{O}H$ and $\dot{O}_2^-$	Linkage with $\dot{O}H$ and $\dot{O}_2^-$ to prevent cleavage of double bonds

Some phenolic phytohormones also have antioxidant properties. Siegel (1962) established that compounds having indole rings, such as IAA (indoleacetic acid), indole, and tryptophan, are antioxidants. They protected the seedlings of the cucumber *Cucumis sativus* against oxidants. IAA has even been shown to decrease levels of lipid peroxidation. In the experiment exposing the etiolated coleoptiles of

Avena (Dhindsa et al., 1984) to a 10^{-6} M concentration, even for 1 minute there was a decrease in the amount of malondialdehyde (an indicator of suppression in lipid peroxidation; see also the reaction of IAA with ozone in section 3.1.).

In recent years, the antioxidant properties of salicylic acid, when subjected to ozone stress, have been studied (Sharma et al., 1996; Rao and Davis, 1999). This substance is a common component of the genus *Salix* and, perhaps due to the property of its taxon, is very tolerant of air pollution and pathogens. Salicylate may, in fact, scavenge superoxide anion radicals. In some cases, this substance may be served as a producer of the radicals that was seen in tobacco cell suspension culture (Kawano and Muto, 2000). Nevertheless, common view relates salicylates to antioxidants.

4.2.2. Tocoferol and ubiquinones

At present, several thousand similar compounds are known as antioxidants, among which the most pronounced antioxidant action is observed from α-tocoferol (vitamin E) and various ubiquinones. Almost the entire pool of tocoferol is connected with cellular membranes, presumably with a hydrocarbon part in unsaturated fatty acids (Men'shchikova and Zenkov, 1993). In chloroplasts, this antioxidant is located in the thylakoid membranes (Hausladen and Alscher, 1993).

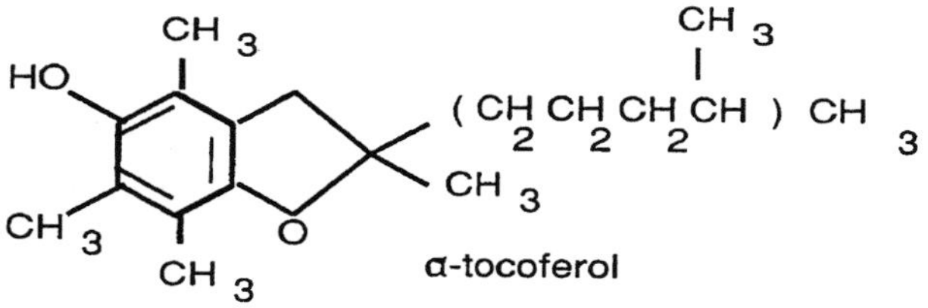

HO, CH_3, CH_3, CH_3, O, CH_3, $(CH_2CH_2CH_2CH(CH_3))CH_3$

α-tocoferol

CH_3O, CH_3O, O, O, CH_3, $(CH_2CH=C(CH_3)CH_2)_nH$

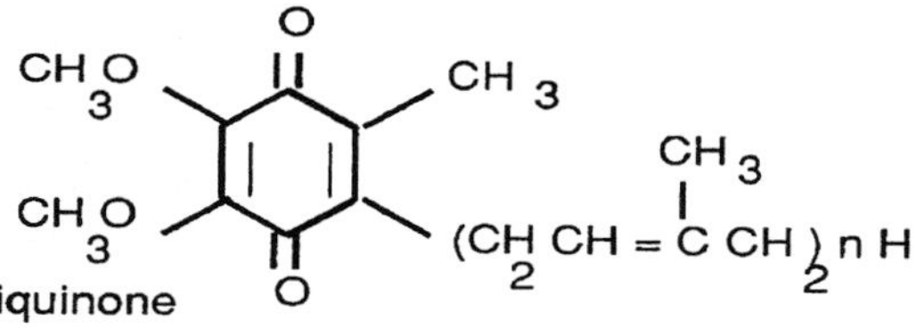

ubiquinone

Ubiquinones play a smaller role in the antioxidant protection of membranes than tocoferol does. The term "ubiquinones" is used to designate substances having the same quinone nucleus as tocoferol but differing in the number of the isoprene residues in the composition of its lateral lipophilic chain. Ubiquinones are, however, the main antioxidant in mitochondria. They effectively inhibit the formation of superoxide, hydroxyl, peroxy and alkoxy radicals.

Like tocoferols, ubiquinones stabilize membranes. Their molecules contain long-chain lateral branches comprised of 16-50 carbon atoms. Being built as lateral chains between unsaturated fatty acids of the phospholipids in membranes, these substances will form complexes with the double bonds of fatty acids. The density of the phospholipid packing in membranes is increased, and this prevents the formation of peroxy radicals from lipids and reduces the total rate of the lipid oxidation.

4.2.3. Ascorbic acid

Ascorbic acid (ascorbate) is γ-lactone. In plant cells, it is located in the cytoplasm and the vacuole. Ascorbate is also found outside of the plasmalemma in the apoplast (Luwe et al., 1993). However, 30-40 % of the total protoplast pool of ascorbic acid is located in chloroplasts (Foyer et al., 1983; Feyer, 1993).

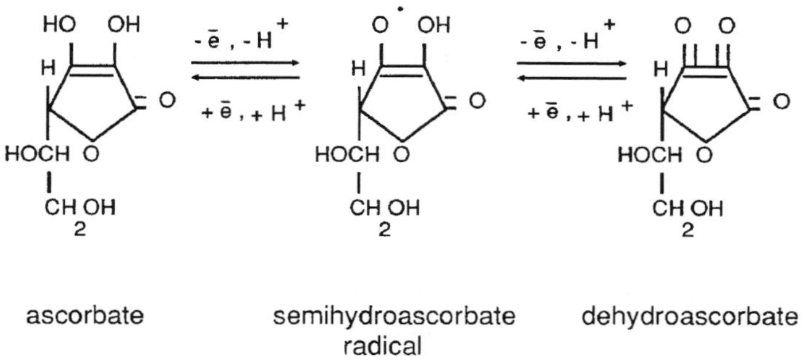

ascorbate ⇌ semihydroascorbate radical ⇌ dehydroascorbate

Figure 35. Transformation of the reduced form of ascorbate into the oxidized form

Ascorbic acid protects cellular components against oxygen radicals by acting (Fig. 35) as a reducing agent (a donor of electrons). Ascorbate quickly reacts with O_2^-, $H\dot{O}_2$, and $\dot{O}H$ by giving them electrons. The semihydroascorbate radical is formed, which can be transformed into dehydroascorbate (Halliwell and Gutteridge, 1985). In the cells of plant and animals tissues, there are mechanisms for transforming the oxidized form of ascorbate back into the reduced one (Fig. 35). The enzyme dehydroascorbate reductase can, for example, help to perform this task.

Ascorbic acid is better able to protect lipids against peroxidation than other antioxidants (Chameides, 1989). The protective effect stems from the fact that it can be oxidized into products—intermediary radicals and molecules—that are less active than the toxic hydroxyl radical (Vladimirov et al., 1991). The stimulative effects of ascorbic acid are well-seen as changes in the autofluorescence of intact pollen and its germination on an artificial nutrient medium (Roshchina and Melnikova, 1998a).

The effects of ascorbic acid on an ozonated plant were investigated in a study of the protoplasts of the rice *Oryza* (Yoshida et al., 1994 a). A decrease in the rate of malondialdehyde formation—an indicator of lipid peroxidation—and the stabilization of ATPase activity, which is 2 times lower in the absence of ascorbate, were found. An enhanced level of the antioxidant in *Glycine soya* was observed under average O_3 daytime concentrations (58 nl/L), and was shown to contribute to the ozone tolerance of this plant species (Robinson and Britz, 2001). However, there are data indicating that, in both normal atmospheric and increased (0.1-4.0 ppm) concentrations of ozone, ascorbic acid does not prevent oxidation of the fluorescing dyes D494 and D283 accumulated in the apoplast of the leaves of the spinach *Spinacia oleracea* (Jacob and Heber, 1998).

4.2.4. Thiol (SH) - compounds

Thiol compounds play an important role in antioxidant protection:

cysteine	$CH_2(SH)CH(NH_2)COOH$
methionine	$CH_2(SCH_3)CH_2CH(NH_2)COOH$
glutathione	$HOOCH(CH_2)_2CONHCHCONHCH_2COOH$ with NH_2 on the first CH and CH_2SH on the second CH

Measurements of the ozone absorption rate of various thiol compounds at pH=7 has shown that this capacity decreases in the following sequence: thiosulfate > cysteine = methionine > glutathione (Kanofsky and Sima, 1995).

In living cells, one of the important thiols is glutathione, which is present in significant quantities in plant leaves (Gilham and Dodge, 1986). According to Hausladen and Alscher (1993), its concentration in chloroplasts can vary from 1 to 4 μM (on the average 66-76 %). In cytoplasm, it is possible to detect high levels of glutathione, which exists mainly in a reduced state. In reaction with free radicals, glutathione (GSH) forms thiol radicals ($G\dot{S}$) with subsequent dimerization to disulfide (GSSG). Glutathione, furthermore, entraps hydroxyl radicals ($\dot{O}H$), as its interactions with $\dot{O}H$ results in the rapid oxidation of this radical:

$$2\,GSH + 2\,\dot{O}H \rightarrow 2\,H_2O + GSSG$$

It is also possible that the reaction produces thiol radicals $G\dot{S}$, which are less chemically active than hydroxyl radicals but which are also capable of oxidizing biologically important molecules, such as alcohols, ascorbate and sugars.

$$GSH + \dot{O}H \rightarrow G\dot{S} + H_2O$$

There are, furthermore, indications (Halliwell and Gutteridge, 1985) that glutathione directly reacts with hydrogen peroxide, and thus protects cells.

4.2.5. *Uric acid, cytokinins and biogenic amines*

For a long time uric acid was considered to be the final product of the metabolism of nucleic acids in mammalia. However, it is now known (Goodwin and Mercer, 1986), to be widespread in both animals and plants. Uric acid is a powerful recipient of singlet oxygen and the hydroxyl radical.

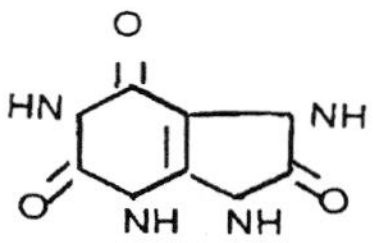

URIC ACID

The reaction

$$\text{Uric acid} + \dot{O}H \rightarrow R\dot{O}_2$$

forms peroxy radicals, which are less reactive than hydroxyl radicals. Uric acid is capable of inhibiting lipid peroxidation. Additionally, its function is probably similar to ascorbate, especially when very little active superoxide dismutase and. catalase are present (Ames et al., 1993) (see also chapter 5., section 2.3.).

Phytohormones cytokinins also appear to be related to antioxidants. They can directly participate in the regulation of free radical processes through the inhibition of purine oxidation and the scavenging of oxygen radicals (Leshem et al., 1981). Benzyl adenine and zeatin effect an improvement in the state of already ozonated plant tissues (Fletcher, 1969) and reduce ozone damage (Pauls and Thompson, 1982).

It is possible that biogenic amines found in plants (Roshchina, 1991; 2001a), such as dopamine, noradrenaline, adrenaline, serotonin and histamine, are also antioxidants. Their antioxidant features are due to the double bonds in the phenolic (catecholamines dopamine, noradrenaline and adrenaline), indole (serotonin) or imidazole structures of their molecules. In any case, they are known to have a protective role against pathogens, UV- and ionizing radiation, all of which give rise to active oxygen species. Plant species enriched in biogenic amines, such as the common nettle *Urtica dioica*, *Laportea moroides*, the entire family of Cactaceae and others (Roshchina, 2001a), are usually very tolerant of environmental changes.

4.2.6. *Multiatomic alcohols, polyamines and polyacetylenes*

Multiatomic alcohols (glucose, ribose, mannitol) are scavengers of free radicals due to the double bonds in their molecules (Zenkov and Men'shchikova, 1999). Especially significant is mannitol, which can react with hydroxyl radical. A plant's first reaction to ozone entails changes in the amount of its carbohydrates (see Chapter 5). The possible role of sugar excretions from many plant cells may also constitute a surface defense against oxidants and, in particular, ozone.

Plant polyamines, such as spermine, also demonstrate antioxidant characteristics (Mahmoud and Melo, 1991). They can eliminate the damaging action of external oxidants.

A group of substances with bactericidic and fungicidic properties, which have both two and three bonds in their molecule, are related to polyacetylenes. They are also met in plant secretory cells and excretions (Roshchina and Roshchina, 1989; 1993) and may be a target for ozone and active oxygen species. Polyacetylene capilline have demonstrated defensive properties in protecting pollen against gamma-irradiation, which forms free radicals (Roshchina et al., 1998b). The potential of polyacetylenes as oxidants represents a subject for future study.

4.2.7. Terpenoids

Terpenoids demonstrate antioxidant activity due to the double bonds in their unsaturated carbon chain. Among the most well known antioxidants are various triterpenes, pigments carotenoids, phytohormone abscisic acid and sesquiterpene lactones (for instance, prozulenes and azulenes).

Carotenoids are triterpenes, that is C_{40} -compounds, the carbon skeleton of which is constructed from eight C_5-isoprene fragments (see Fig.18). They are usually located in the lipophilic regions of a cell. In membranes, they form a complex with proteins. The system of coupled (connected) double bonds makes carotenoids susceptible to oxidizing reactions with oxygen and ozone. Carotenoids can serve as protectors against reactive oxygen species and as agents that decrease the rate of lipid peroxidation (Young and Lowe, 2001). They can directly react with peroxy and alkoxy radicals and prevent the formation of hydroperoxides. Carotenoids are capable of restricting the formation of singlet oxygen, and they even help to eliminate it when it is formed (Halliwell, 1981; Haliwell and Gutteridge, 1985). In connecting with membranes, they stabilize them, preventing the chain reactions of autooxidation. β-carotene is also a major quencher of singlet oxygen (Karnaukhov, 1988; 1990; Olson, 1992).

Other groups of potential antioxidants include triterpenes, in particular carnosic acid, carnosol, rosmanol and epirosmanol from *Rosmarinus officinalis* (Haraguchi et al., 1995). In small concentrations (3-30 μM), the compounds inhibit lipid peroxidation as well as the generation of superoxide anion radical.

Antioxidant features also belong to the terpenoid derivatives, for instance abscisic acid. Treating a plant with abscisic acid reduces ozone injury because it regulates the reactions of the stomata and, thus, prevents O_3 penetration into a leaf (Evans and Ting, 1974). The presence of ozone has also been shown to increase the amount of abscisic acid in the rice *Oryza sativa*. Moreover, quantities of the acid are higher in varieties (grades) tolerant of ozone than in sensitive ones (Evans and Ting, 1974).

The double bonds in sesquiterpene lactones, especially proazulenes and azulenes, allow them also to serve as antioxidants (Konovalov, 1995; Roshchina et al., 1998b). Azulenes (chamazulene, guajazulene, etc), which possess bactericidic features, are found in the secretory cells of glands and trichomes (Konovalov, 1995), as well as in pollen (Roshchina et al., 1995) and the vegetative microspores of horsetail (Roshchina et al., 2001a). They can be discerned not only in covers but in the isolated chloroplasts of red clover *Trifolium repens* and pea *Pisum sativum* (Roshchina, 1999a). According to our observations in field conditions, its blue color, characteristic of the leaves and stems of many plant species from legumes to cereal grasses, appears in spring but is absent in summer from the same species. The cause of the phenomenon appears to be the accumulation of azulenes (not anthocyanins because there is not any change in color at acid pH). Increase in the springtime ultraviolet radiation and the related enhanced ozone formation requires the formation of antioxidants, perhaps including azulenes. The antioxidant properties of

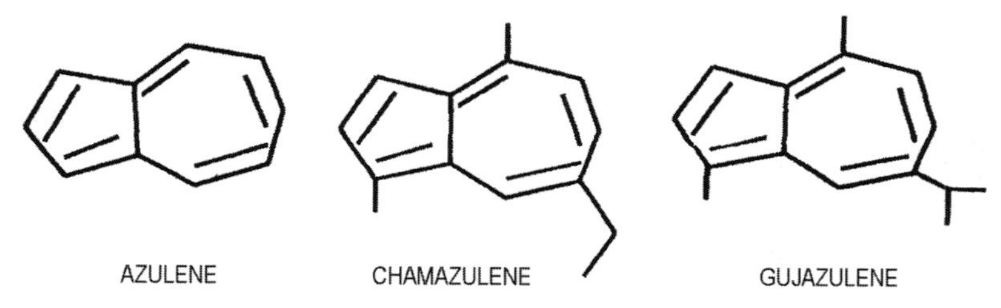

azulenes have not yet been studied, and investigations should be specifically undertaken in this direction.

4.2.8. Carbon dioxide.

Recently, it has been demonstrated (Kogan et al., 1998) that carbon dioxide gas in concentrations of 8.2 and 20 % decreases the generation of superoxide anion radical, not only in multicellular animals but also unicellular (yeast) organisms. The antioxidant influence of CO_2 on the generation of other reactive oxygen species has also been observed. The findings testify to the fact that carbon dioxide is a regulator of free radical homeostasis in living organisms. It has arisen during evolution as a protective mechanism activated in a response to the presence of oxygen in the atmosphere. The manner in which this question applies to plant organisms has not been explored. As the product of respiration, CO_2 may be accumulated within the extracellular and intercellular spaces of plants. High concentrations of the gas (> 20%) have been found in woody plants (Roshchina, 1971; 1973). Perhaps, the accumulation of carbon dioxide is one of the antioxidant mechanisms defending against external oxidants.

4.3. PLANT EXCRETIONS AS NATURAL ANTIOZONANTS

The scavengers of ozone and the intermediates of ozonolysis also include plant excretions, which may contain both high- and low-molecular antioxidants. Different organs of green plants—developing seeds, leaves, flowers, fruits and even roots—release a number of complex organic substances. The substances are constantly present in the air, and are all encountered in the lowest layers of the troposphere. They are especially found in the air surrounding plants in places of plentiful vegetation. In terms of their chemical structures, these air organic inclusions are rather diverse. Among them are various hydrocarbons, mainly terpenes, alcohols, organic acids, complex esters, aldehydes, ketones and other compounds (Roshchina and Roshchina, 1989; 1993). Under favorable conditions, the concentration of volatile organic substances in the troposphere can reach several mg per 1 m^3 of air.

Numerous reactions between ozone and the products of plant excretions occur that results in the instability of the organic compounds in the atmosphere. The reactions have been insufficiently investigated, and only some of them will be given consideration here (also see chapter 1).

Unsaturated compounds with double bonds, for example ethylene, isoprene and terpenoids, have the greatest ability to bond with ozone. In these reactions, O_3 intervenes in the C=C link. As a result, peroxy and ester bonds will be formed, which can easily degrade to form strongly oxidized compounds composed of short carbon chains.

During a normal exchange of substances, vegetative tissues constantly form ethylene. Its concentration in agricultural regions is very small. In the USA, annual ethylene emissions are nearly 1.8 $x10^4$ tons per year. This concentration is 1000 times less than the total amount of organic substances produced by the industry (Altshuller, 1983). The scheme of the ethylene reactions with ozone can be seen in fig. 36. The first stages of the interaction of ethylene with ozone proceed according to the usual Criegee mechanism of attaching to double bonds (Criegee, 1975). Subsequently, the primary ozonide of ethylene will produce biradical Criegee, formaldehyde and the peroxide of methylene as its primary products (Duce et al., 1983). The latter substance will be transformed into a dioxide, and then into a dioxy radical. Short-living, the dioxyradical, via a chain of spontaneous reactions, yields the final stable products: CO, CO_2, H_2, H_2O and formic acid. The largest component of the output is carbon monoxide, followed respectively by water, carbon dioxide and hydrogen. The increased release of ethylene resulting from ozone stress is part of the plant's defense against the oxidants in its environment (Pell et al., 1997).

Isoprene, also a product of the normal metabolism of plants, was found in vegetative excretions by G.A.Sanadze (1964), and later by Rasmussen and Jones (1973). Sanadze et al. (1972) demonstrated that isoprene is one of the products of photosynthesis. Although animals also release isoprene (Roshchina and Roshchina, 1993), the basic sources of this gas in the atmosphere certainly are plants. Tropical plants release a significant amount of isoprene. The greatest quantity of isoprene in the forests of the Amazon is contained in the air under the forest canopy, where it reaches 0.008 ppm (Rasmussen and Khalil, 1988). The Global yield of isoprene is nearly $3.5x10^{14}$ tons of carbon per year. Isoprene is one of the main components in the gas excretions from the foliage of such trees as the oak *Quercus* , the poplar *Populus*, the willow *Salix* and the aspen *Populus tremula* (Finlaysen-Pitts and Pitts, 1986). The reactions of ozone with isoprene have been described in the section 3.1.

The large interest in the release of ethylene and isoprene by plants is connected

ethylene $\rightarrow$ ozonide $\rightarrow$ formic aldehyde + peroxide

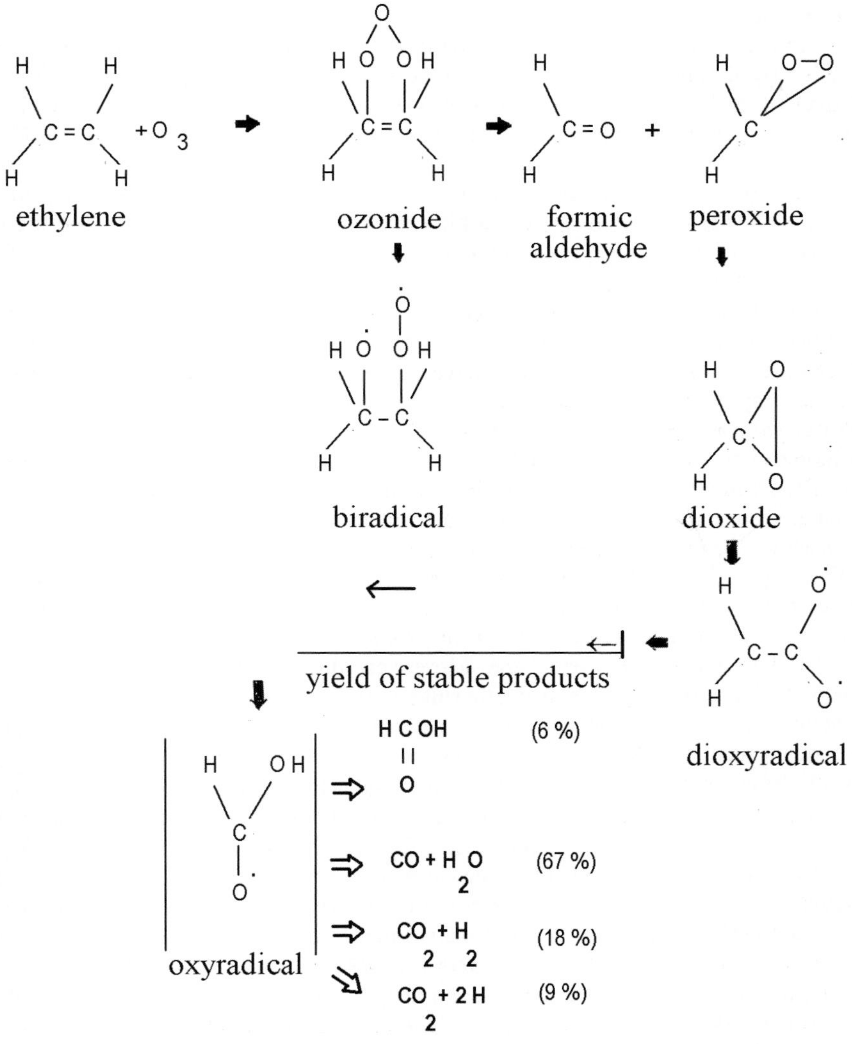

$H_2C{=}CH_2 + O_3$

ozonide $\rightarrow$ biradical

peroxide $\rightarrow$ dioxide $\rightarrow$ dioxyradical

yield of stable products

oxyradical

Product	Yield
$HCOOH$	(6 %)
$CO + H_2O$	(67 %)
$CO_2 + H_2$	(18 %)
$CO + 2H_2$	(9 %)

Figure 36. The ethylene reactions with ozone (Duce et al., 1983).

to the main role of such hydrocarbons in the photochemical formation of the ozone in troposphere (Trainer et al., 1987; Chameides et al., 1988). Moreover, ethylene and isoprene participate in the formation of CO, oxygenated hydrocarbons and organic acids (Jacob and Wofsi, 1988).

Coniferous plants release large quantities of terpenes. According to data assembled by Zimmerman et al. (1978), total production of terpenes is nearly 4.8 x 10^{14} tons of carbon per year. Rasmussen (1972) further suggests that the global contribution of terpenes is roughly the same (230–460 x 10^{12} ton per year). The variety of terpenes is great. Species and varieties of plants differ from each other in accordance with the chemical structure of their terpenes. Regardless of this diversity, ozone reacts with terpenes at high rates (Grimsrud et al., 1975), and these reactions play a paramountly important role in the destruction of monoterpenes. In table 14 lifetimes and the second-order rate constants of similar reactions for monoterpenes of various structure are shown. The briefest lifetimes in an atmosphere containing ozone (about 4.5 s) belong to the conjugated dienes, such as α-phellandrene and α-terpinene. They have the highest rates of reaction with ozone. The acyclic systems and reduced benzolic rings have, accordingly, longer lifetimes and lower values for the rate constants.

Among plant excretions are many compounds with aromatic rings. Here, like in the reaction with ethylene, ozone attaches to double bonds. The ozonides formed in this manner are usually unstable and readily transformed into aldehydes or ketones by emission of a radical. The reaction of ozone with phenols, which constitute a significant part of plant excretions (Roshchina and Roshchina, 1989), mainly involves the estrangement of hydrogen from a hydroxyl group and the opening of the ring. The polycyclic aromatic hydrocarbons formed by plants are more sensitive to ozone than monocyclic ones are. Moreover, the attachment of ozone to an aromatic nucleus protects other aromatic rings against ozonolysis.

The natural concentration of ozone varies from 0.01 up to 0.06 mg/m^3. However, ozone quantities increase prior to thunderstorms. At this time, a plant releases plentiful amounts of moisture and organic substances. During the 2-3 hours preceding a thunderstorm, plant fragrances are amplified, induced by the enhanced gas excretion. Some authors (Dmitriev, 1971; 1972; Roshchina and Roshchina, 1993; Roshchina, 1996) consider the aroma substances of plants as natural antiozonants. These views are based on experimental data indicating that hydrocarbons, phenols, terpenoids and many other compounds encountered in plant excretions could possibly be antiozonants (Chameides et al., 1988; Hewitt et al., 1990; Hiller et al., 1990; Salter and Hewitt, 1992). For instance, the addition of β-pinene to the leaf surface of the apple tree *Malus pumila* decreased ozone damage.

As a result of the reaction of ozone with various compounds in plant excretions, the O_3 concentration in the air is reduced. Similarly, Enders (1992) has shown that the amount of ozone sharply decreases when quantities of α-pinene increase. Plant excretions are also likely detoxicants of ozone itself. This is confirmed by the fact that the volatile products from reactions involving these excretions and O_3 are not so poisonous as ozone is.

Table 14. Lifetimes of terpenes and their rate constants of reactions with ozone (adopted from Grimsrud et al., 1975).

Terpenes	*Name*	*Rate constants, ppm/s*	*Lifetimes*
	Isobutene	147 - 240	3.1 - 3.6
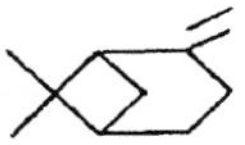	β–pinene	175 - 186	7.5 - 10.3
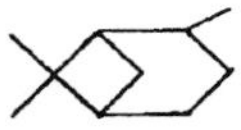	α–pinene	41 - 144	3.2 - 3.9
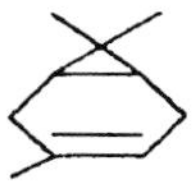	Δ^3– carene	131 - 208	2.0 - 3.8
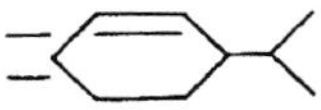	β–phellandrene	49 - 143	3.6 - 4.2
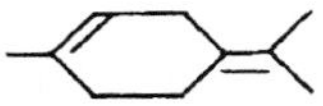	γ– terpinene	66 - 157	5.8 - 7.6
	carvomenthene	29 - 54	1.0 - 1.5
	limonene	23 - 41	1.3 - 1.6
	dihydromyrcene	31 - 73	1.4 - 2.1
	myrcene	29 - 51	2.5
	cis - ocimene	12 - 37	3.8 - 5.9
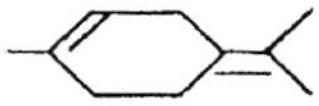	terpinolene	13 - 219	1.9 - 2.5
	α–phellandrene	7.5 - 14	2.1 - 2.8
	α– terpinene	4.5	1.7 - 2.6

The rate constant of (K) was calculated using the formula:

$$K = \frac{\ln[O_3]_0[O_3]_i}{[Hc]_{ave} \quad x \quad ti},$$

where $[O_3]_0$ and $[O_3]_i$ are ozone concentrations measured at the beginning of the experience prior to the interaction of ozone with monoterpenes as well as at the end of it, t is time of measurement; and $[Hc]_{ave}$ represents the speed of the gas flow (mean average from initial and final concentrations of the hydrocarbon participating in the reaction with ozone).

A study of the ozone reactions with volatile terpenoids has shown (Tingey and Taylor, 1982) that the oxidation of terpenoids produces appropriate alcohols, aldehydes and ketones. Depending on the chemical structures involved, reactions between monoterpenes and O_3 proceed at different rates. Demonstration of this point occurred when β-pinene, isoprene, α-pinene, 3-Δ-carene, β-phellandrene, terpinolene and α-terpinene were fumigated with ozone 0.05 - 0.5 ppm for 1 hour (the limit in lifetime of O_3 for any of the terpenes). The percentage of ozone absorption was linearly dependent on the concentration of this gas in the range between 0.045-0.105 ppm. Appreciable absorption of ozone was observed only for α-terpinene (66.71 %), terpinolene (11.75 %) and β-phellandrene (13.5 %). The amount of O_3 absorption by the other terpenes studied was no more than 2 %. In the presence of ethylene, cymene and methylene, the absorption of ozone was the lowest. The rate constants of the ozonolysis of many monoterpenes have been published in various papers (Hull, 1981; Atkinson et al., 1990).

When volatile plant excretions are transformed by ozonolysis, they become a source of new physiologically active agents—free radicals and ozonides—which can play a protective role against pests and be participants in allelopathic relationships (Ostrovskii et al., 1993; Roshchina, 1996). The possibility is not excluded that changes in one or another biochemical reaction are indirectly induced by plant excretion, or by a product of its interaction with ozone (Roshchina, 1996; 2001). In reactions of O_3 with allelochemicals, the latter are transformed into aldehydes. For instance, lunularic acid turns into benzaldehyde, and anethol into anisaldehyde (Fig. 37). After ozonolysis, naturally occurring alcohols, such as cinnamic, are transformed into glycolic aldehyde under normal conditions and into benzaldehyde and formic acid under stress (Fig.38). Geraniol is also converted into formaldehyde and glycolic acid.

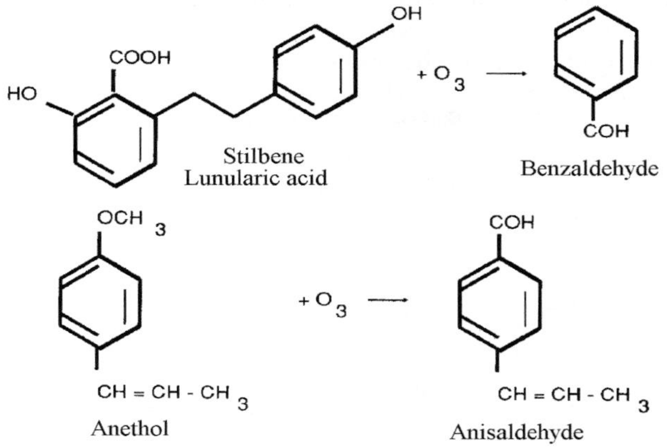

COOH HO OH

Stilbene
Lunularic acid

$+ O_3 \longrightarrow$

COH
Benzaldehyde

OCH_3

$+ O_3 \longrightarrow$

COH

$CH = CH - CH_3$
Anethol

$CH = CH - CH_3$
Anisaldehyde

Figure 37. Formation of aldehydes in reactions of ozone with aromatic compounds.

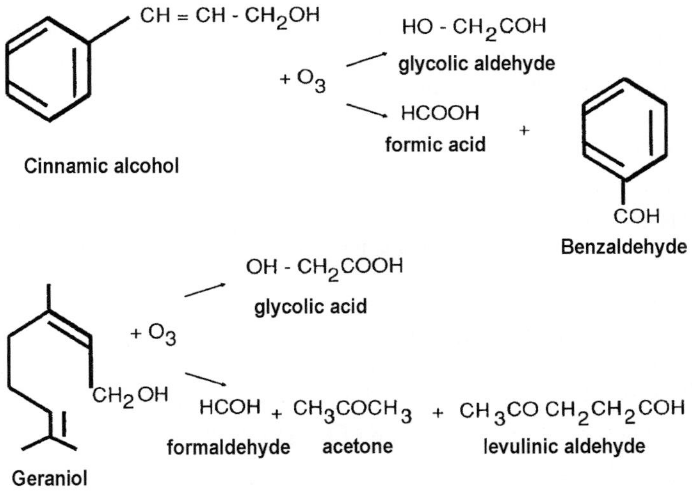

$CH = CH - CH_2OH$

Cinnamic alcohol

$+ O_3$

$HO - CH_2COH$
glycolic aldehyde

HCOOH
formic acid

+

COH
Benzaldehyde

CH_2OH

Geraniol

$+ O_3$

$OH - CH_2COOH$
glycolic acid

$HCOH + CH_3COCH_3 + CH_3CO\,CH_2CH_2COH$

formaldehyde **acetone** **levulinic aldehyde**

Figure 38. Transformation of cinnamic alcohol and geraniol in the reactions of ozonolysis under normal and stressed conditions.

Besides volatile compounds, plants release liquid substances, which may also be antiozonants (Roshchina and Roshchina, 1993; Roshchina, 1996). For instance, nordihydroquariaretic acid, found in the resinous exudate of *Larrea divaricata* and other plants, has antioxidant properties (Halliwell and Gutteridge, 1985). The sugars contained in excretions can absorb free radicals (Asada, 1980). The increase in their quantities can nullify, if only partially, the toxic effects of free radicals produced from ozone. However, this possibility is based more on assumption than established fact.

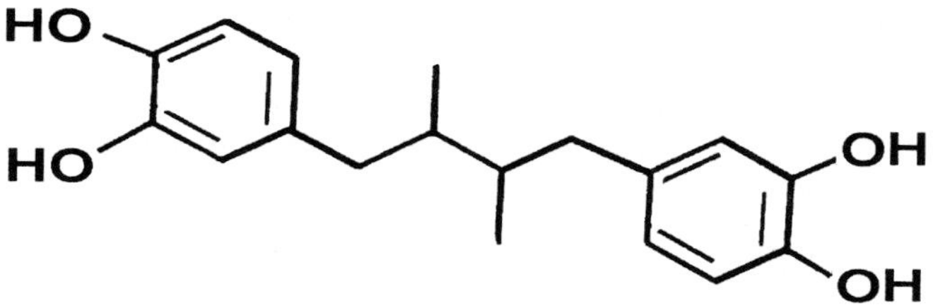

NORDIHYDROQUARIARETIC ACID

Apparently experimental confirmation is still necessary, as we have not found any mention of it in the literature. Of special interest are the excretions of antioxidant enzymes, such as peroxidases and superoxide dismutases, by plant cells (Roshchina and Roshchina, 1993). These enzymes may be concentrated in the apoplast, especially when the organism is under stress.

4.4. INTERACTION OF ANTIOXIDANTS AND THEIR DUAL ROLE IN PLANT LIFE

The above-mentioned information indicates that the protection of living cells from ozone and other reactive oxygen species can occur via different mechanisms that enhance the reliability of biological systems (Table 12).

The activity of each enzyme is specifically directed at the elimination of the initiators or products of lipid peroxidation ($\dot{O}_2^-$; H_2O_2). They are not involved in the detoxification of other reactive oxygen species, although the latter can deactivate the enzymes, as illustrated by the hydroxyl radical's effect on tocoferol and other low molecular inhibitors of free radical processes. It is essential to note that the antioxidants mainly work as a complex and that enzymic systems are specialized so that they are activated at different stages of the oxygen reduction reaction:

$$\dot{O}_2^- \xrightarrow{\text{Superoxide dismutase}} H_2O_2 \xrightarrow{\text{Catalase}} H_2O$$

In contrast, the inhibitors of organic radicals initiate a chain of transformations, by which less active types of radicals are formed. For instance, the ability to link peroxy radical $R\dot{O}_2$ is in a range : α-tocoferol> ascorbic acid> uric acid.

In the living cell, the same low-molecular substances can play a dual role—either as oxidants or antioxidants. For instance, such biogenic amines as dopamine, noradrenaline, adrenaline, serotonin and histamine are all antioxidants. But in certain conditions (in particular at pH > 7.5-8), the oxidation of catecholamines (dopamine, noradrenaline and adrenaline) gives rise to superoxide anion radical and hydrogen peroxide (Roshchina, 1991, 2001a). The rate of the reaction is strongly accelerated in the presence of ions of the transitive metals (Halliwell and Gutteridge, 1985). Moreover, an excess of catecholamines may act on the antioxidant enzymes, in particular peroxidase, like oxidants. Quinones can also form free radicals by interacting with active oxygen species (see chapter 3), or even serve as antioxidants. Azulenes and proazulenes function as antioxidants but, in certain conditions, form free radicals (Bachmann et al., 1994).

Since some low-molecular antioxidants can form toxic products themselves, they are significant for a plant's defense as a whole. In particular, the reaction of ozone with isoprene and other biogenic alkenes produces some hydroperoxides in plant tissue (Melhorn et al., 1988). Ethene (ethylene) and isoprene will form formaldehyde and carbon monooxide, as well as significant amounts of bis(hydromethyl)peroxide in the ratio: ozone/alkene < 1. Among the products formed by the ozonolysis of terpenes are compounds having high biological activity. Some of them demonstrate antimalaria properties (Barbosa et al., 1993), although ozonated water is itself, in some cases, a pesticide (Inness, 1994). In this connection, ozone and the products of ozonolysis also influence the mutual relations of plants with insects (Bolsinger et al., 1992).

The influence of ozone and other active oxygen species on plant excretions leads to changes in the allelopathic interactions between plants in the biocenosis (Roshchina, 1996; 1999; 2001b). The formation of free radicals by ozonolysis can be one of the factors inducing the allelopathic effect (Roshchina, 1996; 2001b). The high concentration of superoxide anion radicals is supposed to kill some weeds (Elstner et al., 1980). Hydrogen peroxide, as the product of a reaction involving the dismutation of the radicals, disrupts seed rest and stimulates germination (Hendricks and Taylorson, 1975).

Ultraviolet irradiance, the main source of ozone and the intermediates of ozonolysis in air, enhances the formation of toxic products from many phenols, which can be antioxidants as well (Aucoin et al., 1992). It also changes the plant state in the biocenosis.

The occurrence of many antioxidant systems is a necessary condition of the defense against oxidative bursts, including those instigated by various reactive oxygen species. For instance, the vegetative organs of plants, such as tobacco and birch, accumulate H_2O_2 in response to ozone (Schraudner et al., 1998; Pellinen et al., 1999), whereas Arabidopsis builds up quantities of both the superoxide anion radical and hydrogen peroxide (Rao and Davis, 1999). O_3 and extracellular amounts of the superoxide radical, but not H_2O_2, induce intracellular accumulation of the same radical (Overmyer et al., 2000). The sites of the resulting accumulation of active oxygen species and visible damage (lesions) are of a distinct size. The injurious effect of the active oxygen species needs to be scavenged, transformed and minimized by the activation of various antioxidants. For instance, microspores, such as pollen and the vegetative spores of horsetail, which produce superoxide radicals but have no superoxide dismutase (Roshchina et al., 2002b; 2003), contain peroxidases, as well as a lot of phenols and carotenoids as antioxidants in their cover (Stanley and Linskens, 1974). Vegetative plant organs synthesize a greater variety of antioxidants.

The antioxidant mechanisms were produced throughout evolution and are part of the generic make up. They prevent development of free radical processes and lipid peroxidation in cells. In the normal conditions of living, the antioxidant systems do not nullify these processes. Their products constantly arise and are present in systems of various complexity. However, due to the functioning of antioxidant systems, the amount of these products is kept at a constantly low level. In extreme conditions, for instance those created by ozone fumigation, the balance is moved in the direction of the free-radical processes and the activation of lipid peroxidation. The consequences of such influence were considered in the chapter 3.

CONCLUSION

The protective cellular reactions to ozone in plants and the formation of reactive oxygen species involves the participation of steady-state high-molecular enzymic systems, such as superoxide dismutase, catalase, peroxidase, glutathione reductase and others, as well as low molecular systems, including, in particular, ascorbate, carotenoids, phenols, etc. Plant excretions contain both volatile and liquid antiozonants, including various antioxidants. In low concentrations of O_3, they can scavenge ozone and the intermediates of ozonolysis in order to prevent the negative effects of these oxidants. Moreover, this phenomenon appears to be favorable for the chemical relations in the biocenosis as a whole. If ozone concentration increases beyond the amount that the pool of steady-state antioxidants can absorb, ozone and its derivatives penetrate into the plant cell or send a signal to the cell to engage further adaptive mechanisms in its antioxidant defense (see chapter 5).

CHAPTER 5

OZONE-INDUCED CHANGES IN PLANT METABOLISM

Chronic ozonation of plants causes adaptive or damaging changes in plant metabolism. The stable quantity of antioxidants in plant cells undergoes an increase, and new antioxidants or other protective compounds are formed. The damage inflicted on the cell by ozone is, like many other factors, first displayed at the biochemical level before visible morphological symptoms of toxicosis can be seen (Dugger and Ting, 1970). Photosynthesis, respiration, nitrogen and lipid exchanges are all affected, along with the activities of the secondary metabolism. Later, visible damage appears, such as necrotic spots and bronze coloration on leaves (see chapter 6). In this chapter, the main changes in plant metabolism occurring in response to ozone fumigation will be considered.

5.1. CARBON METABOLISM

Exposed to ozone, the general carbohydrate metabolism is subject to change. The effects are attributed to changes in the processes of photosynthesis and respiration as a whole, in the glycolytic and the pentose phosphate pathways, as well as in the synthesis and catabolism of starch and cellulose.

5.1.1. Photosynthesis

The effects of ozone on photosynthesis, as a main process of assimilation required for normal plant growth, are different for higher plants, lower plants and micro-organisms (Guderian et al., 1985). They are, however, always dependent on dose. Ozone doses of 0.25-1.0 ppm (in 4-h exposures) stimulated the regenerative capacity of the moss *Funaria hygrometrica*, while a 6-8-h exposure was inhibitory of photosynthetic capacity (Comeau and Le Blanc (1971). Photosynthesis in the lichens *Parmelia sulcata* and *Hypogymnia enteromorpha* was diminished by a 12-h exposure to 0.5 or 0.8 ppm of ozone (Nash and Sigal, 1979). The foliose lichens *Hypogymnia enteromorpha* and *Letharia vulpina* were tolerant of ozone in mountain conditions of up to 150 ppm of ozone per h, and only after doses of 285 ppm per h was there a decline (Sigal and Nash, 1983). Brown and Smirnoff (1978) suggested that photosynthesis in the lichen *Cladonia rangiformis*, determined as $^{14}CO_2$ – fixation, was tolerant of the acute ozone exposures of 2 ppm for 0.5-2.5 h. Photosynthesis in *Euglena gracilis* degraded by 75 % when exposed to 240 ppm of O_3 for 60 min (Chevrier and Sarhan, 1992). It has to be admitted that there is only a small amount of information dealing with lower plants and photosynthesizing microorganisms.

Under the influence of ozone, photosynthesis in higher plants is clearly inhibited (Bennett and Hill, 1974; Osmond et al., 1981). However, the initial rate of this process is, in the majority of cases, restored one day after termination of the treatment (Pell and Brennan, 1973). The age of the leaves plays an essentially important role in these experiences. The restoration of photosynthetic activity is observed in young leaves but not in old ones. Experiments on the leaves of the rice *Orysa sativa* (Nacamura and Saca, 1978) established a threshold concentration of ozone required to induce an effect on the photosynthesis of this plant. After a brief fumigation with ozone, the inhibition of photosynthesis in the lucerne *Medicago* L. was observed at concentrations of 100 ppm of ozone per hour. In other work, the same effect was induced with concentrations of 0.2-0.5 ppm, but required longer periods of fumigation (Darral, 1989). When ozone fumigation occurs for more than one day, the effect is already appreciable at concentrations of 0.035-0.045 ppm. However, lower doses of O_3 (0.02 ppm or 20nl/L for 7-h over midday periods and administered 5 days per week) can affect leaf growth and development without producing long-term effects on photosynthesis, starch accumulation or the partitioning of the photosynthate (Britz and Robinson, 2001). Particular examples of photosynthesis inhibition by ozone are listed in table 15. As can be seen from these data, the action of ozone in woody plants requires higher concentrations or longer durations than it does in grassy plants. Under ozone exposure, the values of the compensatory points for CO_2 and light are increased. Experimentally, this has been demonstrated on the leaves of the poplar *Populus euramericana* (Furukawa and Kadota, 1975). The authors believe that this increase results from the fact that the activity of CO_2-fixing enzymes is inhibited.

The extent to which ozone's effects on photosynthesis depend on stomata conductivity was studied in vivo on leaves of the wheat *Triticum aestivum* (Farage et al., 1991). Completely formed secondary leaves were treated with 0.2 or 0.4 ppm of ozone. There was a decrease in the rate of CO_2 fixation in saturated light intensities, accompanied by a parallel decrease in stomata conductivity. However, this limitation of stomata conductivity lasted only for a few hours after exposure to 0.4 ppm ozone and was not observed under any other conditions. The determining factor for a decrease in CO_2 absorption is believed to be a reduction in the efficiency of the carboxylation of triose phosphates, which were reduced by 21-58 % after an experimental duration of 16 hours (Farage et al., 1991). Thus, these experiments also confirmed that the main cause of the photosynthesis inhibition in vivo is a decrease in carboxylation intensity.

Ozone acts directly on the carboxylases participating in the photosynthetic fixing of CO_2. The ozone-induced granulation of the chloroplast stroma is thought to result from the oxidation of the SH-groups in ribulose bisphosphate carboxylase or RDP-carboxylase for short (Mudd, 1982; Doming and Heath, 1985). The inhibition of the enzyme occurs soon after exposure to ozone. Accordingly, after treating the poplar *Populus trichocarpa* with ozone for 3 days, RDP-carboxylase activity is reduced (Pell et al., 1992). Some information about the inhibition of RDP-carboxylase is shown in table 16. It is necessary to mention that, despite a high percentage of inhibited RDP-carboxylase (sometimes up to 80 %), no visible damage to leaves is

Table 15. The inhibition of photosynthesis with ozone.

Plant species	*Ozone. ppm*	*Exposure. hours*	*% of inhibition*	*Reference*
Acer saccharum	0.5	4.0/2 days[x]	21	Carlson, 1979
Avena sativa	0.4	0.5	33	Hill and Littlefield, 1969
Fraxinus americana	0.5	4.0/2 days[x]	0	Carlson, 1979
Glycine max	0.4	4.0	37	-.//-
Lycopersicon esculentum	0.6	1.0	43	Hill and Littlefield, 1969
Medicago sativa	0.1	1.0	4	Bennet and Hill, 1974
Nicotiana tabacum	0.4	1.5	78	Hill and Littlefield, 1969
Phaseolus vulgaris	0.6	1.0	29	Hill and Littlefield, 1969
Populus euramericana	0.9	1.5	50	Furukawa and Kadota, 1975
Quercus velutina	0.5	4.0/2 days[x]	30	Carlson, 1979

[x] - i.e. by 4 hours per day for 2 days

apparent (Pell and Pearson, 1993). In a paper by Hill (1971), inhibition of photosynthesis is said to have occurred at concentrations of 2.4×10^{-8} M ozone per cm^2 of cellular surface. According to other authors (Ticha and Catsky, 1977), a similar effect is achieved with higher values of ozone (3.2×10^{-6} M cm^{-2}).

RDP-carboxylase is a protein consisting of large and small subunits organized in a uniform structure (Goodwin and Merser, 1986). Experimental data provided by Landry and Pell (1993) indicates that fumigation with ozone causes the enzyme to lose its large subunits, a factor that is, perhaps, the cause of its deactivation, as the enzyme's catalytic activity is connected with its large subunit. The loss of the large55 kDa subunit by RDP-carboxylase was observed in the poplar hybride

Populus maximawizii x trichocarpa when subjected to ozone in concentrations of 0.1 μl/l or ppm for 4 h per day over 1.5 months (Landry and Pell, 1993). According

Table 16. The changes in the activity of RDP-carboxylase under ozone treatment

Plant species	*Ozone concentration, μl/l or ppm*	*Duration of the treatment*	*Inhibition of RDP-carboxylase, %*	*Reference*
Oryza sativa	0.12	2 hours	50	Nakamura and Saka, 1978
Solanum tuberosum	0.06-0.08	4 days	the decrease in the activity	Dann and Pell, 1989
Medicago sativa	0.25	48 hours	36-80	Pell and Pearson, 1983
Triticum vulgare	0.1	8 hours	the decrease in the activity	Lehnherr et al., 1987

to Conklin and Last (1995), ozone also causes a decline in the levels of two chloroplastic antioxidant mRNAs (iron superoxide dismutase and glutathione reductase) and two photosynthetic protein mRNAs (chlorophyll a/b-binding protein and ribulose-1,5-bisphosphate carboxylase/oxygenase small subunit) in *Arabidopsis thaliana*. This decline does not include all mRNAs encoding chloroplast-targeted proteins, since O_3 causes an increase in the amount of mRNA encoding the chloroplast-localized tryptophan biosynthetic enzyme phosphoribosylanthranilate transferase .

In the membranes of chloroplasts, the formation of free radicals is possible due to lipid peroxidation and disturbances in the electron transport (Treshow, 1984). For instance, exposure to ozone increases the production of superoxide radical associated with photosystem 1 (Roulans et al., 1970).

Chloroplasts have a system of radical scavengers; therefore the influence of ozone depends on the ability of the plastids to entrap and to destroy free-radical products. Measurements have been made of the effects of ozone on photosynthesis and some free-radical scavengers in the leaves of 5 clones of the poplar *Populus deltoides*, which were fumigated with 0.18 ppm of ozone for 3 hours (Gupta et al., 1991). During the first 90 minutes of the exposure to ozone, the rate of photosynthesis did not change, whereas the level of the antioxidant glutathione (G) and the activity of superoxide dismutase increased. Later, the rate of photosynthesis began to fall, while the level of glutathione (reduced + oxidized) and the activity of superoxide dismutase continued to grow. The relation GSH/GSSG in the leaves of trees exposed to ozone also decreased from 12.8 to 1.2. The level of the superoxide dismutase in ozonated plants increased twofold. After 4 hours, the rate of photosynthesis had decreased to one half of that in the control, while both the level of glutathione and the activity the superoxide dismutase were higher than in the control. The enhanced level of antioxidants was maintained for 21 hours after the ozone treatment, but the rate of photosynthesis was restored to 75 % of that in the control by this time. It is interesting to note that neither electron transport nor the level of NADPH changed in response to the action of ozone. Hence, an increased antioxidant metabolism due to exposure to ozone can protect the photosynthetic apparatus.

5.1.2. Respiration

Ozone acts on respiration to different degrees, as it can both stimulate and inhibit the process. The type of effect depends on the object of research, concentration and duration of the ozone action. The inhibition of respiration and phosphorylation was found to occur in experiments on isolated mitochondria (Lee, 1967). When leaves of the tobacco *Nicotiana tabacum* were treated with 4 ppm of ozone, their rates of respiration were reduced for one hour, and then increased. In the root systems of such plants as the string bean *Phaseolus vulgaris,* exposure to a 0.15-ppm concentration of ozone for 2 days at 6 hours per day resulted in decreased respiration intensity. A similar inhibition of respiration was also registered in an experiment on the leaves of the tobacco *Nicotiana tabacum* (Mac Dowall, 1965, as well as in research by many other authors). A stimulation of respiration has been noted in still other works (Barnes, 1972; Dugger and Palmer, 1969; Todd, 1958).

In plant leaves, an increase in respiration intensity is frequently observed to occur when there is a simultaneous reduction in photosynthesis (Dizengremel and Citerne, 1988). Both processes are associated with visible damage of leaves. When a callus culture of the tobacco *Nicotiana tabacum* is given a 2-hour exposure to a 0.1-μl/L ozone concentration, respiration appears to be stimulated in both O_3-sensitive and O_3-tolerant plants. However, the dose of ozone causing the increase in respiration in tolerant plant types was twice as much (Anderson and Taylor, 1973).

In comparing the effects of ozone on respiration and photosynthesis, as well as on the appearance of any visible leaf damage (Dizengremel and Citerne, 1988), it should be noted that a 2-26 hour exposure to 0.4 ppm of ozone caused a 20% decrease in the photosynthesis performance of *Phaseolus vulgaris* seedling but a 15-38 % increase in respiration. Visible injuries of leaves were also seen. As for woody species such as *Populus deltoides* and *P. trichocarpa*, as well as *Pinus elliotti, P. serotina, P.taeda* and *P.strobus*, the effects on photosynthesis and respiration were different like those observed above, although there were no visible symptoms of leaf damage from a 15-18 week exposure to 0.12-0.15 ppm of ozone. In seedlings of *Glycine max*, exposure to 0.05-0.13 ppm of O_3 for 52 days caused photosynthesis to decrease by 10-22 %, whereas there were no changes in respiration and visible leaf damage. Respiration is a process less sensitive to ozone than photosynthesis is (Mudd, 1973; Treshow, 1984).

5.1.3. Changes in the pool of metabolites and in the activity of some enzymes.

The biochemical changes instigated by ozone are frequently expressed as infringements of the carbon metabolism (Dugger and Ting, 1970; Koziol et al., 1988). The elements in the carbon metabolism that generally serve as targets of ozone action are shown in fig 39. The changes in the carbohydrate metabolism result to form ozone's effect on enzymic activity. In fig. 39, the sites of O_3 action connected with the modification of appropriate enzymes are shown. Under the influence of ozone, ATPase and the enzymes involved in the transformation of phosphorylated sugars are modified. Ozone suppresses enzymic systems for synthesising cellulose and starch, inhibits the glycolytic pathway of the exchange and stimulates the pentose phosphate pathway.

The change in the carbohydrate exchange can be seen with the help of $^{14}CO_2$ (Wilkinson and Barnes, 1973). Ozone acts to decrease label inclusion in simple soluble sugars and to amplify the inclusion in the phosphates of sugars arising from strengthened respiration in the leaves of *Pinus strobus* and *Pinus taeda.*

After 2-6 hours of low ozone concentrations (0.05 μl/L or ppm), the hydrolysis of starch in the cucumber *Cucumis sativus* L. is inhibited. The same picture is observed in leaves of the string bean *Phaseolus vulgaris* and the cassia *Cassia* L. after 4 hours of ozone at 0.05 ppm. The effect seems to occur as a result of the inhibition of phosphorylase (Hanson and Stewart, 1970). An experiment with homogenates from *Avena* coleoptiles exemplifies the benefits of suppressing the enzymic system of cellulose synthesis (Ordin and Hall, 1967). It was found that an ozone concentration of 116 ppm caused both cellobiose content and the amount of polysaccharides to double. It is assumed that, in such cases, ozone suppresses the synthetases (phosphoglucomutase) necessary for cellulose synthesis. Conversely, cellulase activity is increased by ozonation of the string bean *Phaseolus vulgaris* (Dass and Weaver, 1972). In the first 48 hours of exposure to 0.35 ppm of ozone, not only was cellulase activity increased but, simultaneously, that of peroxidase and, to a less significant degree, of lactate dehydrogenase. Table 17 shows the change in enzyme activity resulting from various periods of fumigation with ozone. In increasing the duration of ozone fumigation, the activity of peroxidase grows 2.2 times, the activity of cellulase 4 times, and the activity of lactate dehydrogenase by 15 %. Rapid increases in cellulase activity under the influence of ozone entails rapid destruction of cell walls. In general, it is possible to conclude that ozone reduces the total rate of photosynthesis, changes the enzyme activity pool of assimilates, and distributes them among plant organs.

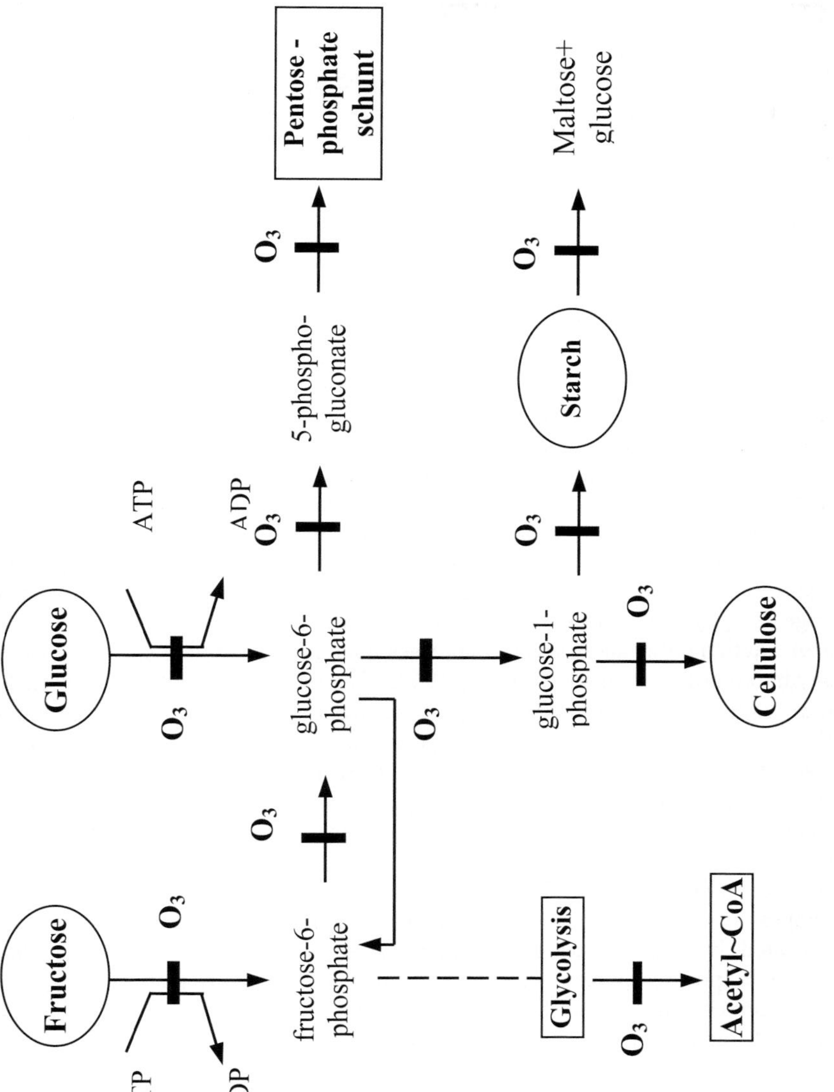

Figure 29. Scheme of the targets for ozone in carbohydrate metabolism

Table 17. Action of ozone on the enzyme activity in intact leaves of the string bean Phaseolus vulgaris (Dass and Weaver, 1972).

Duration of the fumigation with ozone 0.35 ppm	*Enzymic activity after the ozone fumigation (48 hours)*		
	Peroxidase units (units of optical density of guaycol at 470 nm/min g^{-1}fresh mass)	*Celulase (Change of viscosity of carboxymethyl cellulose for 44 hours)*	*Lactate dehydrogenase (units of enzyme/min g^{-1} of fresh mass)*
0	8.55	15.08	170
0.5	14.25	21.63	180
1.0	13.65	49.56	203
1.5	18.00	61.57	175

Under exposure to ozone, the activity of other membrane-bound enzymes is disturbed. Consequently, 2 hours of a 0.5-µl/L (ppm) concentration of ozone causes the activity of glucose-6-phosphate dehydrogenase to be increased by 30 %, whereas the activity of glyceralaldehyde-3- phosphate dehydrogenase of the soya *Glycine soja* L is decreased by 20 % (see the scheme of general carbohydrate exchanges in fig. 39). This effect is attributed to ozone's inhibition of the glycolytic pathway (Dass and Weaver, 1982) and to the stimulation of the pentose phosphate pathway of exchange (Tingey et al., 1975; Tingey et al., 1976a).

Ozone action either increases or decreases quantities of soluble sugars and other carbohydrates in leaves (Dugger et al., 1962à; Wellburn and Wellburn, 1994). For instance, exposure to ozone (0.49 ppm) immediately decreased the amount of soluble sugars in leaves of the soya *Glycine soja* L. (Tingey et al., 1973c) and branches of the pine *Pinus* L. (Wilkinson and Barnes, 1973). Since starch quantities do not change, the reduction in soluble sugars must be due to the suppression of photosynthesis. After a single treatment with ozone, an increase in the amount of sugars follows the initial decrease. This increase can be a result of changes in the translocation of sugars or the inhibition of starch synthesis.

Changes in the translocation of carbohydrates under ozone treatment are demonstrated by the following experimental results for the yellow pine *Pinus ponderosa*. Exposure to ozone resulted in an increase of the soluble sugars and starch in needles and a decrease in their quantities contained in roots (Tingey et al., 1976a,b). Ozone reduces the amount of photoassimilates flowing from the leaves of the pepper *Capsicum annuum* during the period when fruits are maturing (Bennet et al., 1979b). Similarly, ozone reduces assimilate translocation from leaves to roots in some crop species (Oshima et al., 1978) and, therefore, has an effect on the growth of the root top (Tingey, 1977).

5.2. LIPID METABOLISM

Ozone has an influence on the lipid metabolism of plants, mainly by changing the ratio of lipid components to lipid synthesis and the products formed by lipid peroxidation (Table 18). First, there is a decrease in lipid quantities due to inhibition of their biosynthesis (Mudd et al., 1971). Ozone causes the amount of fatty acids to be reduced and their structure to be changed (Frederick and Heath, 1975), both factors that affect a cell's viability. In leaves of the tobacco *Nicotiana tabacum*, ozone effects a decrease in the total quantities of all fatty acids and, to the greatest degree, the quantities of palmetic and linolenic acids (Tomlinson and Rich, 1969).

Under ozone treatment, the total amount of galactolipids and phospholipids in membranes changes. In experiments on *Pharbitis* and the string bean *Phaseolus vulgaris*, Nouchi and Toyama (1988) have, similarly, shown that exposure to ozone (0.15 μl/l for 8 hours) increases the amounts of phospholipids in the first species and decreases it in the second. In addition, the number of glycolipids, unsaturated fatty acids and malondialdehyde varied in both species. It is assumed that the initial stages of ozone action trigger changes in the metabolism of polar lipids, in particular phospholipids.

Though the lipids in cells are protected from oxidation by the presence of many antioxidants (see chapter 4) that may even constitute a sort of protective layer, this protection is not complete. To a certain extent, peroxidation takes place in cells because interaction with ozone does, nevertheless, occur and peroxides and free radicals are, consequently, produced. The nature and basic properties of the free radicals formed from lipids are principally similar to the properties of radicals formed from other classes of organic compounds (Mudd, 1979). Lipid peroxidation produces malondialdehyde, so that the amount of this compound increases to the extent in which process is allowed to progress (see Chapter 3). Experimentally, increased formation of this compound subsequent to ozone exposure has been demonstrated in leaves of the rice *Oryza sativa* (Nakamura and Saka, 1978) and the bean *Phaseolus vulgaris* (Mudd et al., 1971).

Jasmonic acid is also a product of lipid peroxidation, one that also involves phospholipases. This acid and its methyl esters occur in all plant tissues and serve a signalling compound in plant growth and development (Crelman and Mullet, 1997). It accumulates in response to ozone fumigation, participates in the expression of a defense gene and helps to realise the characteristic picture of cell death (Rao and Davis, 1999; 2001). Jasmonic acid and its derivatives serve as antioxidants in reaction to an oxidative burst, in which such active oxygen species as radicals and peroxides are formed (Overmyer et al., 2000) and act as mediators in programmed cell death (Jabs,1999).

The modification of lipids by ozone occurs in the membranes of all organelles— the plasmatic, chloroplast, mitochondrial and others. Work on isolated chloroplasts has demonstrated the direct action of ozone on lipid biosynthesis. Ozone disrupts the ability to metabolize ^{14}C acetate or acetyl~CoA in lipid biosynthesis (Mudd et al.,

Table18. Effects of ozone on the lipid exchange

Ratio of lipid components	*Content of fatty acids*	*Disturbance in the lipid synthesis*	*Formation of products of lipid peroxidation*	*Content of lipophilic components*
Change of the molar ratio of monogalactosyldiacyl glycerol/digalactosyldiacyl glycerol (Wingsle et al., 1992) and sterols (Tomlinson and Rich, 1969	Decrease in total amount of all fatty acids in leaves of *Nicotiana tabacum,* mainly palmetic and linolenic acids (Tomlinson and Rich, 1969).	Inhibition of the glycolipids' synthesis in chloroplasts (Tomlinson and Rich, 1973).	Formation of malondial-dehyde and increase in its amount in leaves of *Oryza sativa* (Nakamura and Saka, 1978), *Phaseolus vulgaris* (Mudd et al., 1971) and *Pinus abies* (Kronfuss et al., 1996). Formation of lipofuscin in seeds (Brooks and Csallany, 1978) and in pollen of various plant species (Roshchina and Karnaukhov, 1999).	Changes in amounts of carotenoid in seedlings of *Pinus silvestris* , xanthophyll, violaxanthin and zeaxanthin (Wingsle et al., 1992)

1971). In chloroplasts, ozone inhibits the synthesis of glycolipids and promotes the formation of malondialdehyde and glutathione (Tomlinson and Rich, 1973). In an ozone concentration consisting of 0.5-μl/L per hour, the number of steroid glycosides grows larger, and the quantity of free sterols is reduced (Tomlinson and Rich, 1973). According to data provided by Sellden et al. (1990), ozone causes greater damage in galactolipids than phospholipids.

At low concentrations of ozone (> 0.5 ppm) over longer durations, there is an increase in the free sterols and a decrease in the number of steroid esters and glycosides (Tomlinson and Rich, 1971 Grunwald and Endress, 1985).

Of special interest to the reader is the exchange of lipophilic pigments, such as chlorophyll and carotenoids. In higher plants, the biosynthesis of chlorophyll is usually decreased by exposure to high concentrations of ozone. But Rosentreter and Ahmadjian (1977) have shown that a one-week exposure to 0.1 ppm of ozone increased chlorophyll quantities in the lichen *Cladonia arbuscula* and phycobiont which includes free-living fungi *Trebouxia* as weel as lichens *Cladonia arbuscula* and *Cladonia stellaris*. Higher ozone concentrations had no effect. Increase of the chlorophyll content in response to the same total dose of ozone (0.015 ppm per day for two weeks) are observed in the vegetative microspores of the horsetail *Equisetum arvense* (Fig.40). The red fluorescence of individual cells and the concentration of chlorophyll increased in the ozonated cells in comparison with the control sample. An especially significant increase was observed upon the ozone concentrations of 0.007-0.03 ppm. The enhanced chlorophyll synthesis instigated by ozone may be explained by the participation of O_3 as an oxidant in the biochemical process. According to Roshchina, (unpublished data), this decrease in chlorophyll concentrations may also result from ozone doses of 0.07-0.1 ppm for 2 hours, although normal cell development is retarded in these cases (Roshchina and Melnikova, 2000). Perhaps, accumulation of chlorophyll plays, in such cases, a role in the antioxidant defense rather than in the normal processes of photosynthesis. If this is so, then many superoxide radicals are formed, and they serve as acceptors for protons and electrons emitted by chlorophyll. If the amount of chlorophyll is insufficient, subsequent destruction of the chloroplast structure results from effects of toxic radicals. Depending on plant taxon, higher concentrations of ozone (from 0.05 up to 12 ppm) decreased quantities of chlorophyll in Golosperms and Angiosperms (Horsman and Wellburn, 1973). In particular, chlorophyll loss from genera *Phaseolus*, *Zea* and *Glycine* occurred in response to exposures of 0.05-0.25 ppm of O_3 for 2-4 hours, from *Pinus* in response to exposures of 0.5 ppm for 9-18 days and from *Lemna* in reaction to exposures of 1 ppm for 4-24 hours. Chronic fumigation of ozone is more dangerous for chlorophyll than an acute exposure (Schreiber et al., 1978; Frage et al., 1991; Wingsle et al., 1992). The chlorophyll heterogeneity was also seen in many species under these conditions (Gaponenko et al., 1988). Chlorophyll "a" is destroyed faster than chlorophyll "*b*" or, at least, to an equal degree (König De and Jegier, 1968a; Beckerson and Hofstra, 1979 c).

Carotenoids are also lipophilic components found in membranes as parts of complexes involving proteins. They may additionally serve an antioxidizing function (see Chapter 4). Exposures to high concentrations of ozone (0.3 ppm for 8 hours per day over 10 days) caused the xantophyll cycle in seedlings of *Pinus*

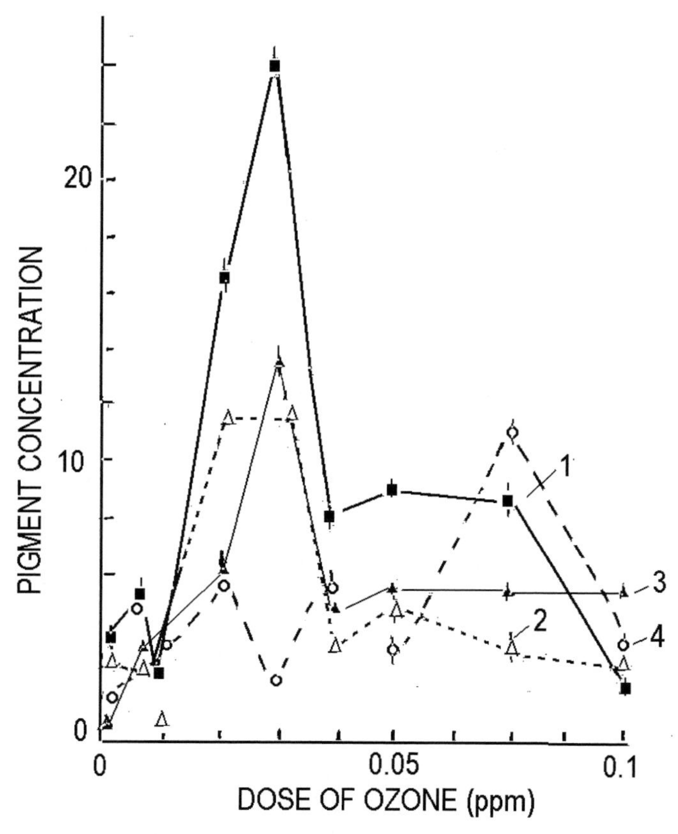

Figure 40. The "dose-effect" concentration curves for the pigment content (mg g^{-1} fresh mass) of the dry vegetative microspores of the horsetail Equisetum arvense after ozone treatment. 1 - chlorophyll a+b; 2 - chlorophyll a; 3 - chlorophyll b; 4 - carotenoids.

sylvestris to be changed—the level of violaxanthine grew by 19%, whereas the amount of zeaxanthin was lowered by 42 % (Wingsle et al., 1992). However, no observed changes occurred in the composition of carotenoids, such as α- and β-carotenes, lutein, and neoxanthin. In the same experiment (Wingsle et al., 1992), the high concentrations of ozone effected an 11% variation in the molar ratio of monogalactosyl diacylglycerol/digalactosyl diacylglycerol. The implication is that the enzyme galactolipid galactosyl transferase, located on the outside of the chloroplast membrane, is stimulated by ozone.

Under extreme ozone treatments, lipofuscin pigments, with a characteristic fluorescence of 420-580 nm in the visible part of a spectrum, are accumulated in animal cells (see section 3.1.). According to widespread views (Vladimirov and Archakov, 1972; Tappel, 1975; Singh etal., 1995), the occurrence of liposoluble fluoresced substances is associated with interactions involving the products of lipid peroxidation (in particular malondialdehyde) and compounds containing, free amino acids. Karnaukhov (1988; 1990) suggested that the formation of this yellow pigment of aging indicated a transformation of the carotenoids. Similar products are also formed in other plants in response to ozone action. For example, pigments of aging, fluorescing in the region 420-440 nm, are found in the lipid factions of the soya *Glycine soja* and the corn *Zea mays*, after these plants had been subjected to ozone fumigation (Brooks and Csallany, 1978). Ozone stimulated the synthesis of lipofuscin by 2.7 times. The pigment is also found in ripening fruits (Maguire and Haard, 1976) and aging leaves (Wilhelm and Wilhelmova, 1981). Furthermore, the formation of lipifuscin-like pigments have been found in the pollen of some plants (Roshchina and Karnaukhov, 1999) after chronic ozonation (0.05 ppm per hour for 100 h at 3 h per day, producing a final dose of 5 ppm). Fig. 22 shows that, after chronic ozonation (0.05 per hour for 100 hours at 3 hours per day, producing a total dose of 5 ppm), changes occurred in the fluorescence spectrum of pollen from the mock-orange *Philadelphus grandiflorus*. These pollen were rich in carotenoids, and the exposure to ozone caused a maximum at 540-550 nm, characteristic of carotenoids, to disappear and a new maximum at 440-480 nm to appear. The analysis of chloroform/ethanol extracts from this pollen indicated the presence of a pigment with maximum fluorescence at 430 nm, which is similar to that of lipofuscins. In objects, in which the concentration of carotenoids is small, that are shielded by other pigments, or in which any pigments are absent (as is the case for the plantain *Plantago major*), lipofuscin is not formed. The pigments of ageing that occur in plants have various natures, and the mechanism of their formation cannot be described within the framework of any uniform mechanism (Merzlyak, 1989).

5.3. METABOLISM OF NITROGEN-CONTAINING COMPOUNDS

The main effects of O_3 on the nitrogen metabolism of plants involve changes in quantities of amino acids and proteins, as well as in the synthesis and activity of enzymes (Table 19). Ozone causes changes in amino acids and proteins, acting on carboxylic and primary amino group (Rowland et al., 1988). Moreover O_3 induces the synthesis of new proteins that serves as the protector mechanism against oxidative stress. This gas also modifies the exchange of adenine nucleotides.

Most significant effect of ozone in plants is changes in contents of amino acids and proteins (Tomlinson and Rich, 1967). For instance, in leaves of beans *Vicia faba* concentration of glutathione falls after the ozone (1 ppm) treatment during 15 min, but the amount of γ - amino butyric acid increases. The fumigation by ozone in a mixture with O_2 sharply increased quantity of free amino acids. It has been shown ozone to strengthen the synthesis of alanine and serine in pines *Pinus strobus* and *Pinus taeda* (Ting and Mukerji, 1971). In the soya *Glycine max,* the 24 h -treatment with O_3 causes the increase in the content of proteins, a fact that also implies enhanced their synthesis. However, protein quantities in other plant species, for example *Ulmus americana*, greatly decreases under the influence of ozone (Constantinidou and Kozlowski, 1979), which means that either destructive processes in proteins are strengthened or their synthesis decreased. In leaves of the bean *Vicia faba*, the reduction in the total amount of proteins occurs even when the ozone dose is 0.08 pm for 1.5 hour/day over 40 days (Rowland et al., 1988). If proteins are bonded to sugars or pigments (in particular, to chlorophyll), they are more stable and less susceptible to damage (Rowland et al., 1988). Exposure to ozone suppresses the pools of metabolites in the tops and roots of plants to different degrees. In the tops of plants, there is no change in the amounts of nitrogen and amino acids, but the level of these metabolites in root systems is increased (Tingey et al.. 1976 a, b).

Table 19. Effects of ozone on the nitrogen exchange

Amino acids	*Amino acids*	*Synthesis of enzymes*	*Synthesis of enzymes*
Increase or decrease in the total amount of amino acids and aminobutyric acid (Tomlinson and Rich, 1967), alanine and serine (Ting and Mukerji, 1971	Decrease or increase of total protein amount or enhancement of biosynthesis of certain proteins (Constantinidou and Kozlowski, 1979; Rowland et al., 1988)	Synthesis of glucanases and chitinases in leaves of tobacco *Nicotiana tabacum* (Schraudner et al., 1993), guajacol peroxydase (Kronfuss et al., 1996).	Decrease in the activity of nitrate- and nitrite reductase in leaves of maize *Zea mays* and soja *Glycine max* (Leffler and Cherry, 1974). Increase in the activity of superoxide dismutase, peroxidase, catalase, glutathione reductase (Tuomainen et al., 1996)

In experiments on leaves of the corn *Zea mays* and the soya *Glycine max* in vivo, ozone's ability to inhibit activities of nitrite and nitrate reductase disrupts the entire nitric metabolism (Tingey et al., 1973c; Leffler and Cherry, 1974). However, a similar phenomenon has not been observed in vitro. It is possible that the natural event is not caused by a direct inhibition of enzyme activity but by an unbalanced formation of the reduced NADP necessary for enzymic activity. Of interest is the fact that the nitrate reductase in chloroplasts is more strongly inhibited than cytoplasmic nitrate reductase (Tingey et al., 1973 c).

The formation of protective proteins occurs under various stress conditions. Ozone causes the gene transcription to be amplified and the synthesis of many enzymes to be strengthened. An example of this process involving the synthesis of glucanases and chitinases has been studied in leaves of the tobacco *Nicotiana tabacum* (Schraudner et al., 199). The sets of enzymes whose replication is stimulated by ozone may vary, a diversity that is illustrated by comparing the consequences of O_3 fumigation and UV irradiation on leaves belonging to two lines (varieties) of the tobacco *Nicotiana tabacum*, one sensitive (BelW3) and the other tolerant (BelB) to ozone (Thalmair et al., 1996). Two days fumigation with ozone (0.16 ppm for 5 hours per day) induced the formation of β-1.3-glucanase in both lines and, moreover, chitinase synthesis in the sensitive strain. After ultra violet irradiation, the enzymes were not formed. If the fumigation by ozone lasted 32 hours, the protective protein R-10, with a molecular weight of 28 kDa, was collected in intracellular liquid (the primary extract). It is a 100 %. homologue of chitinase. After 10 weeks of fumigation by 0.1 ppm of ozone, the activity of guayacol peroxidase (EC 1.11.1.7) in apoplast of 4-year-old seedlings of the fir tree *Picea abies* decreased, and the quantity of malondialdehyde, indicator of lipid peroxidation, increased (Kronfuss et al., 1996).

Antioxidant isoenzymes, which function to eliminate free radicals, are located in several different subcellular components of the plant cell. Despite the theoretical possiblity that superoxide dismutase may participate in the protective mechanism, it appears not to play a key role in establishing ozone tolerance (Melhorn and Wenzel, 1996). In an ozone-sensitive clone of the birch *Betula pendula* after 24 hours of fumigation with ozone (0.15 ppm at 8 hours/day), the activity of some antioxidant enzymes—superoxide dismutase, peroxidase and glutathione reductase—is increased. Additionally, the synthesis of the gene, governing the formation of phenylalanine ammoniumlyase, catalyses the synthesis of phenylpropanoids (Tuomainen et al., 1996).

Such antioxidant enzymes as superoxide dismutase, peroxidase and catalase display high sensitivity to ozone. At lethal and sublethal doses of ozone (accordingly 0.25 and 0,125 ppm), the activity of superoxide dismutase and peroxidase in the spinach *Spinacia oleracea* is greatly augmented (Decleire et al., 1984). Lethal doses of ozone increased catalase activity and sublethal doses decreased it. The increase in peroxidase activity instigated by ozone has been noted in many studies (Tingey et al., 1976a; Curtis et al., 1976; Dass and Weaver, 1982; Decleire et al., 1984 as well as others).

Ozone effects that stimulate gene expression for the antioxidant enzymes have also been demonstrated (Table 20). When *Arabidopsis thaliana* is exposed to ozone, there is an accumulation of mRNAs encoding both cytosolic and chloroplastic antioxidant isoenzymes (Conklin and Last, 1995). The steady-state levels of three mRNAs encoding cytosolic antioxidant isoenzymes (ascorbate peroxidase, copper/zinc superoxide dismutase, and glutathione S-transferase) also increase. The glutathione S-transferase mRNA responds very quickly to the oxidative stress (2-fold increase in 30 min) and is elevated to very high levels, especially in plants grown with a 16-h photoperiod. In contrast, ozone causes a decline in the levels of two chloroplastic antioxidant mRNAs (iron superoxide dismutase and glutathione reductase) and two photosynthetic protein mRNAs (chlorophyll a/b-binding protein

Table 20. Stimulating effects of ozone on gene induction and expression of proteins

Metabolic way	*Protein, coding by RNA*	*Plant species*	*Reference*
Oxidative metabolism	Ascorbate peroxidase (cytosolic)	Transgenic tobacco *Nicotiana tabacum* plants	Örvar and Ellis, 1997
Oxidative metabolism	Ascorbate peroxidase, copper/zinc superoxide dismutase and glutathione S-transferase (cytosolic and chloroplastic)	*Arabidopsis thaliana*	Conklin and Last ,1995
Oxidative metabolism	Glutathione-S-transferase	*Arabidopsis thaliana*	Heidenreich et al., 1999
Lignin biosynthesis	Cinnamoyl alcohol dehydrogenase	Scots pine *Pinus sylvestris*	Wegener et al., 1997
Extracellular protein biosynthesis	Cytosolic ascorbate peroxidase	Transgenic tobacco *Nicotiana tabacum* plants	Örvar and Ellis, 1997
Stilbene biosynthesis	Stilbene synthase	Scots pine *Pinus sylvestris*	Zinser et al., 1998; 2000b
	Stilbene synthase and cinnamoyl alcohol dehydrogenase	Norway spruce *Picea abies*	Galliano et al., 1993
	Resveratrol synthase	Tobacco *Nicotiana tabacum*	Grimming et al., 1997
Pathogenetic	Pathogenesis-related proteins	Parsley *Petroselinum crispum*	Eckey-Karltenbach et al., 1994a,b; 1997b,
		Tobacco *Nicotiana tabacum*	Ernst et al., 1992 a,b; 1997 a,b

(Table 20 to be continued)

Metabolic way	*Protein, coding by RNA*	*Plant species*	*Reference*
Heat-shock	Heat-shock proteins		Eckey-Karltenbach et al., 1994a,b; 1997a,b
Stress-related	Stress-related peptides 12 and 50 kDa	Conifers	Großkopf et al., 1994
Synthesis of glycoproteins and related proteins	β-1,3-glucanase, chitinase	Tobacco *Nicotiana tabacum*	Ernst et al., 1992 a,b; 1997 a,b
	Extensin (glycoprotein)	Norway spruce *Picea abies,* scots pine *Pinus sylvestris,* european beech *Fagus sylvatica*	Schneiderbaur et al., 1995

and a small subunit of ribulose-1,5-bisphosphate carboxylase/oxygenase). This decline does not include all mRNAs, encoding chloroplast-targeted proteins, since O_3 causes an increase in the mRNA, encoding the tryptophan biosynthetic enzymes phosphoribosylanthranilate transferase. In *Arabidopsis thaliana*, mRNAs ,encoding glutathione-S-transferase and several pathogenesis-related proteins, were not induced in plants grown on a mercury-containing medium, and a similar picture was observed for plants exposed to ultra-violet irradiation (Heidenreich et al., 1999). Only ozone fumigation alone temporarily increased the level of all these transcripts. According to data provided by Pino and Mudd (1995), ozone not only modifies proteins, but also changes the manner in which they are synthesized. Exposed to 0.36 μl/L (ppm) of ozone, the synthesis of protein 32 κDa in chloroplasts of the corn *Zea mays* is reduced in accordance with the increase in ozone concentration.

Ozone changes the level of the redox enzymes superoxide dismutase and glutathione reductase (Maccarrone et al., 1997). In response to oxidizing stress, the activity of glutathione reductase changes, as shown by changed level of synthesis of the enzyme in the transgenic lines of the tobacco *Nicotiana tabacum* (Broadbent et al., 1995). Moreover, ozone stress causes the excretion of ethylene to be amplified, and it correlates with accumulated amounts of the enzyme synthetase of 1-amino-cyclopropan-1-carboxylic acid in the tomato *Lycopersicon esculentum* (Bae et al.,

1996). Especially particular to tobacco, ozone also induced a change in the level of the mRNA responsible for synthesizing β-1,3-glucanase , chitinase and protein 1b, all concerned with pathogen-induced products (Ernst, et al., 1992 a).

In an experiment on *Euglena gracilis* (Chevrier and Sarham, 1992) involving the rather high quantities of ozone in the air (240 μl/L or ppm with a rather high rate of diffusion: about 1 μmoles/min), rapid changes occur in the cellular metabolism of the adenine nucleotides. Their total concentration decreased by 15 % after 15 min of ozone action. This change is associated with a reduction in the contents of ATP and ADP, whereas AMP had practically no change. Simultaneously, there were no appreciable changes in the adenylate concentration as a whole and in the ratio ATP/ADP. These results assume that, during the first minutes of fumigation with ozone, the size of the nucleotide pool is regulated to maintain an acceptable energetic balance during ozone stress. On the other hand, the rate of photosynthetic fixing CO_2 does not change during the first 30 min of the fumigation with ozone, indicating that the changed level of adenine nucleotides does not influence CO_2 fixation during this time. However, the rate of the carbon dioxide fixation begins quickly to be reduced after 60 min of exposure to ozone, and at the end of its action (at 120 min) is inhibited to a 75 % level.

After terminating the ozone treatment, the level of adenine nucleotides and the fixation of CO_2 by cells of *Euglena gracilis* during photosynthesis return to normal levels in 5 hours. Consequently, the changes caused by ozone are convertible and the restoration period results in strengthened reparation after the ozone damage (Chevrier and Sarham, 1992). As a whole, the ATP level due to a inhibition of oxidative phosphorylation starts to decrease when exposed to 0.005-0.015 ppm of O_3 for 35-154 day of exposure, or to 1 ppm for 0.5-1 h (Horsman and Wellburn, 1973).

5.4. METABOLISM OF SECONDARY COMPOUNDS.

As shown above, the effects of ozone and, especially, its phytotoxicity is displayed as changes of the primary metabolism, such as photosynthesis, respiration, etc. However, the secondary metabolism is an even more vulnerable target for O_3, since ozone influences activities of genes and enzymes as well as the biosynthesis of secondary products (Pell, 1988; Sanderman et al., 1990; Langebartels et al., 1990; 2000; Galliano et al., 1993 a, b; Sandermann, 1996; Sanderman et al., 1998). Table 21 shows the main secondary metabolites of which the biosynthesis is sensitive to ozone. These mainly include phenols, amines and hydrocarbons, all of which often may serve as antioxidants (see chapter 4). Colored compounds also indicate the damaged parts of the ozonated tissue of leaves and stems, if the concentration of O_3 is rather high.

5.4.1. Phenols

When the steady-state pool of phenolic compounds is exhausted under ozone, the amount of phenols is increased due to the enhanced activity of phenylalanine ammonia lyase, polyphenoloxidase and peroxidase (Tingey et al., 1975a; 1976; Curtis et al., 1976). Phenol production is stimulated even before any visible leaf damage appears and remains at a high level for several days. The mechanism that allows phenols to act as antioxidants against ozone and its intermediates has been described in chapter 4. But we should, at this point, note that phenols function not only as protectors. In some cases, the toxicity of phenolic compounds can be

Table 21. Effects of ozone on secondary metabolism

Phenols	*Amines*	*Hydrocarbons*
Increase in total amount of phenols (Tingey et al., 1975a; 1976; Curtis et al., 1976).	Accumulation of polyamines - putrescine, spermidine and spermine (Reddy et al., 1993)	Formation of stress ethylene (Craker, 1971; Pell, Puente, 1986)..
Increase in enzyme activity - peroxidase (Dass, Weater, 1972; Tingey et al., 1975.), Phenylalanine ammonia lyase, polyphenoloxidase (Tingey et al., 1975a; 1976; Curtis et al., 1976), dehydrogenase of cinnamic alcohol (Galliano et al., 1993 a,b) Accumulation of isoflavonoids, flavon glycosides, furanocoumarins and flavons (Keen and Taylor,1975) and salicylic acid (Rao and Davis, 1999) Induction of stilbene biosynthesis (Rosemann et al., 1991). Formation of anthocyanins - red pigmentation in leaves (Konkol and Dugger, 1967) Induction of catechin biosynthesis (Hiller et al., 1990).	Increase in enzyme activity of arginine decarboxylase (Langebartels et al., 1991).	Increase in enzyme activity of the ethylene biosynthesis

increased by oxidation, particularly due to the formation of quinones and superoxide anion radicals derived from phenols (Averyanov, 1984). Such occurrences are, however, observed mainly when phenols are exposed to doses of ozone high enough to produce visible damage in cells.

O_3 influence not only to promote the liberation and transformation of phenolic compounds, already present in a cell, but also stimulates the phenolic metabolism as a whole, so that newly synthesized phenolic derivatives are generated. Ozonation causes plants to respond to the stress conditions, in particular isoflavonoids to accumulate, by switching on appropriate metabolic processes (Keen and Taylor, 1975). When treated with 0.2 ppm of ozone for 2.5 hours, the lucerne *Medicago sativa* accumulates 4,7-dihydroxyflavone in concentrations that greatly increase as the symptoms of visible damage begin to appear (Hurwitz et al., 1979). Under ozone stress, salicylic acid is accumulated; it then functions as an antioxidant, signaling agent for a hypersensitive cellular response and instigating apoptosis-like cellular damage (Sharma et al., 1996; Rao and Davis, 2001). The compound may also regulate cell death (Rao and Davis, 2001).

Various sorts of stress conditions induce plants to synthesize protective compounds, be they antioxidants, such as flavone glycosides (resulting from ultra-violet irradiation) or furanocoumarin phytoalexins (in response to the elicitors of fungal stress). Flavone glycosides are located mainly in the epidermal layer of a leaf, and furanocoumarins in leaf oil ducts. The similarity between the effects of ozone, radiation and fungal elicitors on protective biosynthetic processes has been suggested by Eckey-Koltenbach et al. (1994) as follows:

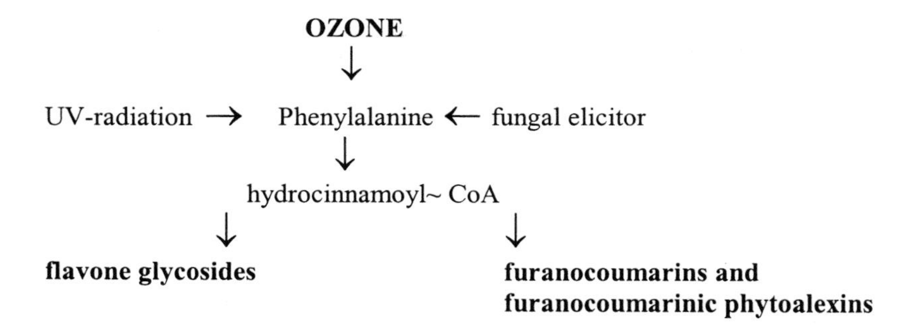

Both pathways (synthesis of flavones and furanocoumarins) were activated in an experiment on *Petroselinum crispum* L. in which the plants were fumigated with 20 nl/L of ozone for 10 hours. However, the maximum amounts of flavine glycosides and furanocooumarins are, respectively, induced after 12 and 24 hours of the treatment (Eckey-Kaltenbach et al., 1994). Thus, ozone acted as a cross-inducer, because the two different biosynthetic processes were simultaneously stimulated.

Ozone stimulates the activity of enzymes participating in the phenolic metabolism. Increases in the activity of phenylalanine ammonia lyase, polyphenoloxidase and peroxidase can stimulate the oxidation of phenols into quinones and cause the products of polymerization to accumulate (Tingey et al. 1976a). Accordingly, the dehydrogenase of cinnamic alcohol is accumulated in the ordinary pine *Picea abies* both in response to ozone and to the action of elicitors involved in a parasitar invasion (Galliano et al., 1993 a, b).

The activity of the enzymes participating in the phenolic metabolism is suddenly suppressed after a two-hour exposure to 0.5 μl/L concentrations of ozone but, when subsequently restored, exceeds the activity level in control plants. Such an increase in peroxidase activity after O_3 treatment has been described in many papers (Dass and Weaver, 1972; Tingey et al., 1975; as well as others). The analysis of the isoenzymes found among the peroxidases in two different grades of ozone sensitivity in the soya *Glycine max* has shown that all isoenzyme activity is increased by ozone in the sensitive strains whereas only a few isoenzymes have enhanced activity in the tolerant line (Curtis et al., 1976). In general, sensitive strains accumulate more phenols than tolerant ones, a tendency that was confirmed in tests on various grades of the arachis *Arachis hypogaea* (Howell, 1974).

Ozone-induced changes in metabolic processes are visually expressed as an increase in pigmentation (Howell and Kramer, 1973; Korhonen et al., 1998). This phenomenon occurs by inducing pigment synthesis or blocking ways of the utilization of the pigment intermediates. The most visible symptoms of ozone damage are the pigmentation on plant leaves. Pigmented sites can contain phenols, for example anthocyanins, or polymerized products of their reaction with ozone.

Phenolic compounds participate in pigment formation when subjected to ozone stress. In particular, anthocyanins can be synthesized and accumulated under normal conditions in all vegetative organs, flowers, leaves, roots and seeds, but under stress their contents increase. An example of this response is provided by the anthocyanin formation (red sites) in *Rumex crispus* (Konkol and Dugger, 1967). The change in the amounts of red pigments (the characteristic color of anthocyanins in an acid medium) after treatment with 2 to 12 μl/L (ppm) ozone for several days can be seen in an experiment on the flowers of the petunia *Petunia hybrida* and the geranium *Pelargonium hortorum*, as well as on the bracts of the poinsettia *Euphorbia pulcherima*, (Craker and Feder, 1972). The total amounts of anthocyanins in extracts (methanol- 1 % HCl) from the above-mentioned items were determined by measuring the absorbance at 525 nm. It was shown (Craker and Feder, 1972) that the anthocyanin content in the flowers of both plants was reduced when the ozone concentration was increased > 6 ppm. The crude weight of the flowers simultaneously grew heavier. Conversely, when the bracts from the poincettia *Euphorbia pulcherima* were exposed to 6-12 ppm of O_3, they displayed an increase in pigment quantity, while their weight and corresponding volume decreased. Thus, a low concentration of ozone can influence blossoms, even though no visible damage is observed in the tissues.

Accumulation of phenols in response to ozone occurs in many species of plants and is connected with some visible signs of leaf damage (Howell and Kramer, 1973). It is assumed that, when ozone enters a cell, oxidizing enzymes react with phenols to produce such compounds as quinones, which then form polymers with proteins and amino acids. As a result, the pigments are formed that are considered as visible symptoms of ozone damage.

The phenols, participating in the stress metabolism, include stilbenes (Fig.41). Ozone in concentrations of 0.15 or up to 0.2 μl/L (ppm) was shown to induce 100-1000 times higher production of stilbenes (natural C6-C2-C6 phenol compounds like pinosylvin and pinosylvin-3-methyl ether) in the primary branches of 6-week-old pine (*Pinus sylvestris*) seedlings. The effect was a consequence of the increased activity of the pinosylvin-forming enzyme, stilbene synthetase (Rosemann et al., 1991). Damage was discernible after no less than 4 hours of ozone exposure. In the

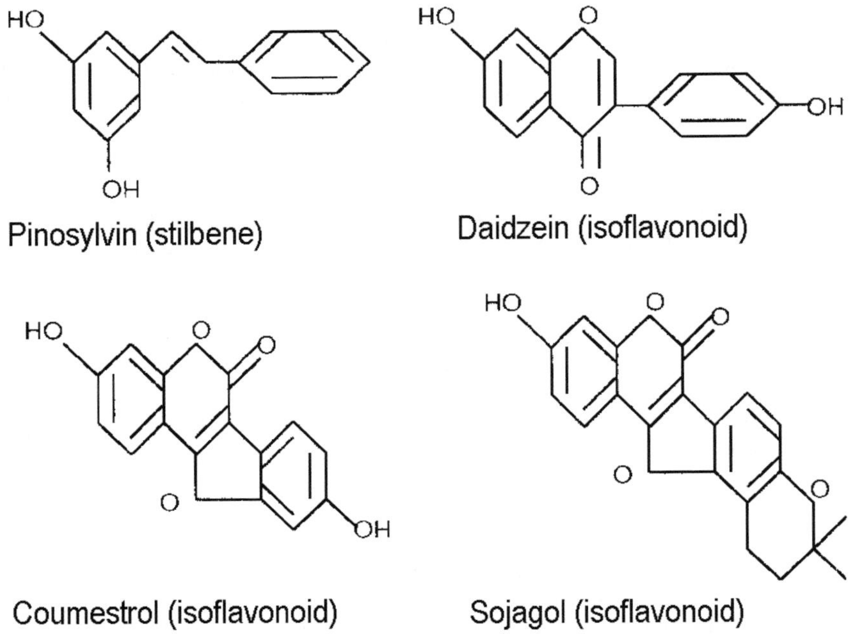

Pinosylvin (stilbene)

Daidzein (isoflavonoid)

Coumestrol (isoflavonoid)

Sojagol (isoflavonoid)

Figure. 41. Stilbenes and flavonoids, formed under ozone stress

same conditions, the enzyme activity of phenylalanine ammonia lyase and chalcone synthetase, both of which participate in the synthesis of flavonoids, was doubled (Rosemann et al., 1991). The stimulation of the gene expression for enzymes of the stilbene biosynthesis is shown in table 20. Ozone is believed to be an abiotic elicitor of plant defense reactions (Sandermann et al.1998).

Taking into account the specificity of the induced stilbene biosynthesis and its dependence on the concentration of O_3, the possibility exists to use the quantity of these substances as a biological marker in ozone stress studies on pine trees. In coniferous plants, such as the pine *Pinus* and the fir tree *Picea excelsa* Link, ozone not only induces the formation of stilbenes but also that of catechins (Hiller et al., 1990).

Pigmented tissue sites other than those produced by anthocyanins contain polymers of amino acids with metals, sugars, phenols, that resemble plant reactions to biogenic pathogens. Such pigmentation also occurs as a result of the ethylene increase in membranes due to stress, and the accumulation of phenols in the resulting damaged tissue (Howell and Kremer, 1973). The increase in the pigmentation is connected with a strengthening of the activity of polyphenoloxidase and phenylalanine ammonia lyase

Special consideration should be given to the data on the phenol formation induced by the mutual influence of ozone exposure and parasitical infestation (Heagle, 1982; Heagle and Strickland, 1972; Heagle and Key, 1973; Heagle et al., 1979). Such combined phenomena are important for the manufacture of grain, where ozone effects can be aggravated by an infection. In field conditions, it is, however, rather difficult to isolate the effects of these factors, as plant respond to both in almost the same way (Heagle and Key, 1973).

5.4.2. Hydrocarbons

Among hydrocarbons excreted during ozone fumigation, the main one is ethylene, released by a plant under both normal and stressed conditions. However, after the homogenization of a tissue or its death, the gas will not be produced. Hence, its formation is peculiar only to living cells. The synthesis of ethylene depends on the degree of damage inflicted on plants by ozone (Craker, 1971; Reddy et al., 1991; Kettunen et al., 1999). The primary process involved in its biosynthesis (Adams and Yang, 1981; Yang and Hoffman, 1984) 1981; Yang and Hoffman, 1984) is the transformation of S-adenosylmethionine into I-aminocyclopropane-I-carboxylic acid (ACC), the process catalyzed by ACC.synthase This latter compound then interacts with the ethylene-forming enzyme to produce ethylene. In this scheme, ACC synthase serves as a target for ozone:

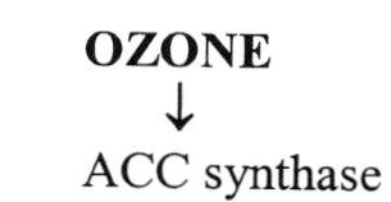

Methionine → S-adenosylmethionine → →1-amino-cyclopropane-1-carboxylic acid

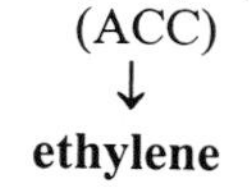

Plant damage instigated by ozone leads to the production of ACC synthase and an invigorated ethylene synthesis. The reaction is not specific to ozone. Mechanical damage, cooling, drought, flooding, pathogens, aging, maturing of fruits all induce production of the enzyme, which then participates in the formation of ethylene. Any stress appears to promote ethylene formation. This process can be considered as a protective reaction by the plant to external influences, as the increase in the ethylene concentration within plant slows down growth and promotes leaf fall.

When the plant is treated with ozone (0.25 μl/L for 2 hours) in a hothouse, the release of ethylene is changed. Stimulation of ethylene release by ozone depends on plant species and can vary from 1.5–4 times that of the untreated control sample. Ethylene synthesis was observed when plants were exposed to air containing 25μl/L (10^{-6} %) ozone for 2 hours (Craker, 1971). In hothouse conditions, the fumigation of plants with 588 μl/m^{-3} of ozone for 3 hours also stimulated the formation of ethylene (Pell and Puente, 1986). The petunia *Petunia nyctaginiflora* was especially sensitive to ozone, because, after fumigation with O_3, a drop in flower kidneys and a reduction in the flower amount were both observed (Posthumus, 1982).

Mehlhorn and Wellburn (1994) believe that free radicals (see section 2.6) are formed when plants are exposed to ozone. Such radicals injure leaves and stimulate the generation of the stress ethylene. By using spin traps (scavengers), it was established that free radicals cause ethylene to be formed from 1-amino-cyclopropane-1-carboxylic acid (Legge et al., 1982). The use of electronic paramagnetic resonance (EPR) made it evident that the formation of free radicals occurred simultaneously with leaf damage in the pea *Pisum sativum* and the string bean *Phaseolus vulgaris*. They clearly represent a cellular response to fumigation with ozone at 70-300 nl/L or 0.07-0.3 ppm (Mehlhorn et al., 1990). Thus, the release of ethylene increased from 2.5 nmole/g of dry leaf weight per hour (in the control) to 6.2-16.5 (in the variant treated with O_3). The formation of the stress ethylene is connected to free radical processes in lipid peroxidation that cause damage in membranes and stimulates excretion of this gas. The toxicity of ozone correlates with the formation of ethylene. If the synthesis of ethylene is inhibited, for instance, by aminoetoxivinylglycerol, the sensitivity of plants to ozone decreases.

The rates of ethylene formation depend on the concentration and duration of the ozone action (Mehlhorn et al., 1991; Kangasjärvi et al., 1997; Tuomainen et al., 1997). For instance, the production of ethylene was increased under the treatment with 50-150 ppm O_3 for 3.5 hours per day, but longer exposure, for 7 hours per day blocked ethylene synthesis (Mehlhorn et al., 1991). The plants growing through long exposures to ozone have lower rates of ethylene formation than those subjected to short exposures. This is probably the case because the short exposures to ozone can be more phytotoxic (Mehlhorn et al., 1991). After short expositions to ozone, the plants of the pea *Pisum sativum* released, however, large amounts of ethylene and had high percentages of leaf damage. In another series of experiments (Mehlhorn et al., 1991) on the tobacco *Nicotiana tabacum*, the release of ethylene increased from 140 picomiles per g of dry leaf weight in the control sample to 230 picomiles per g of dry leaf weight after short exposures to ozone and 1002 picomiles per g of dry leaf weight after average ozone exposures. In the first case, damage was visible on more than 50 % of leaf, and in the second, on more than 80 %. A direct correlation between formation of stress ethylene and sensitivity of a plant to ozone has also been shown (Mehlhorn et al., 1991). Lines of the tobacco *Nicotiana tabacum* tolerant to ozone have less visible damage after O_3 exposure than sensitive lines due to the fact that the former produce ethylene in smaller amounts. The difference in ethylene production in both strains grown in pure air is not very large. The lines of tobacco sensitive to ozone lose its sensitivity during flowering, distribution of pollen and formation of seeds. In all these conditions, the sensitive strain released large quantities of stress ethylene (Mehlhorn et al., 1991).

The release of other hydrocarbons in response to ozone fumigation has not yet been studied. These include such volatile products of lipid peroxidation as ethane, pentane, isoprene and isoprenoid-related terpenes (see chapter 3).

5.4.3. Polyamines

One of the symptoms of ozone stress is the accumulation of polyamines such as putrescine, spermidine and spermine (Fig.42). The precursor to their synthesis is the

Putrescine $NH_2 -(CH_2)_4 - NH_2$

Spermidine $NH_2 -(CH_2)_4 -NH - (CH_2)_3 - NH_2$

Spermine $NH_2 - (CH_2)_3 - NH - (CH_2)_4 - NH -(CH_2)_3 -NH_2$

Figure 42. Polyamines formed under ozone stress.

amino acid arginine, from which urea and ornitine are formed in the presence of the enzyme arginase. Putrescine (tetramethylenediamine) is synthesized in the enzymic decarboxilation of ornithine. Spermidine (N-(3-aminopropyl)-tetramethylene diamine) or spermine (N, N'-bis (3-aminopropyl) tetramethylenediamine) are formed in the reaction of putrescine with methionine.

Ozone induces the biosynthesis of polyamines (Reddy et al., 1993). Several authors (Elstner et al., 1985; Heath, 1988; Federico and Angelini, 1986) have shown that the amounts of spermidine in the barley *Hordeum* L., the wheat *Triticum* and the fir tree *Picea* Link change when these plants are exposed to ozone.

Under the influence of ozone, plants of the tobacco *Nicotiana tabacum* quickly increased their levels of free and conjugated putrescine (Langebartels et al., 1991). The induction of putrescine and spermidine conjugates occurs when the doses of ozone are great enough to induce damage. Under conditions of stress caused by ozone, the more significant role belongs to putrescine and the less significant one to spermidine (Langebartels et al., 1991).

Plants tolerant and sensitive to ozone involve different mechanisms for inducing polyamines. In the presence of ozone, O_3-sensitive strains, such as the Bel W3 of the tobacco *Nicotiana tabacum*, display an increase in the activity of arginine decarboxylase, and a rapid accumulation of putrescine, reaching maximum levels after 8 to 12 hours. In such cases, visible leaf damage is not observed. In the common biochemical response of plants to ozone, the induction of putrescine induction is an early reaction. A tolerant strain of tobacco, such as Bel B, only manifests a temporary increase in the activity of arginine decarboxylase, and the high level of putrescine is maintained for 3 days after the ozone treatment. The second type of the putrescine induction is slower and correlates with the appearance of leaf damage. Unlike tolerant strains, ozone-sensitive plants are characterized by the accelerated formation of ethylene and high levels of I-aminocyclopropane-I-carboxylic acid, an intermediate of ethylene synthesis (Langebartels et al., 1991).

Besides ozone, the polyamine accumulation is induced by other stress factors. In the horse bean *Vicia faba*, for instance, an NaCl surplus or a K^+ deficiency also increases putrescine production (Smith, 1985). The accumulation of the polyamine spermidine in the seedlings of the cucumber *Cucumis sativus* has also been shown to occur upon cooling (Wang, 1987). As putrescine is formed from the decarboxylation of ornithine in the presence of the decarboxylases of amino acids under normal conditions, the same enzymes appear to be activated under stress.

The polyamines resulting from ozone action can have more than one function. They are supposed (Sandermann, 1996; Sandermann et al., 1989; Langebartels et al., 1991; 1997) to protect plants from O_3 damage by reducing the production of the ethylene formed in response to ozone activity (see the previous section). In addition, polyamines are effective scavengers of the oxyradicals produced during ozonation. Monocaffeylputrescine, feruloylputrescine and coumaroylputrescine, derivatives of putrescine that are found in plants, are also scavengers of free oxyradicals. They have been observed in the apoplast liquid in leaves of the tolerant strain of tobacco, and their amounts were seen to increase in proportion to the exposure to ozone. The formation of polyamine conjugates also occurs when a virus or fungal pathogens invade mozaic tobacco. These conjugates are accumulated in the living cells surrounding necrotic spots. Accounts indicate that, if the cell wall constitutes about 50 % of the fresh weight of a cell, and consists of 60 % water, the concentration of caffeylputrescine in the apoplastic liquid is nearly $2x10^{-5}$ M in the control sample and $4x10^{-4}$M- in plants treated with ozone (Langebartels et al., 1991).

Providing nutrition to the tomato *Lycopersion esculentum* and the tobacco *Nicotiana tabacum* complete with exogenous polyamines appreciably suppresses the leaf damage discernible by pigmentation (Langebartels et al., 1991). Polyamines also eliminate root damage (Bors et al., 1989). Specific inhibitors (for instance, of arginine decarboxylase - difluoromethylarginine) increase the ozone damage in leaves by preventing polyamine synthesis. Thus, initiating processes of polyamine and ethylene formation can contribute to a plant 's ability to resist ozone.

CONCLUSION

Ozone induces changes in cellular metabolism that are either adaptive or damaging. The first type of change stems from the reparations that occur after the initial change in metabolism has taken place. In such cases, the presence of O_3 causes defensive metabolites, mainly low and high molecular antioxidants, to be accumulated or synthesized de novo. Usually these substances are protective proteins (enzymes or stress-related proteins) and phenolic compounds. The damaging type of metabolistic modification is seen when injuries are inflicted to some tissues, even to the extent of causing cell death. Necrotic processes may also be retarded by the synthesis of the protective substances in surrounding parts. Changes in cellular metabolism (enhanced release of ethylene, accumulation of polyamines, phenols and peroxidases) can, consequently, serve as markers of increased ozone concentration.

CHAPTER 6

CELLULAR MONITORING OF OZONE

Nowadays, the effects of ozone on plants is an important problem in the field of environmental monitoring (Feder, 1978; Feder and Manning, 1979; Treshow and Anderson, 1989; Sandermann, 1996; Sandermann et al., 1997; Kley et al., 1999; Langebartels et al., 2000). Earlier publications, summarized in several reviews and general monographs, concentrated on the analysis of the plant damage caused by high concentrations of O_3 and the diagnostic methods for assessing such impacts. This monitoring involved issues concerning both plant communities (natural biocenosis, population, succession or agricultural plantations) and individual plant organisms. Such controls required that the parameters of normal growth and development be estimated, as well as the degree of foliage damage. References to the main publications and Internet files along these lines are included in appendices 1 and 2. The subject of this chapter mainly concerns the possibilities for monitoring ozone at the cellular level, and involves considerations of plant cell sensitivity to ozone, search of potential cellular models and bioindicator processes for detecting ozone effects.

6.1. SENSITIVITY OF PLANT CELLS TO OZONE

The effects of ozone on plants can be sharp (the so-called acute consequences of exposure to high concentrations of the gas for short time intervals) or chronic (resulting from exposure to low concentrations over long periods). The chronic ozonation of plants results in the occurrence of either adaptive or harmful changes. Thus, various plant parameters—at the level of the individual cell, tissue or entire organism—can be employed to analyze ozone influence. In general, plants are rather highly sensitive to ozone, although the sensitivity of a plant will depend on the effective ozone concentration or dose. The ability of particular species and varieties to be tolerant or sensitive to O_3 is determined either by the genome as a whole or by separate genes, in combination with other external and internal environmental factors.

The O_3 amount of air determines the effectiveness of ozone action. Low quantities of ozone constantly exist in nature, and the compound only becomes a pollutant when concentrations exceed a certain limit (Treshow, 1984). Under natural conditions, this level can vary as a consequence of many factors, but ozone concentrations usually range from 0.03 to 200 ppm. A biologically effective aacumulated dose of O_3 is usually estimated as one that visibly injures the plant. Historically, this approach is used to evaluate ozone pollution when the concentration of the gas is high, as well as any resulting forest decline (Sanderman, 1996; Sandermann et al., 1997). However, studies have also revealed stimulant effects of small doses of O_3, for instance, on the growth and photosynthesis of lichens (Nash and Sigal, 1979), the regenerative capacity of moss (Comeau and Le Blanc, 1971) and pollen germination in general (Roshchina and Melnikova, 2000, 2001). As a consequence, it has become necessary to distinguish favorable and detrimental ozone concentrations.

6.1.1. Dependence of plant sensitivity on the ozone dose

As mentioned above, it has been conventional to determine biologically effective ozone doses on the basis of the visible symptoms of plant damage (Guderian, 1985). In estimating the effects of ozone concentration and exposure duration on living objects, it is usual to employ the notion of a dose. The minimum dose capable of causing effects is referred to as the threshold. If this threshold amount is quite small, then the plant involved is considered to be especially sensitive to ozone. Sensitive, intermediate and tolerant kinds of organisms are therefore distinguishable.

The interrelation between concentration and duration of ozone exposure does not permit precise definition of a single threshold dose, because frequently the same degree of ozone damage can be caused by exposures to different concentrations for different lengths of time. For example, the same effects may arise in response to an acute (sharp) high dose (e.g. 0.1 ppm or 196 $\mu g/м^3$ for 1 hour), or a chronic but small concentration of the gas (0.05 ppm or 98 $\mu g/м^3$ for 10 hours; see "Diagnosing Vegetation Injury Caused by Air Pollution, 1978" for more details). Determing a minimum threshold dose that causes visible damage in sensitive plant species is, nevertheless, a very important undertaking (Feder, 1978). The cultivar Bel-W3 (sort, variety, clone) of the tobacco *Nicotiana tabacum* is sensitive to ozone, and its leaves are damaged by a threshold dose of 0.05 ppm for 4 hours (2×10^{-9} mol/l for 4 hours; Heggelstad and Menser, 1962). Though this threshold can vary depending on external environmental conditions, it is accepted as the basic minimum value for a hazardous dose. A scale has been constructed indicating the extent that damage is inflicted on an injured leaf (% leaf surface injured) depending on the duration of the ozone fumigation. The linearity in a curve "dose-effect"for the monthly exposure to O_3 is observed only in the area within a radius ¼ mile around the relavant plant. Practically such curves mainly demonstrate a nonlinear character.

The threshold doses for the majority of sensitive plant species are relatively similar (Table 22). For instance, damage is observed in Monocotyledoneae (Monocotyledons) plants with narrow leaves, such as the corn *Zea mays* and the oat *Avena sativa* when they are exposed to 0.10-0.12 ppm of ozone for 2 hours. Icn contrast, similar damage occurs to sensitive strains of Dicotyledoneae (Dicotyledons), such as the Bel W3 cultivar of the tobacco *Nicotiana tabacum* or the white pine *Pinus album*, when fumigated by 0.05-0.07 ppm of the gas. The sensitivity threshold also depends on individual peculiarities of the plant organism (Karnosky, 1976). As a whole, very sensitive plants are damaged by ozone concentrations of 0.05-0.1 ppm. According to international standards, normal levels of O_3 in the air should constitute a dose of 0.08 ppm (157 μg/м^3) for 1 hour once a year. In urban regions it is, however, much higher.

Table 22. The threshold ozone dozes that cause damage in plants of different ozone sensitivities (1 ppm = 1960 μg/м^3) (Heggelstad and Heck, 1971).

Duration of the action (hours)	*Concentration of ozone, which induces damage in*		
	sensitive plants	*intermediate plants*	*tolerant plants*
< 0.5	0.15-0.30	0.25-0.60	
1.0	0.10-0.25	0.20-0.40	< 0.35
2.0	0.07-0.20	0.15-0.30	< 0.25
4.0	0.05-0.15	0.10-0.25	< 0.20
8.0	0.03-0.10	0.08-0.20	< 0.15

Today, the term " effective dose" is not only applied to the amount required to inflict leaf damage, but also to the quantity resulting in diminished growth and development of both the plant as a whole and individual cells in particular. The expression specifically refers to doses that affect the cellular state, cell division, any processes leading to cell death, or any other identifiable cellular reactions (Sandermann et al., 1997; Rao and Davis, 2001; Roshchina and Melnikova, 2001). Low (< 0.05-0.1 ppm or μl/L) and high doses of ozone act differently on plant cells

Small doses. There are only a few studies dealing with the influence of small O_3 doses on plants. In low concentrations (< 0.05-0.1 ppm), this gas does not suppress plant reactions but stimulates them. For example, exposure to 0.25-1.0 ppm ozone for 4 h stimulated the regenerative capacity of the moss *Funaria hygrometrica*, while a 6- 8 h exposure inhibited this capacity (Comeau and Le Blanc, 1971). Combined with abundant nitrogen nutritients, ozone can increase yields of the tobacco *Nicotiana tabacum* (Cowing and Koziol, 1982). Chronic exposure to low doses of the oxidant (nearly 4 hours per day for several days resulting in a total dose smaller than 0.05 ppm or 2 x 10^{-9}-4 x 10^{-8} M/L) stimulated in-vitro pollen germination (Roshchina and Melnikova, 2000). It is our opinion that more studies on the effects of low doses of O_3 need to be undertaken.

Large doses. Increase in the size of the ozone dose to a level corresponding to the concentrations that result from urban pollution of the environment or that appear to arise during continuous thunder-storms (more than 1 μl/L or ppm, or 4x10^{-8} M/L) induces visible damage in cellular systems. These injuries can be determined by analyzing the external state of the entire plant, any changes in the absorbance or fluorescent spectra of the plant or individual intact cells,as well as any changes in biochemical processes (see chapters 2 and 3). Harmful effects become visible only at high concentrations of atmospheric ozone and involve leaf damage, along with the decline of forest and agricultural plantings as a whole. Guderian (1985, 1997),

Sandermann (1996) and Sandermann et al. (1997) have all described this problem in detail.

6.1.1.1. Dose-dependent sensitivity of plant cellular reactions

On the cellular level, various reactions have different sensitivities to ozone, and a summary of them is provided in table 23.

In comparing the cellular responses to O_3 concentration and duration of fumigation, it is also possible to discern differences in the effects of acute and chronic ozone treatments. In the first case, the cells in plants that are treated with high concentrations of ozone for several minutes become increasingly more permeable to ions, their essential components are oxidized, and the activity of their enzymes affected. In the second case, the changes develop much more slowly, and only several reactions (autofluorescence of intact cells, pollen germination, level of disulfides, activity of some enzymes) show high sensitivity to chronic fumigation by low doses of ozone. Moreover, these ozone-sensitive reactions are observed only in the sensitive cells of some organs or sensitive plant species. Such cells or plant species could serve as indicators of ozone hazards (see sections 6.2.and 6.3).

The changes in cellular reactions caused by ozone may not only be negative. It is well known that molecular oxygen is necessary for pollen germination (Stanley and Linskens, 1974). Like O_2, O_3 in small concentrations (< 0.05 ppm) also stimulates this process, and therefore may be favorable for normal cell development. Moreover, in response to chronic low doses of ozone, antioxidant synthesis is activated, so that any plant tissues already altered or even injured may be repaired. For instance, the characteristic green-yellow fluorescence on the topside of the leaves of *Zea mays, Raphanus sativus, Hippeastrum hybridum* and *Plantago major* appears after fumigation with 1.5-0.45 ppm of ozone, and before any visible damage is seen (Roshchina and Melnikova, 2000). Following exposure to O_3, reparation of the leaf plate occurs in the first three plant species, even when this fluorescence has disappeared, perhaps due to the synthesis of antioxidants or activation of the antioxidant enzymes superoxide dismutase and peroxidase in the species (when total doses of ozone > 0.12 ppm). Only in *Plantago major* did similar conditions result in necrotic features developing on the sites of the green-yellow fluorescing spots, and after at least 30-day exposure, the seedlings were dead. Consequently, this plant species is very sensitive to ozone as a whole because its pollen germination diminishes even at lower concentrations of ozone (0.05 ppm). Visible damage, involving the destruction of chlorophyll (see symptoms of chlorosis and bleaching in the diagnostics of damage symptoms shown in table 24 and appendix 2) and simultaneous decrease in photosynthesis, is seen in sensitive organisms at concentrations < 0.05 ppm. However, pigmentation (stipples), small dot-like lesions or flecks and necrotic spots on leaves are, as a rule, observed after fumigation with O_3 in concentrations 10-20 times higher.

The ozone-induced activation of genes and the related accumulation of defensive antioxidants as well as the activity of some enzymes can occur even at ozone.

Table 23. Sensitivity of cellular processes in plants to ozone

Dose of ozone (ppm)	*Duration of exposure to ozone*	*Process or level of metabolite*	
		Stimulation or increase	*Inhibition or decrease, or appearance of new products and other changes or damage*
0.012-0.03	3 h per day for 1-7 days	Pollen germination *in vitro* (*Philadelphus grandiflorus, Hippeastrum hybridum*)	Pollen germination *in vitro* (*Plantago major)*
0.012-0.20	106 days		Fertilization (estimated amount of flowers and formation of fruits)
0.005-0.025	3 h	Level of disulphides (*Phaseolus*)	
0.12-0.20	106 days		Pollen germination (*Capsicum annuum)*
0.0125	52 days	Respiration (*Populus*)	Photosynthesis (*Populus*)
0.013-0.05	30 min	Peroxidase (*Phaseolus*)	Lactate dehydrogenase (*Phaseolus*)
0.015	2 h	Ribonuclease (*Phaseolus)*	
0.015	8-18 weeks	Respiration (*Pinus*)	Photosynthesis (*Pinus*)
0.015	8-10 weeks		Destruction of chlorophyll on the topside of a leaf in plants with non-differenciated mesophyll (Conifers and grassy plants)
0.02	10-12 h	Biosynthesis of flavine glycosides and furanocoumarins (*Petroselinum*)	

Table 23 (Contd)

Dose of ozone (ppm)	*Duration of exposure to ozone*	*Process or level of metabolite*	
		Stimulation or increase	*Inhibition or decrease, or appearance of new products and other changes or damage*
0.02-0.1	30-50 days		Abberations and damage of chromosomes (*Picea abies*)
0.025	3 h	Level of sterol glycosides (*Phaseolus)*	Level of free sterols (*Phaseolus)*
0.03-0.04	0.5-3 h	Respiration (*Phaseolus*)	Photosynthesis (*Phaseolus*)
0.03-0.05	5.5 h per day for 60 days		Pollen germination (*Zea mays*)
0.04	2.5 h		Level of ascorbic acid (*Petunia*)
>0.04	1 h	Activity and then accumulation of enzymes of stilbene synthase and cinnamyl alcohol dehydrogenase (*Pinus sylvestris*)	
0.05	1 h	Autofluorescence in leaf secretory hairs (*Lycopersicon)*	Appearence of green-yellow fluorescence in vegetative microspores (*Equisetum arvense*)
0.05-0.1	30 min - several hours		Stomata closing
0.05	2 h		Level of soluble sugars and reduced sugars and nitrate reductase (*Glycine*) and chlorophyll (*Phaseolus*)
0.05	2-6 h		Hydrolysis of starch (*Cucumis, Mimulus, Phaseolus, Cassia* , etc*)*

Table 23 (Contd)

Dose of ozone (ppm)	*Duration of exposure in ozone*	*Process or level of metabolite*	
		Stimulation or increase	*Inhibition or decrease, or appearance of new products and other changes or damage*
0.07	Several weeks		Levels of secondary alcohols, diols, and fatty acids in needles (*Picea rubens)*
0.062-0.088	4 h		Level of chlorophyll and nitrate reductase (*Zea*, Glycine)
0.08	1.5 h per day for 40 days		Protein content
0.1	1-12 h	Induction of gene activity of β-glucuronidase (Transgenic tobacco *Nicotiana*)	Abortion of seeds (*Brassica napus*)
0.1	5.5 h		Pollen germination (*Nicotiana tabacum*, cultivar Bel W3)
>0.1	> 100 h		Level of anthocyanin in flowers (*Petunia* and *Pelargonium*)
0.125-0.25	Several hours	Activity of superoxide dismutase and peroxidases	
0.14-0.15	1.5 h		Passive efflux of calcium ions from cells
0.15	2-3 h		Level of RNA (*Phaseolus*), Oxidation of β-carotene
1.5	5 h	Level of the mRNA of β-1,3-glucanase (*Nicotiana tabacum*)	
1.5	100 h	Formation of lipofuscin in seeds (*Glycine soja*)	

Table 23 (Contd)

Dose of ozone (ppm)	*Duration of exposure in ozone*	*Process or level of metabolite*	
		Stimulation or increase	*Inhibition or decrease, or appearance of new products and other changes or damage*
0.15-0.25	8 h	Cell membrane permeability for ions (*Citrus*)	
0.15-0.45	3-9 h		Appearence of green-yellow fluorescence in leaf surface (*Plantago major, Raphanus sativus,* etc)
0.15-0.20	Several hours	Increase in stilbene content	
0.2	2.5 h	Accumulation of dihydroflavones (*Medicago*)	
0.2	3-10 h	Level of mRNA for the genes of pathogenesis related proteins and elicitor-induced protein, and accumulation of furanocoumarin phytoalexins (*Petroselinum crispum*)	
0.2-1.0	1 h		NADH level, chlorophyll content
0.23-0.35	20-35 min		Amount of polysomes and RNA level in chloroplasts (*Phaseolus*)
0.25	3 h	Formation of malondialdehyde	
0.25	6 h	Accumulation of superoxide anion radical and formation of lesions (damage)	
0.25	2 h	Excretion of stress ethylene. Formation of malon-dialdehyde *Phaseolus*)	

Table 23 (Contd)

Dose of ozone (ppm)	*Duration of exposure in ozone*	*Process or level of metabolite*	
		Stimulation or increase	*Inhibition or decrease, or appearance of new products and other changes or damage*
0.25-0.3	3 h	ATP level (*Phaseolus*)	
0.28-2.0	>2 h		Pollen germination (*Lilium longifolium*)
0.35	48 h	Activity of cellulase and peroxidase and lactate dehydrogenase	Appearance of symptoms of leaf damage (chlorosis, necrosis, etc)
>0.36	1-2 days	Permeability of plasmalemma for ions	Protein synthesis
0.4-0.6			Photosynthesis (*Pinus ponderosa*)
0.4-0.9	30-120 days		Photosynthesis and transpiration
0.41-0.5	9-18 days	Formation of formic acid (*Spinacea*)	Level of chlorophyll
0.5	2 h	Activity of enzymes of phenolic metabolism	
0.5	3 h		Pollen germination (*Petunia hybrida*)
> 0.5	Several hours	Content of free sterols	Activity of catalase and peroxidases, as well as respiration
0.5-0.8	1 h	Level of free amino acids (*Gossipium*)	Photosynthesis
0.6-0.75	0.05 ppm at 3 h per day for 30 days		Necrotic sports on leaves (*Plantago major*)
0.6-1.0	1 h	Respiration (*Nicotiana*, *Phaseolus*)	Oxidative phosphorylation and ATP level
0.8-1 ppm	30min -1 h	Respiration (*Euglena*). Content of α-aminobutyric acid	Content of SH - groups and of glutamine
1.0	15 min	Levels of γ-aminobutyric acid, alanine, glutamine	Activity of acetylcholinesterase *in vitro*

Table 23 (Contd)

Dose of ozone (ppm)	*Duration of exposure in ozone*	*Process or level of metabolite*	
		Stimulation or increase	*Inhibition or decrease, or appearance of new products and other changes or damage*
1.0	0.5-1 h		Level of saturated fatty acids (*Nicotiana*)
1.0	4-24 h		Level of chlorophyll (*Lemna*)
>1.0	30 min-1 h		Injury of plasmalemma
1.8-5	20-100 h	Formation of lipofuscin in carotenoid-containing pollen	Changes in autofluorescence of pollen
2-12	Several days	Content of anthocyanins in flowers and bracts (*Petunia, Geranium, Euphorbia pulcherima*)	Lesions, then necrosis
>2.5	60-90 min	The seed germination (*Hippophae*)	
20-240	2-3 min	Permeability of plasmalemma for ions	
300-1000	4-10 min	Oxidation of amino acids, unsaturated fatty acids, nicotinamide and lecithin	Activity of phosphoglucomutase and ribonuclease

This table is composed of information included in previous chapters and also various other sources (Hill et al., 1961; Feder, 1968; Mumford et al., 1972; Mudd, 1973; Taylor, 1973; Nakada et al., 1976; Brooks and Csallany, 1978; Heath, 1980; Feder et al., 1982; Horsman and Wellburn, 1982; Mudd, 1984; Dizengremel and Citerne, 1988; Chevrier et al., 1990; Ernst et al., 1992; 1996; Eckey-Kaltenbach et al., 1994; Gordon et al., 1981; Mudd, 1994; Pearcy et al., 1994; Schneiderbauer et al., 1995; Castillo and Heath, 1996; Muller et al., 1996; Stewart et al., 1996; Grimming et al., 1997; Clayton et al., 1999; Overmyer et al., 2000; Zinser et al., 2000;Roshchina and Melnikova,2000; 2001).

concentrations < 0.04-0.05 ppm, but only in more higher doses is it regularly observed.

Acute high doses acted for a few minutes cause sharp shifts in membrane permeability, but the same concentrations for 30-90 minutes may stimulate seed germination. Ozone induces mainly non-specific reactions, which respond to any conditions of stress. Stimulation of seed germination is observed only for seeds that require stratification, and may have relevant applications in agricultural practice (Rezchikov et al., 1998). Unlike acute exposures, chronic fumigation with O_3 may induce various responses depending on the sensitivity of the plant species involved.

Sensitivity appears to be generally determined by the genetic properties of the plant species and also by the structures and functions of some cells in certain sensitive organs, especially those involved in regeneration and reproduction. Short-term chronic experiments (of a few hours in duration) have demonstrated the sensitivity of the leaves of genera *Phaseolus, Nicotiana* and *Pinus* to ozone, as well as the leaves and pollen of *Plantago major* (Table 23). Inhibition of gene activity is observed after a minimum 1-hour exposure to low concentrations of O_3.

Biochemical changes are seen earlier than visible leaf damage and result in diminished plant growth and development. The sensitivity of plants to ozone can be defined not only by the visible symptoms of leaf-plate damage, but also by these biochemical reactions, which differ in sensitive and tolerant plants. According to Treshow (1984), most distinct biochemical activities induced by ozone involve the following: formation of free radicals, formation of SH-groups, change of lipid synthesis, and shifts in the reactions of photosynthesis. However, various factors may intervene so that ozone instigates precisely the opposite group of effects. Therefore, it can increase/decrease stomata conductivity, inhibit/stimulate the activity or synthesis of enzymes, increase/decrease the quantities of free amino acids, stimulate/inhibit carbohydrate synthesis. Unequivocally, ozone does not directly act on electron transport and little is known of its reaction with the cell wall. The sensitivity of various biochemical reactions to ozone also depends on the interactions with the secondary products of ozonolysis. Ozone is ten times more soluble in water than oxygen is, and this dissolution, accompanied by interactions between O_3 and water, generates such free radicals as the peroxy radical, the superoxide anion radical and the hydroxyl radical (see sections 3 .1 and 3.2.). Very reactive oxygen species may affect the sensitivity of biochemical processes throughout the whole plant (Treshow, 1984). Usually the concentration of free radicals in tissues is insignificant, though they are always present, but in the presence of high ozone concentrations, they cannot be effectively neutralized by protective systems. Consequently, a surplus of free radicals joins the assault on cellular components, especially the thylakoids of chloroplasts, with the result that the photosynthetic processes, especially photophosphorylation, are affected. The occurrence of this or that biochemical reaction in sensitive plant species will, ultimately, depend on the activity of the protective antioxidant systems. It is possible also to define plant sensitivity to ozone-induced stress by discerning the level of antioxidants present.

According to Mudd (1973), the biochemical reactions of plants to ozone have different sensitivity thresholds in vivo. Low concentrations of ozone (0.05 ppm and below) are already sufficient to inhibit starch hydrolysis. Any increased doses of this gas (from 0.2-1.0 ppm) also effect a corresponding inhibition of photosynthesis, reduction in the concentration of chlorophylla *b*, decrease in the level of NAD(P)H and ATP, depression of respiration, stimulation of oxygen uptake, and an increase in the amount of free amino acids. Fumigation by 0.25 ppm of ozone for 3 h hours leads to the formation of malondialdehyde in the string bean *Phaseolus vulgaris*, after other damage symptoms have appeared on the leaf, though in other plant species, the same reaction requires higher concentrations of the gas (0.8-1.0 ppm for 5 hours).

A plant cell's specific response to ozone itself is difficult to determine. An oxidative burst (formation of reactive oxygen species) is the earliest plant reaction to the presence of a microbial pathogen and is an integral component in any hypersensitive response leading to cell death (Rao and Davis, 2001).

Individual plant cells may have different sensitivities to ozone (Table 23), although this topic has not yet been sufficiently studied. The types of individual cells that should be considered in such a study include unicellular chlorophyll-containing organisms, vegetative (from plants using spores to reproduce) and generative (the pollen of higher plants) microspores, as well as cellular tissue cultures. There is little information about unicellular organisms, except for the unicellular algae *Chlorella sorokiniana* (Hearth et al., 1982) and a few other microorganisms (Guderian, 1985). These cells are sensitive to high concentrations of ozone (> 0.1-0.8 ppm). Sensitivity of unicellular secretory hairs, as indicated by their autofluorescence, has also been observed (Chapter 2, Figs.10, 12). They were sensitive to lower concentrations of ozone (0.15-0.45 ppm).

Ozone can also influence pollen tube germination from pollen microspores or the formation of thallus cells and rhizoid formation from vegetative microspores (see sections 6.1.1.2 and 6.2. for more detail). The sensitivity of these cells to the stimulative action of O_3 is observed at low concentrations of the gas (0.012-0.03 ppm), whereas negative effects manifest themselves at doses > 0.1 ppm (Table 23). Consequently, the cellular systems could be considered as cellular model-biosensors (section 6.2). Fertilization and flowering also react to concentrations of O_3 ranging from 0.15-0.45 ppm.

At higher concentrations > 0.05 ppm, stomata cells begin to close, preventing gas penetration into the leaf and stem. Increase in the ozone doses up to 0.1-0.2 ppm leads to serious changes in cellular metabolism, such as the hydrolysis of starch, the decomposition of some proteins and others. At the same doses of O_3, the genes coding the biosynthesis of defensive proteins and enzymes are activated. Visible symptoms of damage are usually seen when exposure to ozone reaches 0.5—1.0 ppm. Perhaps, endogenous antioxidant systems such as catalase and peroxidase will be inactivated in these cases, although inhibition of the oxygen uptake by mitochondria is reversed by exogenous ascorbic acid and glutathione (Mudd, 1973). The highest concentrations of ozone (from 5-100 ppm and even up to 1000 ppm in 1-10 min exposures) induced sharp oxidation of amino acids, unsaturated fatty acids and nicotine amide (Mudd, 1973).

A review of earlier diagnostics will require us to survey both sensitive biochemical reactions and sensitive cells and plant species. Some of them will be considered below in sections 6.1.3. and 6.2.

6.1.1.2. Dose-dependent sensitivity of whole plant growth and development.

As mentioned in the previous section, growth and development of individual plant cells under the influence of ozone are dose-dependent. Table 24 shows that the same integral processes of whole plants also depend on the dose of O_3. See also the fundamental publications by Guderian (1985; 1995).

Table 24. Sensitivity of plant growth and development to ozone

Dose of ozone (ppm)	*Duration of exposure in ozone*	*Process*	
		Stimulation or increase	*Inhibition or decrease, or appearance of new products and other changes or damage*
0.012-0.20	106 days		Fertilization (estimated amount of flowers and formation of fruits)
0.015-0.1	8-10 weeks	Leaf damage as part of the beginning of chlorosis in Conifers.	Seed formation (*Brassica napus*)
0.05-0.1	6-7 h daily during 64-68 days1-12	14-25 % leaf damage (*Zea mays*)	30-50 % yield of seeds (*Zea mays*) and shoot dry weight (*Medicago sativa*)
0.05-	8h daily for 5 days/week, 5 weeks		31% plant biomass, 54% of root fresh weight (*Raphanus sativus*)
0.05-0.09	24 h daily, 53 days		30% of fresh weight of flowers (*Petunia hybrida, Dianthus caryophyllus)*
0.06	40 days, 5 days/week		26 % height growth , 505 root dry weight (*Phaseolus vulgaris*)
0.06-0.13	7 h daily, 37 days	38-68 % leaf necrosis (*Spinacia oleracea*)	37 % of fresh weight (*Spinacia oleracea)*
0.064-0.094	9 h daily, 55 days		31-56 % seed yield (*Glycine max)*
0.07-0.01	9.5 h daily, 90 days	Leaf chlorosis, leaf drop	50 % flowering (*Pelargonium hortorum*)
0.08	6 weeks		17 % leaf dry weight (*Festuca sp.)*
0.09	8 h daily, 6 weeks		36% yield of dry weight of biomass (*Lolium multiflorum*)
0.09	3-4 h daily, 5 days/week, 5 weeks	21 % leaf chlorosis (*Dactylis glomerata, Lolium perenne*)	

Table 24 (Contd)

Dose of ozone, ppm	*Duration of exposure in ozone*	*Process*	
		Stimulation or increase	*Inhibition or decrease, or appearance of new products and other changes or damage*
0.1-0.13	7 h daily, 54 days		33 % seed yield (*Triticum aestivum*)
>0.1	> 100 h		Level of anthocyanin in flowers (*Petunia* and *Pelargonium)_*
0.1-0.3	6-12 h daily, 10-30 days or 6-20 weeks	20 % foliar injury in pines (*Pinus ponderosa*)	59 % of total growth (*Populus deltoides, Liriodendron tulipifera* and poplar hybrids)
0.2	6 x 3 h biweekly		30-54 % of tuber number and tuber weight (*Solanum tuberosum*)
0.41-0.5	9-18 days	Leaf chlorosis (*Spinacea*)	
2-12	Several days	Leaf necrosis, content of anthocyanins in flowers and bracts (*Petunia, Geranium, Euphorbia pulcherima*)	
>2.5	60-90 min	Seed germination (*Hippophae*)	

Sources (Feder, 1968; 1973; Mudd, 1973; Guderian et al., 1985; Taylor, 1973;rezchikov et al., 1998; Koch et al., 2000).

It is well known that chronic ozone influence suppresses processes in nature by changing the balance between respiration and photosynthesis, altering the pools of secondary metabolites (including growth regulators), and finally damaging main tissues (for diagnostics, see section 6.1.2). Plants may compensate for these unfavorable effects by initiating reparative processes (synthesis of antioxidants, as seen in chapters 4 and 5, biosynthesis of main cellular components *de novo*, etc), so that the plants can continue with their normal course of development. In some cases, plant species are so sensitive to ozone that ozonation causes them to decrease their seed yield, total biomass and amount of flowers. Furthermore, high O_3 doses may also produce leaf damage, which usually begins as chlorosis (loss of green color due to chlorophyll drop) and then develops into a wide necrosis of plant tissues (Table 24).

The primary way in which ozone penetrates into a leaf is via the stomata through which gas exchanges with the environment normally occur. Visible ozone damage will depend on the fact that they are open and allow gas to enter the mesophyll. At high concentrations of ozone (> 0.1-0.2 ppm), significant reductions in seed yield, total biomass and dry weight are accompanied by such leaf injuries as chlorosis and even necrosis. Leaf damage is a great problem that contributes to forest decline and urban inhabitation (Guderian et al., 1995; Sandermann et al., 1997).

Especially sensitive inhabitants of fields, meadowland and forests are the grassy genera *Festuca, Lolium* and *Dactylis* and the woody species *Pinus ponderosa, Populus deltoides* and *Liriodendron tulipifera*. Among the common cultivated plant species of which the development was analyzed, the corn *Zea mays*, the lucerne *Medicago sativa*, the pinto bean *Phaseolus vulgaris,* the winter wheat *Triticum aestivum,* and the spinach *Spinacea oleracea* were all sensitive to low doses (0.03-0.06 ppm) at the lengthy ozone exposures. As a whole, chronic exposure to ozone (i.e. treatment with 0.1-0.13 ppm for more than 1 day) decreased the seed yield of crops by 10-133% (Heagle, 1979; Guderian, 1985).

The effects of ozone on the growth and development of individual plants and plant communities has also been studied. However, the main concern of such research has been the problem of air pollution, and therefore only high doses of the gas have been examined (Feder, 1973; Laurence and Weinstein, 1981; Treshow and Anderson, 1989; Smith, 1990; Kangasjarvi et al., 1994; Heath and Taylor, 1997; Sandermann et al., 1997; Kley et al., 1999). The principle consideration of such inquiries is economic, as air pollutants decrease crop yields (Heagle, 1989). This phenomenon does not, however merely result from the direct effects of increased O_3 levels on plant functions but also its influence on the plant's ability to resist parasites (Heagle, 1973) and diseases (Manning and Tiedemann, 1995). The consequences of ozone in combination with other air pollutants have also been specifically studied (Feder, 1973; Reinert, 1984).

The breeding of plants is always sensitive to any stress, including ozone-induced stress. This sensitivity manifests itself in stages of flowering and fertilization, before fruit and seed production. The most evident symptoms involve the formation of flowers and fruits during chronic exposures to ozone. Until maturity, plant growth is suppressed in the presence of 8-10 ppm of ozone, whereas flowering may be more prolonged than in a control (Feder, 1973). It has been known for some time that ozone alters the ratio between flower formation, fruit production and leaves. For example, when plants of the pepper *Capsicum annuum* are fumigated with ozone (0.12-0.20 ppm during 106 days) the formation of fruits is reduced by 54 % (Bennett et al.. 1979b). Accordingly, the dry weight of the plant is also changed. Such decreases in the amount of flowers and fruits on different plants has already been described elsewhere (Bonte, 1982). Long-term chronic exposures to ozone during the flowering phase of plant development reduces the quantity of flowers in the petunia *Petunia*, the geranium *Geranium*, and the pink *Dianthus* (Feder and Campbell, 1968; Feder et al., 1969a; Feder. Sullevan, 1969; Feder, 1970). Experiments in which various blossoming plants of the cabbage *Brassica* were repeatedly exposed to ozone (0.1 ppm) have shown that there were no visible effects for *B.campestris* whereas seeds were aborted in the plants of *B.napus* (Stewart et al., 1996). The first species probably has compensatory mechanisms, preventing premature fruit fall.

Unlike leaves, stems and flowers, seeds are largely unaffected ozone action, as the structure of a seed cover helps them to resist ozonation. Using special industrial installations to administer doses of ozone higher than 2.5 ppm for various short-term durations (from 5 minutes to 2 hours), the germination of seeds taken from a number of agricultural plants were seldom completely blocked (Rezchikov et al., 1998). Moreover, not only was any disruption of the process rarely seen to result from the high doses of ozone and ozonated water, but often seed germination was stimulated, as it was in the wheat *Triticum* , the pea *Pisum sativum*, the potato *Solanum tuberosum*, the sea buckthorn *Hippophae rhamnoides* and the yellow goat's-beard *Tragopogon pratensis* (Gavrilova et al.,1999). Certain doses of ozone even improved respiration, energy, rates and time of seed germination, plant growth and productivity. The fact that seeds remain unaffected by short-term treatments with high ozone doses and, moreover, that seed germination is even stimulated in such conditions indicates that there may be prospective applications in industrial agricultural practice as either as desinfectant or a stimulator of seed germination (Rezchikov et al., 1998). Unfortunately, publications devoted to this subject are wanting.

As seen above, the sensitivity of different plants to ozone varies widely and depends on many factors. However, the possibility exists to select specific species capable of providing rough indications of the overall ozone damage occurring in a given district under certain conditions.

6.1.1.3. Dose-dependent sensitivity of plant species and cultivars to ozone

The sensitivity of species and even of cultivars of the same plant species to ozone, as indicated by vegetative cell damage, is extremely various. As a rule, the majority of plants are tolerant to ozone, and damage occurs only in cases of severe atmospheric pollution. However, some plant species are very sensitive and are injured by O_3 concentrations only slightly above normal levels. These plants are precisely the ones that have the most value as biological indicators (Heck, 1966; Feder and Manning, 1979; Posthumus, 1988).

In England, the sensitivity of 150 grassy plant species to ozone was investigated (Ashmore et al., 1988). Some families, in particular Fabaceae (Papillionaceae), contained a large number of species that were rather sensitive to ozone. Conversely, few species belonging to the Asteraceae (Compositae) family amount displayed any such ozone sensitivity. The general tendency is evident: evolutionarily more primitive species have greater sensitivity to ozone.

Some lower plants, especially when grown under intensive ultra-violet irradiation, appear to possess adaptive features allowing them to withstand ozone exposure. In particular, the foliose lichens *Hypogymnia enteromorpha* and *Letharia vulpina* were tolerant to ozone exposures equivalent to those occurring in mountain conditions (up to 150 ppm per h), and only suffered damage after doses of 285 ppm per h were administered (Sigal and Nash, 1983).

For various species of plants, the critical concentration of ozone in air causing serious dysfunction in plants varies over a wide range. For sensitive species of plants this critical dose is near 0.05-0.1 ppm of ozone for 2-4 hours. The corn *Zea mays* and the oat *Avena sativa* have critical values ranging from 0.1-0.12 ppm when the duration of exposure is 2 hours. Acute leaf malfunction in the tobacco *Nicotiana tabacum* occurs when exposed to 0.05 ppm concentration of ozone for 4 hours (Hicks, 1978). Most authors, according to Hicks (1978), believe that any plant that suffers functional difficulties when the atmospheric O_3 concentration does not exceed 0.3 ppm should be considered as ozone sensitive. Conversely, plant species are tolerant if functional breakdowns do not occur until ozone concentrations surpass 0.4 ppm. A list of the plant species that are most sensitive to ozone is provided in appendix 2.

In many countries of the world, the legislation governing pollution of the environment by antropogenic ozone is based on the successful implementation of bioindicators. Bioindicators are used to determine the effect of O_3 on test plant species (see section 6.3 and appendix 2).

6.1.2. Dependence of the plant sensitivity to ozone on external and internal factors

The sensitivity of plants to ozone depends on many external factors: edaphitic, climatic, atmospheric and biotic (Table 25). Additionally, such internal factors as phases of growth and development also have roles to play (Table 26).

External factors. Important among these factors is moisture content; the sensitivity of plants to ozone is higher when they are saturated with water (Table 25) than when they are subjected to drought (Wilhour, 1970). For example, ozone fumigation (25 ppm during 2-3 weeks by 4 hours daily) of *Fraxinus americana* seedlings in growth chambers caused the greater damage when humidity was 85 % than it did when humidity was 65 % (Wilhour, 1970). Ozone damage in the plants of the tobacco *Nicotiana tabacum* and the bean *Phaseolus vulgaris* was also facilitated by increased humidity (Otto and Daines, 1969) The extent of injury correlated well with porometric parameters. Increases in humidity cause turgor in closing stomata cells so that ozone conductivity is allowed to increase (Wilhour, 1970 a, b). In arid regions, a large resistance to ozone is observed. The sensitivity of red pine and poplar leaves to ozone especially grows when humidity increases (Fuentes et al., 1994a). In contrast, any small or moderate water deficit inhibits ozone damage (Heugle, 1989).

Soil condition and mineral nutrition also influence plant sensitivity to ozone (Table 25). The connection between sensitivity to ozone and plant habitat has also been revealed (Ashmore et al., 1988). Species grown in calcium-enriched soils are more sensitive to ozone than those develop in regions with acid soils. The elements of mineral nutrition may strengthen ozone damage symptoms, which are more strongly manifested in the presence of high-levels of nitrogen than in areas of low-nitrogen content (Cowing and Koziol, 1982). When nitrogen nutrients are plentiful, ozone effects appear more strongly. The damage induced by ozone is not observed in plants that are deficient in phosphorus. Sulfur increases resistance to ozone, apparently, due to an increase in the amount of the compounds containing sulfhydryl

Table 25. External factors enhancing plant sensitivity to ozone

Factor	*Sensitive plant process*
Moisture	Water exchange
Higher humidity (Wilhour, 1970a,b).	
Soil state	
Calcium carbonate-rich soils (Ashmore et al., 1988).	Mineral nutrition
Optimally high amounts of nitrogen and calcium (Cowing and Koziol, 1982).	
Aerated, light soils (Lacasse and Treshow, 1978).	
Light	
High intensity for short periods or low intensity for long ones (Juhren et al., 1957; Heck et al., 1965).	Photosynthesis, growth and development
Total exposure and short photoperiod (Ting and Dugger, 1968; Juhren et al., 1957	Development
Quality. Damage occurs in the presence of light at wavelengths 420 and 480 nm	Development.
Temperature	
Temperature, especially increases from 3 to 30° C	Photosynthesis, growth and development

groups. In general, mixtures of nitrogen, phosphorus and potassium nutrients reduce the degree to which plants suffer ozone damage. Plants grown in soils with high sodium content display appreciably greater amounts of damage. Role of a nutrition in the tree defense against ozone as very important problem is specially discussed (Polle et al., 2000). Soil structure also influences sensitivity to ozone, as species growing in well aerated, porous soils, are more sensitive to ozone, while those growing in heavy, poorly aerated soils are more tolerant (see Diagnosing Vegetation Injury Caused by Air Pollution, 1978).

Light is an important factor for plant sensitivity to O_3, as the duration of light and dark portions of the day (photoperiod) influence ozone's effects (Table 25). The damage caused by ozone occurs more often during periods of light, when stomata are open. However, if a plant, such as *Gossypium*, is placed in darkness one day before treatment with ozone, the sensitivity to the gas is decreased (Ting and Dugger, 1968). The same effect was occurred when young Virginian pine, *Pinus virginiana*, was used as the experimental plant (Davis, 1970). The length of the photoperiod intervals also played a significant role (see Diagnosing Vegetation Injury Caused by Air Pollution, 1978). Annual plant *Poa annua* was more sensitive to ozone when photoperiods were 8 hours long than when they were 16 hours long (Juhren et al., 1957). The spectral characteristics of light also influence the ozone damage, as Dugger et al. (1963) have shown how the effects of ozone depend on the light wavelengths at which the plants are exposed to ozone fumigation. The greatest damage occurs at 420 and 480 nm. Interestingly, this region in the spectrum of light corresponds to the absorption bands of carotenoids, compounds that are already known to react quickly with atomic oxygen. Apparently, there are general antioxidizing mechanisms operating in the latter case. In any event, lengthy periods of illumination enhance the defeat induced by ozone .

Sensitivity to ozone becomes greater when temperature increases from 3 to 30° C (Dunning and Heck, 1974). Woody plants are especially sensitive to such temperature increases, whereas grassy ones, for instance the tobacco cultivar Bel W3, suffer minimal damage in these conditions (Schenone, 1993). However, contrasting examples also exist of some grasses that are more tolerant to ozone in the warm season of the year than in the cold season (Richards et al., 1980). If, prior to fumigation with ozone, plants are grown at the high temperatures (27-32° C), they are more sensitive to the toxic gas than if they were maintained at low temperatures (10-16° C). However, the opposite phenomenon is observed during the fumigation itself: greater damage occurs at low temperatures (see Diagnosing Vegetation Injury Caused by Air Pollution, 1978).

It is important to consider the regions in which the plants grow. The original geographical status of the plant species inhabitation is also an important factor affecting ozone sensitivity. For instance, the clones of poplars taken from places polluted with ozone have higher resistances to ozone than clones taken from unpolluted regions (Gressel and Galum, 1994).

Internal factors. The ways in which internal factors influence plant sensitivity to ozone are shown in Table 26. The general sensitivity of plant growth and development to ozone depends on the phase of growth, the physiological state of the plant and other conditions. A large degree of variability in ozone tolerance also exists. Of 410 plants of the string bean *Phaseolus vulgaris*, taken from different sites and subjected to sharp ozone stress (0.6 ppm in 2 hours), deformations arose in 23 samples, whereas 17 were extremely tolerant (Gressel and Galum, 1994).

The fact that resistance to ozone is dependent on the age of the plant has been demonstrated in an experiment on the sequoja *Sequoidendron giganteum*. Several plants of this species, of varied ages, were exposed to different concentrations of atmospheric ozone (Grulke and Miller, 1994). 5-year old sequoja plants were the most sensitive. Trees aged from 20-125 years displayed minimal ozone damage. The plants became more tolerant to ozone when they developed protective mechanisms, like low stomata conductivity, highly efficient water use and mesophyll compactness.

The age of the leaf and the phase of its development determine the sensitivity of leaves to ozone. Research into the effects on *Acer saccharum* of total ozone doses ranging from 0.05-0.90 ppm administered in exposures lasting 1, 2 and 3 hours and doses of 0.05-0.50 ppm lasting 5-40 days has shown that trees are not injured if the short-term dose is lower than < 0.30 ppm. Furthermore, if the chronic dose is below

Table 26. Dependence of plant sensitivity to ozone on internal factors

Factor	*Plant*	*Sensitive*	*Tolerant*
Age of plant	*Sequoidendron giganteum* (Grülke and Miller, 1994)	Young 5- year old plants	Old plant (20-120 years)
	Grassy plants (Richards et al., 1980).	9-14 day old seedlings	66-71 day old plants
Age of leaf	*Acer saccharum* (Hibben, 1969).	Before the final leaf size is formed	Young leaves
Size of leaf	*Pinus ponderosa* (Davis and Coppolino, 1974a).	5-7-weeks old needles (length 60-90 cm), grown by extensity	Needles (length > 100-120cm)
Circle of shoots	Adult grassy plants, in particular *Nicotiana tabacum* (Glater et al., 1962; Hill et al., 1970).	Leaves of 2-4 circles	Young leaves of 0-1cycles or old leaves of 6 cycles.
Phase of development	*Raphanus sativum* in phase of extensive growth	Hypocotyls of 14 day old seedlings	7- and 21 day-old seedlings

0.10 ppm, only weak ozone damage is found. Young leaves are more tolerant to ozone than older ones (Hibben, 1969), but leaf stability is compromised before the final size of the leaves is attained. In this phase, intercellular spaces and stomata are formed, a process that favors ozone penetration into the leaf.

The needles of the yellow pine *Pinus ponderosa* show the greatest sensitivity to ozone in the age of 5-7 weeks (lengths of 60-90 mm, when they are undergoing a great deal of growth (Davis and Coppolino, 1974a). When the needles exceed lengths of 100-120 mm, they lose the sensitivity. In early spring, young needles are injured in regions where there is a high content of ozone in air, whereas mature tissues are only damage by chronic exposures to O_3.

9-14 day old seedlings of grassy plants are more sensitive to ozone exposures in concentrations of 0.3 to 0.5 ppm lasting 3 hours than 66-71 day old ones (Richards et al., 1980). Ozone decreases the hypocotyl growth of 14-day old radish (*Raphanus sativus*) seedlings undergoing a phase of rapid growth. In such cases, growth is reduced by about 37 %, whereas 7 and 21-day-old seedlings have reduced growth of only 25 and 15 %, respectively.

The ozone sensitivity of leaves belonging to the same plant can effectively differ. For example, the leaves on a sensitive cultivar of tobacco are only subject to ozone damage when they have an average age of 2-4 cycles, whereas younger (0-1 cycle) or older (6 cycles) leaves remain uninjured. There are, furthermore, more sensitive sites on damaged leaves. The base of leaves of the 4th cycle is mainly is injured in 87 % of the cases. In young leaves, the leaf tip is sometimes damaged (Glater et al., 1962; Hill et al., 1970). In the final year, resistance to ozone is shown to be under genetic control (Gressel and Galum, 1994). However, the genetic peculiarities of this phenomenon have only been investigated in tobacco and legumes (Gressel and Galum, 1994).

Throughout the flowering phase, a plant can show certain specific reactions to ozone. In many respects, these depend on whether sharp (acute) or chronic fumigation treatments were employed. Long-term chronic exposures to ozone result in reduced flower formation in blossoming plants (Feder and Cambell, 1968; Feder et al., 1962a; Feder, 1976). Low ozone doses (from 6-8 to 10-12 μl/l for 6 hours a day and 5 days a week) decrease the amounts of anthocyanin in the flowers of *Petunia hybrida* Vilmorin (grade Blue Lagoon) and *Pelargonium hortorum* Bailey (grade Red Perfection) by as much as 30 %. The diameters of the flowers are not increased by these doses; however their fresh weight accrues by 10-12 % (Craker and Feder, 1972). This findings contrast with the data received for the brightly red colored bracts from the Paul Mikkelson grade of the blossoming poinsettia *Euphorbia pulcherrina* Wildenow. These structures have the nearly 40% increase in the amount of anthocyanin in the same conditions. There is, however, a corresponding decrease in the area of their leaf plate (Craker and Feder, 1972).

Sensitivity to ozone is also determined by genetic factors, a point that is especially demonstrated by various grades of tobacco (see Diagnosing Vegetation Injury Caused by Air Pollution, 1978). Other plants, such as string bean, pinto bean, tomato, sunflower, poplar, soya bean, petunia, etc., have cultivars with different sensitivity to ozone despite the fact that they belong to the same species. Such possibilities are directly related to difference in gene expression (see the book edited by Sandermann et al., 1997; Zinger et al., 2000).

As can be seen from the above, the sensitivity to ozone of different plants varies widely and depends on many factors. However, opportunities always exist to select species that can be used to provide a rough indication of the potential ozone damage in certain conditions.

6.2. DIAGNOSTICS OF PLANT CELL SENSITIVITY TO OZONE

The sensitivity of plant cell to ozone could be determined on the basis of certain criteria that make early symptoms of a plant's reaction to O_3 readily detectable. Such criteria would be derived from observations made of cellular processes, including biochemical and physiological reactions, as well as the analyzed state of the damaged cell.

6.2.1. Diagnostics involving cellular reactions

Individual cells such as microspores have been chosen to illustrate the diagnostics of ozone damage. Microspores are termed separate cells—whether they are generative (the pollen of higher plants) or vegetative (the spores of spore-breeding plants). Their development is sensitive to reactions with ozone (see section 6.1.2).

Pollen. The ability of pollen (male gametophyte) to fertilize an egg cell in the pistil (female gametophyte) is susceptible to all form of stress, some of which may be induced by ozone. Pollen usually germinates, *in vivo*, on the pistil stigma and forms pollen tubes which include gametes (spermia). Pollen tubes grow inside the pistil until they reach the egg cell, allowing one of the spermia to merge with the egg cell as the final part of the process of fertilization. Pollen germination may also take place outside of the pistil on an artificial nutritional medium, i.e. *in vitro*. Ozone can influence pistil and pollen fertility, i.e. the capacity for fertilization that could occur after plants have been pollinated. Thus, sensitivity of male and female organs to O_3 is very important for breeding.

Ozone effects the pollen germination of various plant species differently *in vitro* and *in vivo* (Feder, 1968; Feder et al., 1982; Mumford et al., 1972; Nakada et al., 1976). Depending on the concentration of ozone, this process is often inhibited, and the growth of pollen tube *in vitro* is suppressed. For instance, the growth of pollen tubes on agar disks with 10 % saccharose is completely blocked if the development of pollen occurs during 5 hours of fumigation by 1 ppm of ozone (Feder, 1968). Lower concentrations of ozone (0.1 ppm) caused a 40-50 % decrease in germination and a 50 % reduction in the elongation of pollen tubes in the ozone-sensitive Bel W3 cultivar of tobacco.

Within various plant species, there is a wide diversity in the extent to which pollen is sensitive to ozone. Exposure to 0.1 ppm of ozone for 5.5 hours reduces pollen generation and pollen tube elongation in the tobacco *Nicotiana tabacum* by 40-50 % (Feder, 1968; Fujii and Novales, 1968). Similar results were also obtained for pollen of the corn *Zea mays* (Mumford et al., 1972). According to Benoit et al. (1983), 0.1 ppm of ozone (4-8 h/day before blossoming) decreased pollen germination in *Pinus strobus* only in wet conditions (more than 15-20 % in a sensitive cultivar). Increased ozone concentration also sharply reduced the percent of germinated pollen grains. Elongation of pollen tubes in *Lilium longifolium* grown *in vitro* (on agar plates with 10 % saccharose) was inhibited by 12 % when plants were exposed to 2.09 ppm of ozone for less, than 2 hours. Longer exposures (2 to 5 hours) and higher concentrations (0.28ppm) decreased this pollen elongation by 18 and 20-28 %, respectively (Nakada et al., 1976). At even higher concentrations of ozone, the effect was not strengthened any further. The pollen of ozone sensitive and tolerant clones of *Petunia hybrida* displayed a rather low rate of germination when treated with 0.5 ppm of ozone for 3 hours (Harrison and Feder, 1974). However, differences in the ultrastructure of the pollen grains of these clones appeared after fumigation with this gas. About 50 % of the pollen from the ozone-sensitive clone had a peripheral ring (strip) of cytoplasm in which all organelles were absent except ribosomes. The tolerant clones showed similar phenomenon only in a few pollen grains. Organelles were supposed to have migrated away from the plasmalemma in response to ozone that, in its turn, suppresses the development of pollen tubes as a whole and, especially, modifies the cell wall structure.

Roshchina and Melnikova (2001) investigated the relationship of pollen ozone sensitivity to the pigmentation of pollen grains in studies of the role played by different exine structures in pollen germination (Table 27; Fig.43).

Table 27. Effects of low doses of ozone on pollen germination shown as % of control (Roshchina and Melnikova, 2000 and unpublished data of Roshchina V.V.).

Plant species	*Concentration of ozone, ppm*				
	0.012-0.015	0.02	0.03	0.04	0.05
Hippeastrum hybridum	250	112	132	100	100
Philadelphus grandiflorus	100	110	175	130	136
Plantago major	100-112	100	92	96	80

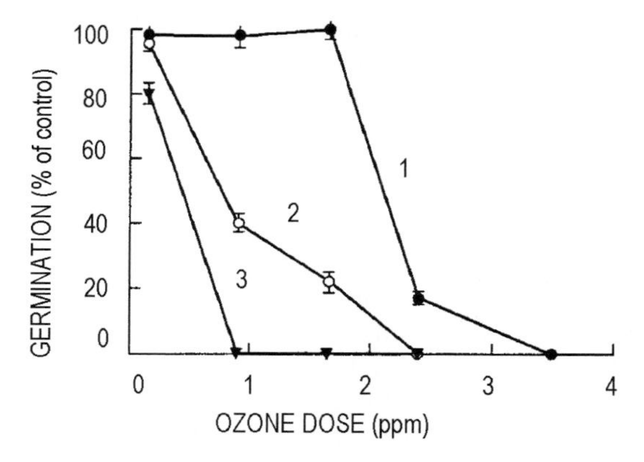

Figure 43. The effects of ozone on the germination index of pollen from different species of plants, shown as % of the control (Roshchina and Melnikova, 2001). A bar—average error at n=4. 1- Philadelphus grandiflorus; 2 - Hippeastrum hybridum; 3- Plantago major.

The pollen of the hippeastrum *Hippeastrum hybridum,* the mock orange *Philadelphus grandiflorus* and the plantain *Plantago major* readily germinated in vitro but contained different pigments or did not contain any at all. The effects of ozone depend on dose and the plant species from which the pollen grains are taken. Concentrations of $O_3 < 0.05$ ppm may stimulate pollen germination in the first two species (Table 27) but not in *P. major*. Concentrations of ozone > 0.05 ppm (Fig.43) begin to inhibit pollen germination. The carotenoid enriched pollen of *Ph. grandiflorus* retained its ability to germinate much longer than the pollen of *H. hybridum* did, which does not contain such pigments. In the first case, this ability was lost only after fumigation with doses of ozone amounting to 2.4 ppm. Furthermore, an increase in the ozone dose led to a corresponding smoother reduction in the germination index than that which occurred in the pollen from either *H. hybridum* or *P. major*. Apparently, the pollen of *Ph. grandiflorus* has more antioxidants, such as carotenoids, which are absent in the other investigated species (Roshchina et al., 1997). The pollen of *H. hybridum* contains phenols, which may also be antioxidants. The sensitivity of this process also depends on the duration of pollen storage; this factor accounts for striking differences in the response of antioxidant systems, which are more active in fresh than in aging pollen.

The significant factors were investigated in an experiment by Nakada et al. (1976), which dealt with the influence of pistil stigma excretions on the ozonation of pollen. In the ozone-sensitive tomato cultivar *Lycopersicon esculentum*, pollen tube growth on the pistil stigma was delayed by fumigating both pollen and pistil with a low dose of ozone (0.15 ppm) for as little as 15 minutes. When only the dry pollen was exposed to ozone, the effect was reduced or completely eliminated even at high doses of the oxidant (0.8 ppm). The same reduction was also observed in another version of the experiment, when the stigma surface was alone fumigated with ozone before the dry pollen was placed on it. However, the growth of the pollen tube was significantly inhibited if the pollen was first kept in a wet medium while it was exposed to ozone for 2 hours *in vitro*, and after that placed on the pistil stigma of the plant, which was then exposed to 0.5-0.8 ppm of ozone for 6 hours. The experiments suggest that ozone appreciably influences the relations between an already germinated pollen tube and a pistil (Feder et al., 1982). Apparently, the effect results from the presence of low molecular antioxidants or antioxidant enzyme systems both in the pistil and the pollen.

The mechanism of the ozone action on pollen is still unknown. It is possible that the gas penetrates into cells only poorly at low concentrations and, therefore, interacts with superficial components to provide a sort of chemosignal. The information received outside of the cells is transferred into the cytoplasm and to other organelles by the help of the secondary messengers in the sporopollenin (the main material of the pollen cover) or in the plasmalemma. It also cannot be excluded that the active matter for inducing ozone effects on pollen tube growth is not ozone itself but the free radicals and ozonides-peroxides produced by ozonolysis (see Chapter 3). Free radicals are formed on the surface of pollen (Dodd and Ebert 1971; Priestley et al., 1985). The interaction of ozone and peroxides with the surface components of the sporopollenin have been modeled in artificial systems (Roshchina and Mel'nikova, 2001). and Mel'nikova, 2001). High concentrations of ozone oxidize the surface components and even penetrate into the cell.

As a test-reaction, the autofluorescence of pollen was measured in the manner previously discussed (Roshchina and Melnikova, 1996; Roshchina., 1996, 1997, Melnikova et al., 1997) and shown to depend on the chemical structure of the sporopollenin components. The fumigation with ozone sharply alters the light emission intensity and position of maxima in the fluorescence spectra (Fig. 13). The maximum of fluorescence after the treatment with ozone was moved into the short-wavelength region of the spectrum, while the intensity of the lightening greatly increased. Thus, the ozone damage of pollen can be expressed as a change in the spectrum and intensity of luminescence. Both these parameters depend on both plant species and ozone concentration. It should however be noted that the fluorescence intensity of pollen collected from self-compatible and self-incompatible lines could differ. In the data that we collected on the *Petunia hybrida*, this parameter was 3-times higher in the latter case. Such marked changes appear to be the earliest indications of ozone damage.

Vegetative microspores. The development of microspores produced by such non-seed plants as the common (field) horsetail belonging to the *Equisetaceae* family is sensitive to ozone (Roshchina and Melnikova, 2000). A study has been made of microspores, from the beginning of their germination to the tallus, where generative cells develop into female gametophyte—archegonia—and male gametophyte—antheridia. The vegetative microspores of the common horsetail *Equisetum arvense* have proven to be very sensitive to ozone (Table 28).

Table 28. Effects of ozone on development of common horsetail Equisetum arvense. (Roshchina and Mel'nikova, 2000)

Dose of ozone, ppm	*Time after the moistening of microspores with water (no. of days)*		
	Blue->red shift in the fluorescence of microspores	*Formation of tallus*	*Formation of archegonia and antheridia*
0	1-7	14-28	60-63
0,2	1-7	21-28	no
0,4	1-14	21-38	no
0,6	1-21	21-42	no
0,8	7-21	21-51	no
1,5	7-14	17-no, *	no,*
2,5	7-14	17-no, *	no,*
5	7-no, *	no, *	no, *

no - is not present, *Limits of development, both the beginning (when the changes in individual cells arise) and the end (basic mass of cells transformed into a new stage) of the process. The asterisk * indicates that main cells died.

Three basic stages of development were studied under the fluorescent microscope that reveals changes in: (1) the initial development of unicellular microspores after moistening by water: (2) the formation of the multicellular vegetative body of the gametophyte (protallus and then thallus); (3) the formation of the generative organs, the archegonia and the antheridia (Roshchina et al., 2002a). Initial microspore development is marked by a well-seen transition of fluorescence from the light-blue region (the luminescence of phenolic compounds, mainly at 440-480 nm) to the red region of the spectrum (indicating intensive formation of chlorophyll, which has a luminescent maximum at 680 nm). The generative organs did not develop under the influence of even the smallest ozone doses. As shown in table 28, even the first stage of the microspore germination (in which their fluorescence changes from light-blue area to red) is slowed after a preliminary fumigation of dry microspores with ozone in low concentration 0.4 ppm. When the O_3 dose was 5 ppm, only a few individual microspores underwent germination. The time required for the following stage of gametophyte formation (involving cell division and congestion, and fluorescing in the red area of a spectrum) also becomes longer. After a 1.5 ppm dose of O_3, there are a few cells that have red-region luminescence, but the complete gametophyte is not formed, and many cells die. A dose of 5 ppm of ozone entirely blocks gametophyte formation. Finally, antheridia and archegonia are not formed at all when exposed to doses of ozone > 0.05 ppm.

6.2.2. Early diagnostics using biochemical and physiological criteria

For early diagnosis, a search of sensitive biochemical reactions and sensitive cells and plant species is necessary, some of which will be considered below.

Analyzing the information included in Chapters 2-5, found in books edited by Treshow (1984a) and Sandermann et al. (1997), or contained in some monographs and reviews (Roshchina and Roshchina, 1993; Roshchina, 2001a,b) on the diagnostics of biochemical processes, it is possible to select the following diagnostic criteria:

1. Plant excretions:

Release of hydrocarbons, mainly stress ethylene and monoterpenes; release of the products of lipid peroxidation; activity of excreted antioxidant enzymes, such as peroxidases and superoxide dismutases; and components of signaling systems, such as cholinesterase. The excretions accrue and also act on the cell surface and outside the cell in the extracellular space (mainly in the apoplast).

2 . Cell surface and extracellular space :

Changes in the redox state of the cover and apoplast ingradients which can be seen as a shifts in the autofluorescence of cell cover and transmitted cellular components, active oxygen species as components of oxidative burst, level of products of oxidation and antioxidants (for instance, ascorbate).

3. Plasmatic membrane:

Permeability of the membrane; efflux and uptake of ions and low-molecular components; state of ion channels and ion pump systems.

4. Components of the signal chain:

Level of signaling molecules, such as ethylene; neurotransmitters, such as catecholamines, serotonin and histamine; pools of secondary messengers such as the cyclic nucleotides cAMP, cGMP, inositol triphosphate, ions of calcium, adenylate and guanylate cyclases; protein kinases; and phosphatases.

5. Organelles:

Chloroplasts (level of pigments, total photosynthesis, RUBISCO level, chlorophyll fluorescence) and nuclei (amounts of stress and protective proteins as well as enzymes).

According to Treshow (1984a, b), the most distinct biochemical effects involve: the formation of free radicals, the formation of SH-groups, changes in lipid synthesis, and shifts in the reactions of photosynthesis. However, there is a group of effects in which, depending on many factors, ozone induces such a strong response that a completely opposite profile develops. Ozone may, as a result, increase/decrease stomata conductivity, inhibit/stimulate activity or synthesis of enzymes, increase/decrease the quantities of free amino acids, and stimulate/inhibit carbohydrate synthesis. Unequivocally, ozone does not directly act on electron transport, and little is know of its reactions with the cell wall.

The cell damage caused by ozone, as well as by many other factors, is displayed first of all on the biochemical and physiological levels. These effects include alterations of enzymic activity, changes in amounts of pigment, increases in ethylene formation, decreases in ATP biosynthesis, reduction of the photosynthesis rate and disturbances in water exchange. Such reactions are often observed prior to any visible damage appearing on leaves. The possible sequence of physiological and biochemical effects, according to Tingey (1977), is indicated in Fig. 44.

OZONE

⇓

Increase in membrane permeability and leaching of ions.
Disturbance of water exchange.

⇓

Stimulated formation of stress ethylene

⇓

Reduction in CO_2 fixation during photosynthesis and decrease in ATP level

⇓

Changes in chlorophyll fluorescence and the ratio of the chlorophyll forms

⇓

Change in enzyme activity (often deactivation)

⇓

Changes in metabolic pools

Figure 44. Sequence of the changes in biochemical and physiological processes caused by ozone.

For many months before any damage manifests itself in such external symptoms as chlorosis and necrosis (see below section 6.2.3), it is possible to find physiological and biochemical changes in the affected plants. Consequently, it is necessary to analyze any pathologies of or changes to the cell or organism involving the above-specified biochemical and physiological criteria (Guderian and Reidl, 1982). Some examples of the criteria will be considered below.

6.2.2.1.Stomata opening and gas exchange of leaves.

The basic way in which ozone penetrates into a leaf is via the stomata through which gas exchange occurs. Stomata function is the plant mechanism that is most directly affected by ozone. The occurrence of any visible ozone damage will depend on the fact that they open and permit the gas to enter the mesophyll.

Thus, the first indication of potentially harmful quantities of ozone is provided by the state of the stomata. The stomata are also directly related to the occurrence of visible damage, which can only occur if the stomata are not injured or remain open under the influence of ozone. As shown in section 2.1, fumigation with ozone disrupts the function of the stomata. The degree of the ozone effect on the stomata opening depends on many factors, in particular on the age of the leaf and the level of light to which it is exposed. Young leaves have higher stomata conductivity than old ones do (Reich, 1987). The treatment of 6-9 and 19-21-day-old leaves with ozone increases stomata conductivity, whereas the treatment of 12-14-day-old leaves decrease it. The effect also depends on light exposure. Under weak light, the conductivity of old leaves was more strongly affected by ozone than it was when exposed to strong light. On the topside of the leaf, the decrease in the stomata conductivity was homogeneously distributed, but on the bottom side of a leaf, its increase is most noticeable in leaves aged from 4 to 10 days. In older leaves, conductivity progressively decreased. In conditions of water stress, there were no observable distinctions in stomata conductivity on the top and bottom sides of a leaf. Studies on the Norwegian fir tree *Picea abies* (Wallin et al. 1990) indicate that stomata conductivity does not directly depend on the concentration of O_3, but is connected to changes in the intensity of photosynthesis and in the intercellular concentration of CO_2. Probably, interactions with ozone cause free radicals to be formed, which are actively toxic substances. In any case, the peak of sensitivity to ozone is not connected to the number of stomata and the strength of their resistance on both surfaces of a leaf.

Any sharp decrease in stomata conductivity may be a direct effect of ozone interaction with the proteins in membranes leading to enhanced permeability of the plasmalemma and increased ion output from the guard cells. This point of view is confirmed by an experiment on the string bean *Phaseolus vulgaris*. After 3-4 hours exposure to rather high concentrations of ozone ($0.37\text{-}0.60 \times 10^{-6}$ M), there was a reduced ability of the stomata guard cells and the cells of the mesophyll to restore a volume of water after turgor (Sober and Anu, 1992). This reduced capacity could be a consequence of the increased rigidity of the cell walls and/or the enhanced permeability of the plasmalemma. Similarly, appreciable ozone influence was also observed after lengthy fumigations with smaller concentrations of ozone (1-2 days with $0.16\text{-}0.30 \times 10^{-6}$ M). Disturbances in the regulative functions of the stomata as a result of ozone influence can serve as an indicator of the ozone stress long before the occurrence of any damage symptoms are visible on leaves (Shabala and Voinov, 1994).

The changes in gas exchange are also viewed as shifts in net photosynthesis and respiration because the stomata cells regulate these processes. Photosynthesis is more ozone sensitive than respiration is, and has greater applications insofar as diagnostic practices are concerned (see section 6.2.2.4).

6.2.2.2. Permeability of membranes and changes in membrane-related processes.

Membranes are the cell components most vulnerable to ozone action (Section 2.2). Especially valuable for diagnosing cell damage is the measuring of plasmalemma permeability. The efflux of ion and organic substances from cells can be determined after 2-10 minutes of acute ozone treatment (Table 24). There are significant changes in the permeability of membranes to water, glucose and ions. Ozone increases the permeability of plasmic membrane to the ions K^+, Ca^{2+}, Mg^{2+}, and their concentration in the extracellular medium, is, consequently, increased. The dynamics of membranous permeability is characterized by high rate constants.

Proteins and lipids are the membrane constituents that are the most sensitive to ozone damage. Upon deacetylation of membranous phospholipids and galactolipids, fatty acids are liberated, which then can become the targets of ozone. The products of their oxidation include the hydroperoxides of fatty acids, as well as carbonic and other oxygenated compounds. The presence of these substances is an early marker of cell damage. The other membrane constituents subject to damage are the protein components included in the receptory, catalytic and other complexes. As a consequence of the adverse action of O_3, changes to the enzymic system occur involving energetic and metabolic reactions, along with the receptors regulating ion permeability. Although these reactions have not yet been sufficiently studied, they may have future diagnostic applications.

One of the membrane processes most sensitive to ozone damage is phosphorylation, which occurs in chloroplast and mitochondria membranes. The reduction of the ATP levels in a cell undermines the barrier function of these membranes.

6.2.2.3. Formation of stress ethylene

Since the visual leaf damage caused by ozone correlates with the formation of stress ethylene, it was assumed that stress ethylene could be used as an indicator of ozone pollution in regions containing mainly coniferous plants and, in particular, *Picea excelsa*. (Wolfenden et al., 1988; Sandermann et al., 1997). Exposure of Angiosperm plants to ozone causes stress ethylene to be formed (Wellburn and Chen, 1990), the release of which can increase by 1.5-4 times. For instance, the tomato *Lyicopersicum esculentum*, after being treated with ozone (0.2 μl/L), had a 3-times higher rate of ethylene release, acting, apparently on the synthetase of I-aminocyclopropane-I-carboxylic acid (Bae and Nakajima et al., 1996). Roberts and Osborne (1981) believe that the biosynthesis of ethylene is increased before any membranes are injured or any compounds leached. It seems that membrane integrity is necessary for synthesis of ethylene. When the available stressor (in this case ozone) is present in the medium, Vick and Zimmerman (1987) suggest that there are a number of reactions leading up to the formation of the stress ethylene, and these occur in the sequence shown in fig. 45. As a stressor, ozone primarily functions to activate phosphorylase in response to membrane damage. Such injuries cause unsaturated fatty acids to be freed from the membranes and the process of lipid peroxidation to occur and to produce, among other things, the free superoxide anion radical. This radical cooperates directly with the precursors of ethylene formation and, in turn, generates their free radicals.

OZONE as stressor

⇓

Activation of phosphorylase

⇓

Liberation of unsaturated fatty acids

⇓

Formation of the hydroperoxy derivatives of fatty acids, along with the appearance of the free superoxide anion radical

⇓

Formation of free radicals of I-aminocyclopropane-I-carboxylic acid

⇓

Ethylene

Figure 45. The scheme of the ethylene formation in response to ozone stress

The use of aminoethoxyvinylglycine, the inhibitor of the enzyme catalyzing a conversion of 1-aminocyclopropane-1-carboxylic acid into ethylene, caused a decrease in ethylene excretion, as well as in visible leaf damage. Quantities of 1-aminocyclopropane-1-carboxylic acid as well as the above-mentioned enzyme may also be a marker of the stress induced by O_3. At certain concentrations of O_3, the rate of the ethylene release is slowed down, but is replaced by the ethane excretion, which is an indicator of the damage caused by free radicals (Wellburn and Chen, 1990).

6.2.2.4. Changes in photosynthesis and chlorophyll content

Photosynthesis can be used as a test process for early diagnosis of ozone-induced stress, when O_3 concentrations are as low as 3×10^{-7} M. As a rule, this process is sensitive to many stress factors, and does not react to the presence of ozone in any specific way. The observation of any changes in photosynthesis can only be recognized as one possible symptom of ozone stress, and the diagnosis need to be confirmed by checking on other factors, such as a modified rate of photoassimilation, an altered fluorescence intensity for chlorophyll and a changed ratio of *a/b* chlorophyll forms.

Inhibition of photoassimilation. The photoassimilation of CO_2 has proven to be a sensitive test on ozone stress (see section 5.1.1). Usually, the level of fall of photosynthesis is insignificant. For instance, a yellow pine *Pinus ponderasa* fumigated with maximal concentrations of ozone (up to 0.15 ppm) for 30 days did not show any visible symptoms of needle damage, despite the fact that the net photoassimilation decreased by 10 % (Guderian et al., 1985). The same finding was observed when citron trees were treated in the same manner. Statistically, a discernible decrease in CO_2 uptake upon light has often been noted to occur in many species before any visible symptoms of damage are detected. At high concentrations of ozone (0.3-0.4 µl/L), rapid suppression of photoassimilation is observed.

Changes in the synthesis of chlorophyll and the ratio of *a/b* chlorophyll forms. With an increase in ozone stress, the ratio of chlorophyll *"a"* to chlorophyll *"b"* is usually reduced. This change probably results from the greater sensitivity of chlorophyll *"a"* to ozone or from the inhibition of pigment synthesis (Beckerson and Hofstra, 1979c; Reiling and Davison, 1994). O_3 action on leaves results in the increased formation of the superoxide anion radical in chloroplasts, which in turn causes the destruction of chlorophyll. It is now recognized that even very low concentration of the radical (10^{-8}-10^{-7}M) initiate this destructive process (Asada et al., 1977).

The loss of green pigment indicates that a plant's normal ability to live has been disturbed. Chlorophyll is destroyed due to the gradual loss of green color, and all parts of a plant become chlorotic or yellowed. Ozone activity, which manifests itself as a number of deterioration in the main physiological processes (total rate of photosynthesis, biosynthesis of pigments, ratio of the chlorophyll forms, intensity and character of chlorophyll fluorescence) can be detected by observing degradations in the above processes with a sufficient degree of reliability (Knudson et al., 1977). Changes in pigment and its biosynthesis in chloroplasts is the most immediate response to changes in the state of chlorophyll in leaves, especially, in its fluorescence in the red region of the spectrum (Kooten van and Howe van, 1988).

Any induction of the chlorophyll fluorescence (so-called Kautsky effect) appears to relate to a stress factor. Ozone also stimulates chlorophyll fluorescence (Shimazaki, 1988; Clark et al., 1999). For instance, the phenomenon is observed in leaves of the string bean *Phaseolus vulgaris* treated with ozone (0.3-0.5 µl/L) at least 20 hours before any visible leaf necrosis appears (Schreiber et al., 1978). The first changes in the fluorescence appear due to a damage of enzymes involved in the photosynthetic water decomposition. Subsequently, inhibition of electron transport between photosystems is observed.

Ozone also acts on decelerated leaf luminescence (this process is additionally known as photosynthetic luminescence or photoremission). The phenomenon of such luminescence occurs due to photochemiluminescence of chlorophyll, which is reversal of a primary photochemical stage of photosynthesis (Veselovskii and Veselova, 1990). Ozone effect on the reaction occurs in two stages. In the first stage (during 10-30 minutes of fumigation with ozone), photosynthetic activity, estimated as the amount of quenching of the decelerated luminescence, was approximately reduced by half. Thus, by measuring the decelerated luminescence, it is readily possible to establish if the studied object has reacted to ozone.

6.2.2.5. Autofluorescence of cell surfaces.

The method of measuring fluorescence is fast as well as simple, and does not require tissue destruction. It is widely used in diagnosing the influences that pollutants have on plants (Schreiber et al., 1978; Kooten and Howe, 1988). Earlier, it was only used to measure chlorophyll fluorescence (see section 6.2.2.5).

In the Microspectral Analysis and Cellular Monitoring of Environment Laboratory (Institute of Cell Biophysics at the Russian Academy of Sciences), a method of microspectrofluorimetry has been used to record the fluorescence spectra of individual cells (Karnaukhov, 1978; 2001). Besides the changes in chlorophyll fluorescence effected by ozone, this method permits researchers to measure the shifts in the maxima emitted by cell surfaces in the light-blue and orange regions of the spectrum, as well as the occurrence blue-green fluorescing pots on a leaf's surface (see fig. 9, see section 2.1).

They can be connected both with a change in the composition of phenolic and terpenoid compounds, as both have double bonds susceptible to oxidation, along with the formation of the pigment of aging, named lipofuscin (see section 3.1). Thus, the analysis of the light emissions from plant surfaces and, therefore, unrelated to chlorophyll can also serve as an early indicator of ozone damage.

Recently, changes in the autofluorescence of pollen have also been considered as a sensitive indication of ozone activity (see section 2.1). It has been demonstrated that the pigments contained in the surface of a pollen grain determine its ability to resist O_3 (Roshchina and Karnaukhov, 1999; Roshchina and Melnikova, 2001). Pollen, collected from natural plant, for instance pines and meadow grasses, as well as from artificial decorative species (Roshchina et al., 1998d), may change the color of its fluorescence under conditions of chronic ozone fumigation. This change is especially marked when the fluorescence maximum shifts from the yellow and orange regions of the spectrum to shorter wavelength regions, an example of which is shown in fig.46 (see also appendix 3).

Shifts of the maxima in the fluorescence spectra of pollen, as registered by the method of microspectrofluorimetry indicate that a degree of damage is being caused by ozone. Pollen, lacking the appropriate pigments in it surfaces, is injured more severely (see section 6.2.1).

The autofluorescence of secretory hairs is also changed in response to ozone treatment (Fig. 10 and 12, Chapter 2). In particular, this modification is readily visible in secretory cells containing lipid-like secretions that can be stained with Sudan III. Clearly, luminescent microscopy and microspectrofluorimetry can help in the cellular monitoring of ozone.

Recently confocal microscopy have been used for the analysis of the ozone effects on plant cells, in particular of vegetative microspores of *Equisetum arvense*..

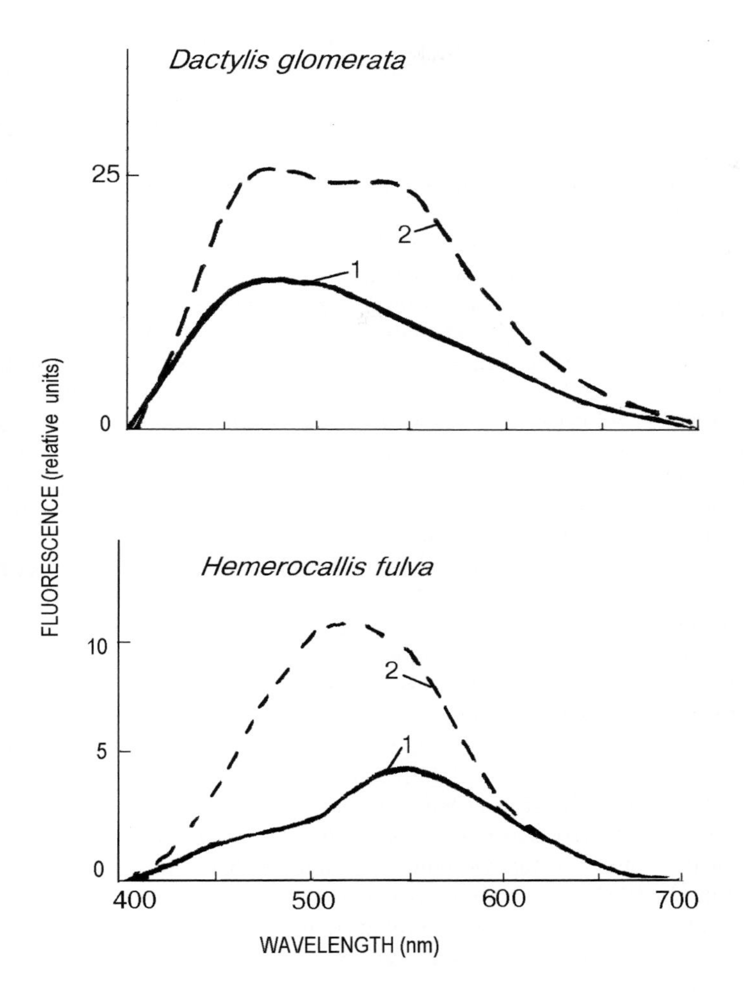

Figure 46. The fluorescence spectra of intact pollen grains collected in nature with (2) and without (1) treatment with 2.4 ppm of ozone.

The changes in their cover autofluorescence excited by ultraviolet ,yellow or red light were seen even after short exposure at 0.03 ppm ozone (Roshchina et al., 2003b). More shifts in the light emission appeared if the microspores were treated with higher concentration of O_3 (0.25 ppm). These data confirm that main cellular reactions with the low and average ozone concentrations occur on the cell surface. In future confocal microscopy could be one of the perspective technique for the cellular monitoring of ozone.

6.2.2.6. Changes in activity and biosynthesis of enzymes. Peroxidases as indicators of ozone stress.

Enzymes, just like many proteins, quickly react with ozone (section 3.0) and, as a result, undergo changes in their activity. Lengthy exposures to ozone lead to changes at the genetic level, as the biosynthesis of enzymes can be affected (Sandermann et al., 1997). The formation of some enzymes is repressed, while others are activated. For instance, the formation of the enzymes participating in cellulose and starch synthesis is suppressed, but the synthesis and activity of cellulase and amylase, enzymes that participate in the desintegration of cell wall, is increased (see chapter 5). Frequently, enzymes that function as oxidases, such as peroxidase, are activated to an appreciable extent. In fact, an increase in peroxidase activity and/or the enhanced biosynthesis of its various isoforms can be the most discernible plant response to ozone, as well as an accurate indicator of ozone stress (Keller, 1974).

The increase in peroxidase activity resulting from exposure to O_3 has been discussed in section 3.3 and chapter 4. As reported in these discussions, the change in this activity occurs prior to the appearance of any external symptoms of damage. In addition, many reactions catalyzed by peroxidase are associated with the formation of heavily pigmented compounds, whose presence greatly facilitates detection of any activity involving this enzyme. Therefore, peroxidases activity can serve as an early indicator of ozone damage.

Some forms of peroxidase can be released into the free space of a cell, as they are capable of transgressing membrane barriers. Acidic forms of the enzyme are even found in the cell wall. Since the free space of a cell is the first compartment into which ozone enters before reacting with organic compounds (see section 2.1.3), the analysis of the liquid leached into this space can provide some information about the primary metabolic changes that a cell undergoes in response to ozone (Tingey et al., 1976a; Curtis et al., 1976). The high sensitivity to ozone of extracellular peroxidases is comparable to that of intracellular peroxidases, as Castillo et al. (1987) have shown to be the case in the needles of the fir tree *Picea excelsa*. The extracellular activity of peroxidases in *Sedum album* are increased threefold when exposed to 0.4 ppm of ozone for two hours, an increase that is furthermore easy to detect (Ogier et al., 1991). The activity of peroxidase in extracellular space can be estimated by comparing it to the total peroxidase activity of a cell, a method that enables the researcher to characterize the extracellular enzyme activity in a quantitative manner (Castillo et al., 1987).

However, an increase in peroxidase activity is not, in itself, a certain sign that the damage was caused by ozone, as many stress factors could result in the same effect. It is important, in addition, to analyze the isoenzymic structure of the proteins. According to Savich (1989), the influence of ozone and other stress factors on a cell causes a unique stress set of isoperoxidases to be formed. Experiments by Podleskis et al. (1984) have provided the most significant support for this view. They employed the electrophoresis method to detect altered peroxidase composition at the 3-4 leaf stage of 20 hybrids of the corn *Zea mays* after exposing the plants to sharp doses of ozone (492 $\mu g/m^3$ or 0.025 ppm for 3 hours). A study was also made of the connection between sensitivity to ozone and isoenzyme composition of peroxidases.

It was demonstrated that determining the number of protein strips associated with peroxidase helps to reveal plant sensitivity to ozone. The cultivar most sensitive to ozone had the least number of such strips, just 14 of them.

But the activity of isoperoxidases in sensitive varieties can also differ from those in tolerant cultivars. For instance, Curtis et al. (1976) cultivated 2 strains of the soya *Glycine max* differing in their sensitivity to 0.035 ppm of ozone administered for 2 hours. In the sensitive cultivar *"Wyu"*, activity of all isoenzymes was increased, whereas only some of the enzymes in the tolerant cultivar *"York"* displayed increased activity, out of a total of 22. Thus, the isoenzymic set of peroxidases can also be indicators of ozone stress.

Studies of other enzymes, mainly antioxidants, are only beginning to be undertaken, and a connection between ozone stress and changes in their activity or pools in cells is yet to be made (see Chapter 5). In future, such diagnostics will have also to be given due consideration.

6.2.2.7. Changes in metabolic pools

Finally, the pools of certain metabolites, along with the means of translocating them, are also altered by ozone-induced changes in enzyme activity, biosynthesis rates, strength of the catabolic processes as a whole, and gene activation or repression (see sections 5.1.1; 5.1.2; 5.1.3; 5.3; 7.4). The resulting accumulation of some metabolites and the decrease in the levels of others are noticeable before any symptoms of plant damage become visible.

Accordingly, the pools of soluble sugars are widely taken to be good indicators of ozone stress. Ozone suppresses the enzymic system of cellulose and starch synthesis, and therefore reduces the amounts of polysaccharides. Conversely, the cellulase and amylase activities are increased, leading to the rapid destruction of the cellular cover and stored starch, relatively. For instance, the activity of cellulase can be increased up to 400 %, while the pool of soluble sugars is reduced due to a decrease in photoassimilation and an increase in respiration intensity. If the pool of soluble sugars is reduced, the amount of free amino acids is increased, as ozone exposure promotes protein destruction, while it inhibits protein synthesis (see section 5.3) .

The lipid pool also changes under ozone treatment (see chapters 3.2.6.7). The decrease in lipid quantities usually results from the inhibition of their synthesis. Although the ratio of the saturated and unsaturated fatty acids can be an indicator of ozone-induced toxicosis, the products of lipid peroxidation provide a precise indication of ozone stress. In this case, the quantity of unsaturated fatty acids decreases, and malondialdehyde is accumulated.

Changes in the pools of the secondary metabolites, especially monoterpenes, phenols and polyamines (see chapters 4 and 6) provide the most specific indication that ozone, alone, is the source of the plant damage. The secondary metabolism is the most vulnerable cellular target for O_3, as ozone activates the genes regulating the biosynthesis of these secondary compounds. Quantities of the secondary substances are, consequently, changed in response to ozone exposure, as the corresponding enzymes are increasingly activated. The most frequent indication of pollution by oxidants in the yellowing needles of such gymnosperm plants as *Pinus abies* is that results from a change in the concentration of monoterpenes (Juttner, 1988). Similarly, ozone stimulates the biosynthesis and activity of enzymes participating in the phenol metabolism (see section 5.4.1). The accumulation or disappearance of certain phenols can be a specific indicator of environmental pollution by oxidants. The accumulation of polyamines, such as putrescine, spermine and spermidine, is also an indicator of ozone stress (see section 5.4.3).

An especially interesting problem is posed by the ozone-induced accumulation of the agents participating in programmed cell death, for instance the products of lipid peroxidation like jasmonic acid, its derivatives and hydrocarbons as well as salicylic acid (Rao and Davis, 2001). Salicylic acid is a phenolic product forming under normal conditions in some plant species, for example the genera *Salix*, but also having defensive functions in response to ozonation (Rao and Davis, 2001). This acid participates in the salicylate-stimulated respiration pathway (Straeten et al., 1995), enhances the activity involving the alternative pathway of respiration in tobacco leaves, and induces thermogenicity. The accumulation of jasmonates and

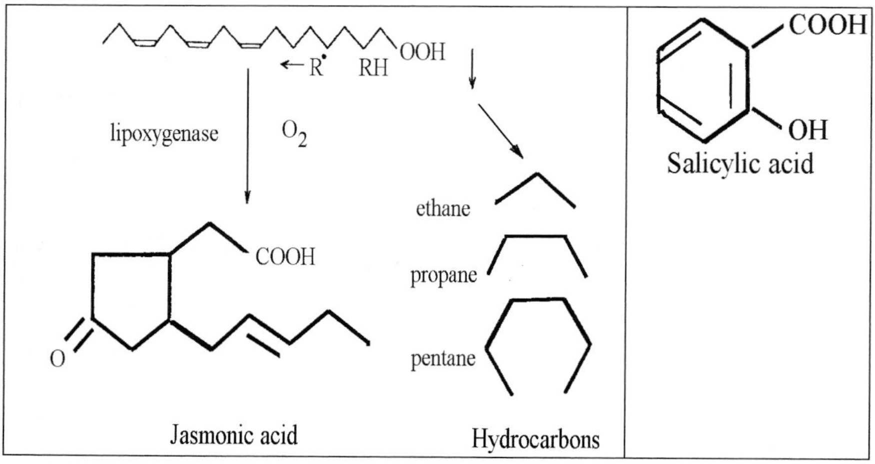

OOH
$\leftarrow R^{\bullet}$ RH
lipoxygenase
O_2
COOH
O
ethane
propane
pentane
Jasmonic acid
Hydrocarbons
COOH
OH
Salicylic acid

salicylates is also a marker of any early damage induced by ozone.

The biochemical reactions of plants to ozone have different thresholds of sensitivity in vivo (Mudd, 1973). Lowest concentrations of ozone (0.05 ppm and below), inhibit the starch hydrolysis. At 0.2-1.0 ppm of the gas, there is also inhibition of photosynthesis, reduced concentrations of chlorophylla *b*, decreases in the level of NAD(P)H and phosphorylation, inhibition of respiration or stimulation of oxygen uptake, increases in the amount of free amino acids, and formation of malondialdehyde.

6.2.2.8. Activation and expression of the genes encoding protective low-molecular and high-molecular metabolites

In laboratories, which have the technique for molecular cloning and genetic engineering, can analyze the activation and expression of the genes encoding protective low-molecular and high-molecular metabolites. Pertinent information is presented in table 20 (chapter 5). As diagnostics, it is possible to use the RNAs coding ascorbate peroxidase (Örvar and Ellis, 1997), glutathione-S-transferase (Heidenreich et al., 1999), stilbene synthase (Zinser et al., 1998; 2000b), and cinnam(o)yl alcohol dehydrogenase (Wegener et al., 1997). Pathogenesis-related proteins, stress proteins, and various glycoproteins (Ernst et al., 1992a, b; 1997a, b; Eckey-Karltenbach et al., 1994a, b; 1997b; Groβkopf et al., 1994) may also serve diagnostic purposes. Currently, transgenic plants, consisting mainly of tobacco clones, are being used to analyze such susceptibility to ozone injury (Örvar and Ellis, 1997).

6.2.3. Diagnostics of cell damage.

Special attention is always paid to the damaged plant cells that occur on leaf or pollen surfaces when exposed to chronically low, middle or acutely high doses of ozone. Among such injuries are apoptosis (requeiring a special microscopic analysis) and visual types of damage (chlorosis, dots-lesions, pigmentation and necrosis). The appearance of such damage is used to define the sensitivity of a whole plant species to ozone (see below in section 6.1.2).

Apoptosis. Apoptosis (programmed cell death) is genetically controlled ablation of cells during development. It is particularly well illustrated by cancer cell or cells, subject to genetic transformation. The form of apoptosis induced by ozone consists in hypersensitive cell death (Rao et al., 2000a, b; Rao and Davis, 2001). Moreover, O_3 is considered as a tool for probing the processes involved in programmed cell death in plants (Rao et al., 2000 a). Apoptosis leads to the appearance of lesions and necrosis in plant leaves (Rao and Davis, 1999).

Unlike necrotic cells, apoptotic cells are characterized by: (1) relative membrane integrity; and (2) alteration in chromatin structure due to the degradation of protein and nucleic acids. In order to distinguish living cells from late apoptotic cells, methods of histochemical staining with fluorescent dyes have been developed.

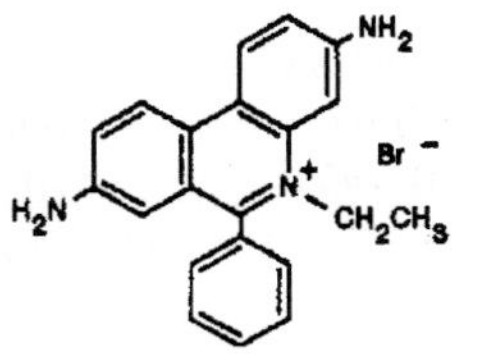

NH_2, H_2N, Br^-, N^+, CH_2CH_3

ETHIDIUM BROMIDE

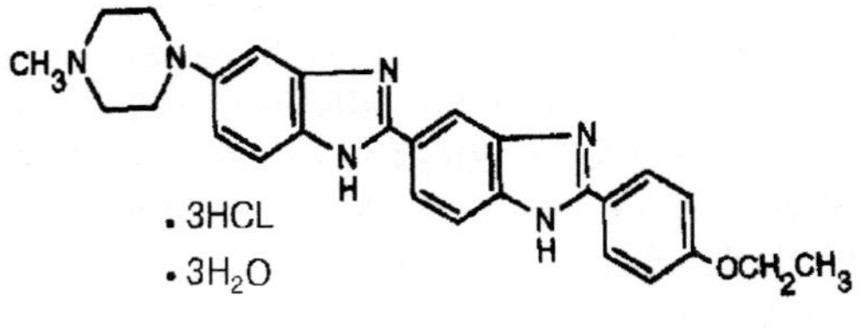

CH_3N, N, N, H, N, H, . 3HCL, . $3H_2O$, OCH_2CH_3

Hoechst No. 33342

(Haugland, 1996). For example, early apoptosis can be assayed by staining DNA with Hoechst 33342, while late apoptotic cells have membranes permeable to ethidium bromide, another fluorescent dye. Table 29 shows the results obtained by staining ozone-treated pollen with above-mentioned dyes

The strenghthened fluorescence of the DNA-Hoechst 33342 complex is characteristic of chromatin damage (Roshchina et al., 1998b, c). Increased amount of cells, permeable to ethidium bromide indicates membrane damage. At the highest concentration of ozone, pollen becomes non-viable and apoptosis of cells appears to be a frequent occurrence. It is seen as the increased fluorescence intensity of the

Table 29. Analysis of results obtained by staining the ozone-treated pollen of Hippeastrum hybridum *with fluorescent dyes Hoechst 33342 and ethidium bromide.*

Ozone dose, ppm	*Amount of germinated pollen grains (% of control)*	*The fluorescence intensity of Hoechst-stained cells at 460 nm (% of control)*	*Amount of orange-fluorescing cells after staining with ethidium bromide (% of control)*
0.05	100±2	95± 4	0
0.10	100±4	97±3	0
1.0	40±4	115±6	34±3
2.4	0	125±5	100±1
5.0	0	136±10	100±1

DNA-Hoechst 33343 complex and as the 100% -staining of cells by ethidium bromide.

Visible damage. Among the effects that ozone has on plants, there are a number of symptoms that can be seen on the external view of a plant. The same damages may also be analyzed under the microscope or by using spectrophotometry, fluorimetry and microspectroflurimetry. These diagnostic techniques should be considered as methods of estimating leaf damage as well as plant growth and development in order to ascertain sensitive and tolerant plant species. By such means, necrosis, chlorosis and other symptoms are observed (Table 30). Ozone mainly attacks the palisade cells of the mesophyll, injury to which is a characteristic attribute of the ozone syndrome. A histological study of leaves from the soya bean *Glycine max* has shown that ozone treatment (590 μg/м^3 for 0.4-24 hours) of cells located in veins are the first ones injured (Pell and Weisberger, 1976). It becomes more widely distributed into the cells of the sponge and the palisade parenchyma, the cells of which are more strongly damaged. In injured cells, the protoplast was compressed and, simultaneously, cells destroyed. In other cases, such damage was not seen, and the membranes of chloroplasts, the endoplasmic network and mitochondria still remain intact. Thus, the epidermal cells were uninjured. Higher concentrations of ozone (0.05; 0.075; 0.1 μl/l) caused an increase in the air space of the mesophyll. Cells of the epidermis the leaves of *Betula pendula* were frown (wrinkled), their stomata apertures diminished, exuded drops appeared as swelled bubbles on the cellular walls of the mesophyll, and the mesophyll was generally degraded (Matyssek et al..

1991). Ozone damage is also discernible in fruit. For instance, tissue in the fruit of the apple tree *Malus domestica* displayed rised lenticels and general collapses (visible sites of dead cells) around them (Miller and Rich, 1968).

Leaf damage is the most conspicuous ozone effect because it involves a pigment change in the leaf plate (Elstner, 1985; Langebartels et al., 1990; Pandey et al., 1991). Under the influence of ozone, four types of leaf pigment damages can be distinguished: dotted pigment damage, bleaching, chlorosis, and bilateral necrosis (Hill et al., 1970). The first three types of damage are characteristic responses to chronic doses of ozone. Bilateral necrosis is mainly observed after short-term exposures to high ozone concentrations. Descriptions of the various forms of the plant pigment damage instigated by O_3 are found in the review written by Guderian et al. (1985) and in the book edited by Lacasse and Treshow (1978). The latter is a survey of the diagnostics used to evaluate atmospheric pollution based on symptoms manifested by the state of vegetation (Diagnosing Vegetation Injury Caused by Air Pollution. 1978). Color photos of the most visible signs of damage to sensitive plants can be seen on various internet sites (see addresses in appendix 1).

Dot damage. The most frequently encountered types dotted damage are seen in the foliage of trees, bushes (shrubs) and grasses, and found mainly on the topside of leaves. As shown in table 30, the distinctly configured dot or stain formations can be dark-brown, black, red or purple in color. This color is caused by ozone-induced biosynthesis of specific anthocyanins or polyphenols (Konkol and Dugger, 1967; Howell, 1970; 1974). Pigments accumulated in the injured sites of some plant species can diffuse into contiguous tissues, boarded by the smallest veins. The epidermis, covering the damaged palisade parenchyma, remains intact. These symptoms are readily seen at small magnifications of the microscope in transverse light and were described in leaves of the grape *Vitis* as early as 1958 (Richards et al., 1958).

Bleaching. Like dot damage, bleaching is mainly observed on the top surfaces of leaves. The palisade parenchyma is most severely injured, but the epidermis can also be damaged. Numerous small, necrotic, non-pigmented sites or white sites appear on the leaf plate so that the leaf seems to be dotted. If the cells are in collapse, they normally remain connected to the healthy cells above and below them. The resulting spaces fill with air and give the leaf a light-gray, milky-white or red-brown shade. In many Dicotyledon plants (the tobacco *Nicotiana tabacum*, the bean *Phaseolus sp*., the soya bean *Glycine max*, the plantain *Platanus sp*), this symptom is observed in its classical form. In other species having little or no differentiated mesophyll tissue, the damage can also occur on the surface of the leaf.

Chlorosis. The symptoms of color loss are observed to occur in their most extreme form in the palisade parenchyma on the topside of a leaf. The primary damage is usually limited to small groups of palisade cells. The upper epidermal cells and the lower-lying cells of the spongy mesophylla remain intact. The extent of the damage may vary from something that affects just a few cells to dysfunctions that create spots of 1 mm in diameter. With chronic or repeated exposures to ozone, these

Table 30. The visible symptoms of the leaf damage induced by ozone (Guderian et al., 1985; Lacasse and Treshow, 1978; Elstner, 1985; Langebartels et al., 1990; Pandey et al., 1991; Günthardt-Goerg et al., 1999; 2000; Günthardt-Goerg, 2001)

Type of damage	*Location of damage*	*Symptoms of damage*	*Mechanism of damage*	*Method of the determination*
Dot damage	In the foliage of trees, bushes (shrubs) and grasses.	The distinct dot damage or stain damage can be dark-brown, black, red or purple in color, and are mainly found on the topside of a leaf (Richards et al., 1958)	This color is caused by the ozone-induced biosynthesis of specific anthocyanins or polyphenols (Konkol and Dugger, 1967; Howell. 1970; 1974)	Visible symptoms are readily seen at small magnifications in transverse light
Bleaching	On the top surfaces of leaves	Numerous small (fine). necrotic non-pigmented sites or white sites in the leaf plate give the appearance of the leaf being dotted. The spaces filled with air give the leaf a light-gray, milky-white or red-brown shade.	Destruction of chlorophyll	Visible symptoms: palisade parenchyma is severely injured; epidermis can also be damaged.

Table 30. (Contd)

Type of damage	*Location of damage*	*Symptoms of damage*	*Mechanism of damage*	*Method of the determination*
Chlorosis.	A common in many coniferous plants, particularly pines, and in grassy plants or plants with non-differentiated mesophyll	Loss of color only on the top of a leaf in the palisade parenchyma. For objects with non-differentiated mesophyll, a bright chlorotic stain usually develops between the vessels on both leaf surfaces (Hill et al., 1961).	Destruction of chlorophyll	Visible symptoms: the majority of chlorotic (colorless) cells still remain alive, but chloroplasts are destroyed, and the amount of chlorophyll is reduced. Effects can be seen under a microscope (Hill et al., 1961).
Bilateral necrosis.	Between upper and lower epidermis; depending on the plant species, the color of the necrosis site can be changed.	Necrosis is seen when the mesophyll tissue between upper and lower epidermis is destroyed.	Destruction of the mesophyll between the upper and lower epidermis	Visible symptom: under the microscope, both layers of the epidermis appear close to each other and seem to form a paper-thin layer of injured cells. The color of this layer can vary from white to red - orange.

injured sites can merge, giving a leaf a spotty appearance. Microscopic analysis shows that the majority of chlorotic (colorless) cells remain alive, but their chloroplasts have been destroyed, and their chlorophyll content is reduced (Hill et al.. 1961). Chlorotic stains are a common symptom in many coniferous plants, particularly pines. In other species, they are, however, rarer than dot damage or bleaching. In grassy plants or plants with non-differentiated mesophyll, bright chlorotic stain usually develop between the vessels on both leaf surfaces. The outside cells of the spongy mesophyll contiguous to a vessel bundle and the cells, lying above fine vessels are always amazed, and chlorosis can be more distributed inside of a leaf (Hill et al., 1961). In response to long-term constant exposures to ozone or a few days after an exposure to high concentrations of the gas, leaves fall.
Bilateral necrosis. Bilateral necrosis occurs when the mesophyll tissue between the upper and lower epidermis is destroyed. Both layers of epidermis are, consequently, placed close to each other and form thin (like a paper) layer of injured cells. Depending on the plant species, the color of this layer can vary from white to red - orange. Bilateral necrosis and the other above-mentioned forms of damage can be observed to occur in the same leaf.
Revealing necrotic sites by spectral methods. The influence of increased concentrations of ozone can lead to the formation of necrotic stains, which can be analyzed by using spectral methods. With the help of spectrophotometry, the spectral leaf reflection in the seedlings of the pine *Pinus taeda* L. can be measured in the visible (400-720 nm) and infrared (720-2500 nm) regions of the spectrum (Carter et al., 1992a, b). Usually, the ozone-induced destruction of chlorophyll, which absorbs visible light, increases a leaf reflection. The water released from cells that have been ozone damaged compensates for the reduced absorption in the visible frequencies, as water also absorbs light. There is, as a result, an increased reflection of infrared radiation (1300-2500 nm), while visible light is primarily absorbed by water instead of chlorophyll.. When the summer wheat *Triticum aestivum*, the white clover *Trifolium repens* and the corn *Zea mays* were exposed to ozone (in concentrations of 180-240 $\mu g/m^3$ or 0.086-0.11 ppm) for 8-12 hours /day, visible changes to leaves correlated with their altered spectral reflections (450-950 nm). There was an evident reduction in the infrared reflection of the leaves of the clover *Trifolium*, though not all leaves had visible symptoms of damage. When the wheat *Triticum* and the clover *Trifolium* displayed visible symptoms of injury, the reflection or radiation in the 400-700 nm range was increased, while the infrared reflection of clover greatly decreased. No changes were observed in the corn. Experiments on the pine *Pinus* also indicated that an increase in the reflection of needles was accompanied by an increase in the fluorescence of chlorophyll (Meinander et al., 1996).

Necrotic spots also can also be analyzed by subjecting them to fluorescent methods. The presence of ozone in the bottom layers of troposphere is frequently caused by emissions from motor transport and industry. Therefore, bronze necrotic stains are often found on the leaves of such plants as poplars or common nettles growing near parking facilities or areas of automobile congestion. The analysis of their fluorescence spectra (Fig.47) shows, that the necrotic stain on leaves of the

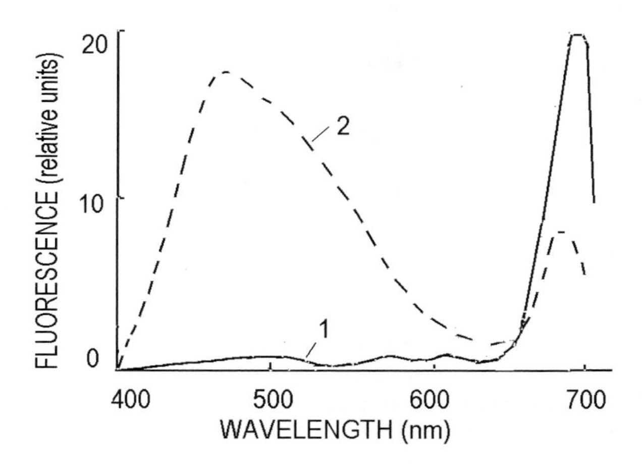

Figure. 47. The fluorescence spectra of intact(1) and necrotic(2) cells of a poplar Populus balsamifera (Roshchina, unpublished data).

poplar *Populus balsamifera* has a strongly fluorescent maximum at 460-465 nm. Destroyed chlorophyll has a weak fluorescence at 680 nm.

6.3. PLANT CELLS AS BIOSENSORS AND BIOINDICATORS OF OZONE

Indication of damages is fundamentally important for predicting the ability of cells to survive increases or decreases in ozone quantities (Feder, 1973, 1978). Plant cells and plants as a whole have rather high sensitivity to ozone (see section 6.1). They can therefore be used as indicators of O_3 presence and may even be used to determine its levels in the air (Lacasse and Moroz, 1969; Feder and Manning, 1979). For this purpose, it is necessary that discernible ozone effects on plants are identified, and the results of experiments readily replicated. In addition, the reactions specific to ozone should differ from any effects caused by other pollutants in the atmosphere. According to Feder (1978), plants serve as collectors and natural indicators of air pollution. Symptoms of leaf damage, modified growth and reproduction, changes in crop yield and efficiency, and any deviations from the usual distribution of plant species in the phytocenosis, either alone or in combination with other methods of the ozone monitoring can serve as diagnostics of ozone pollution. Genetic variability among the plant species is revealed by the wide complex of attributes connected with their resistance or sensitivity to ozone. This diversity enables the cultivation of specific indicator plants on sites for the expressed purpose of monitoring the atmosphere.

In laboratory conditions, a useful test for atmospheric ozone pollution could involve the germination of pollen from plants sensitive to and tolerant of ozone. As the information in section 6.1 shows, the viability of pollen from some plant species is sharply reduced under the chronic influence of small doses of ozone.

Below, the possibility of using individual plant species as bioindicators of ozone effects will be discussed.

6.3.1. Plants as bioindicators of ozone pollution

Different countries often use various plant species belonging to the gymnosperms or angiosperms, as well as lichens and mosses to function as environmental bioindicators (see Appendix 2). To monitor ozone, the tobacco *Nicotiana tabacum*, the petunia *Petunia hybrida*, and the string bean *Phaseolus vulgaris* are frequently planted. As constant indicators of ozone, observations about the state of coniferous trees, in particular, the pine *Pinus strobus*, are made, although only individual varieties of the same species may be susceptible to ozone stress.

The specificity of such bioindicators represents a problem that has attracted a great deal of attention (Taylor, 1973; Mudd, 1973; Feder and Manning, 1983). To dissolve this problem, it is necessary to analyze specific plant responses and cumulative effects of other air pollutants (SO_2, NO, NO_2, and organic wastes). Several studies have compared the separate reactions of the various atmospheric pollutants as well as plant responses to mixtures of the pollutants. For more information on this subject, we refer the reader to the above-mentioned publications.

The effects of ozone on plants can be characterized as acute (resulting from exposures to high concentrations for short intervals) or chronic (subject to low concentrations for long periods). Thus, different parameters of damage to plants can be used. Among the visual negative effects are the fall of plant organs, various kinds of chlorosis and necrosis, as well as decline in plant growth and development. These are represented in table 31.

Table 31. The responses of whole plants to acute and chronic treatments with ozone

Acute treatment	*Chronic treatment*
Fall of leaves, fruits, and sepals of flowers. Leaf wilting and curvature (crook) of stems.	Delay or termination (discontinuance) of growth and development. Acceleration of fall of leaves and fruits before the term of maturing. Chlorosis and necrosis of leaf tops. Slow fading (withering)

Ozone contamination causes the leaf plate to be damaged. Using this symptom, it is possible to show how species can have varied sensitivity to ozone (Ashmore et al., 1988). Plants in which more than 70 % of a leaf plate is injured can be identified as very sensitive species. In intermediate species, the leaf damage does not exceed 45-

70 %, and the leaves of tolerant species do not display more than 30-45 % damage. The particular representatives of these groups of plants are listed in appendix 2.

High doses of ozone cause necrotic spots to appear in the sensitive tissue of a leaf (Taylor, 1973). Chronic exposures of plants to lower concentrations of ozone lead, however, to cumulative effects (Feder, 1973). In such cases, necrotic staining and loss of chlorophyll "*b*" in leaves may result from fumigation with ozone or ozonated hexenes. This phenomenon may be demonstrated by the prolonged ozonation (8-10 ppm for 5.5 hours/day over 60 days) of the seedlings of the corn *Zea mays,* the geranium *Pelargonium* or the poinsettia *Euphorbia pulcherrima.*

Despite the appeal and the simplicity of this method, it should be noted that use of plants as indicators of ozone stress is not always effective. Unfortunately, the reaction of plants to a variety of stresses are often more or less identical. Only by comparing data about the ozone influence on plants obtained both in the laboratory conditions and under natural conditions will allow a researcher to ascertain if chronic symptoms are O_3 related.

Only organisms that provide a convenient means of observing the effects of ozone pollution can be used as reliable bioindicators. Such indicators should include plants that are not just sensitive to ozone but that also display the rather characteristic signs of visible damages. In addition, they should be rather easily susceptible to measurement (Table 32). The list of sensitive plant species that could be used for such purpose is also included in appendix 2 (see also appendix 1 lists the addresses of relevant internet sites).

Table 32. Plant species that are usually damaged by ozone and typical symptoms of the injuries suffered.

Plant species	*Typical symptom of damage*
Allium (Onion).	White stains, colorless ends
Citrullus (Watermelon)	Grey, metal shade of a stain
Cucumis (Cucumber)	White dots
Ipomea (Ipomea	Brown stains, chlorosis
Nicotiana (Tobacco)	White-gray stains.
Phaseolus (String Bean)	Bronze color, chlorosis
Pinus (Pine)	The ends of needles have yellowish-brown color, speckled needles
Solanum (Potato)	Gray-white, metallic colored stains.
Vitis (Grapes)	Dots ranging from red-brown to black

For practical reasons, only a small number of plant species have been used as indicators of ozone air pollution (Feder and Manning, 1985). The first indicators of ozone were certain species of tobacco *Nicotiana tabacum,* in particular the supersensitive Bel W3 cultivar (Heggestad, 1969). Increases in the concentration of ozone to more than 25-50 ppm caused necrotic stains to appear on the leaves of these plants (Elstner et al., 1985; Langebartels et al., 1990). The extent of the injury to the plant could be established on the basis of the leaf part area covered by the necrotic pots. There is no such damage in tolerant cultivars (grades) exposed to low concentrations of the gas 0.08-0.1 ppm, whereas about 41 % of the leaves in a supersensitive strain were damaged. High ozone concentrations (e.g. 0.42 ppm) caused 98 % of the leaves to be injured in the sensitive plants, and the area of the damaged surface of the leaf to exceed 75 %. The same test for determining the sharp influence of ozone was conducted in Holland, England and other European countries, as well as in Israel and North America. In Holland the spinach *Spinacia oleracea* var. Subito Dynamo was used to indicate ozone pollution. The monitored symptom of damage was the appearance of necrotic stains on the upper side of a leaf (Posthumus, 1976). A good detector of ozone is the plantain *Plantago major*, the growth of which is greatly slowed by exposures to high concentrations of ozone (Reiling and Davison, 1994). Treatment with a 70 μl/L (ppm) concentration of ozone changed both the chlorophyll fluorescence and ratio of O_2 /CO_2 gases.

In Japan, ozone damage symptoms in pine species, such as *Pinus thunbergii* and *P. densiflora* were analyzed (Matsunako, 1973; Shiratori, 1973). In Germany, some species of lichens, and some varieties of the genera *Gladiolus*, *Tulipa*, *Medicago*, *Petunia* and other plants have been used for this purpose (Heck, 1966; Feder and Manning, 1979; Shabala and Voinov, 1994).

In practice, the most convenient indicators are the leaves of the pea *Pisum sativum* and the radish *Raphanus sativus*, both of which become chlorotic when exposed to concentrations of ozone up to 0.14 ppb or 0.00014 ppm for one month (Roshchina and Melnikova, 2000). Recently, a general reduction in populations of the plantain *Plantago major* has been observed near an area in Pushchino where automobiles stop. A dense population of the strawberry *Fragaria viridis* replaced this species. Other potential indicator species could be represented by plants that develop poorly in response to ozone treatments administered in the laboratory (in particular, plantain *Plantago major* - see chapter 2.2). The table 33 contains unpublished data on the extent to which the growth of 14-day-old seedlings of plantain *Plantago major* depend on the dose of ozone fumigated for 30-34 days. As indicated, the control sample grew normally and had no chlorosis of leaves, while the increased dose of ozone caused irreversible changes in the seedlings of the experimental sample. Other tested plants, such as *Raphanus sativus* and *Zea mays*, al have shown the intensive green-yellow fluorescence on the leaf upper surface, but there was no subsequent necrosis. Thus, the plantain *Plantago major* may serve as an ozone indicator.

On the whole, woody plants are more tolerant of ozone than grassy ones (Treshow, 1988). 2-5 year old seedlings of 16 species of foliage trees from a north eastern part of the USA demonstrated symptoms of damage when exposed to the gas

Table 33. The effect of ozone on 10 day-old seedlings development of some species.

Plant species	*Control*	*Dose of ozone, ppm*			
		0.1	*0.15*	*1.0*	*1.5*
Plantago major	Normal growth for 30 days	Green-yellow fluorescence on leaf upper surface and then chlorosis of leaves.	Chlorosis of leaves	Death of some seedlings	Death of majority of seedlings
Raphanus sativus	"-"	Green-yellow fluorescence on leaf surface	Normal growth for 30 days		
Zea mays	"-"	Green-yellow fluorescence on leaf upper surface	Normal growth for 30 days		

concentrations as small as 0.25 ppm (for 8 hours, at 21°C, and 75 % humidity). Necrotic sites, common chlorosis and collapses of the mesophyll occurred on the top surface of the leaves (Wood, 1970). Among species such as *Fraxinus lanceolata, Fraxinus americana* and *Quercus niger*, the percentage of plants sensitive to ozone was from 19 up to 92 % (Wood, 1970; Korhonen et al., 1998). Among coniferous plants, the eastern white pine or Weymouth pine *Pinus strobus* is especially sensitive to ozone. Some exemplars of this species are damaged in the presence of 0.05 ppm O_3. Plantings of this pine show symptoms of damage when fumigated by only a 0.070 ppm concentration of ozone for 4 hour (Hicks, 1978). More information dealt with this problem reader could find in some special publications (Günthardt-Goerg et al., 1997;1999; 2000; Günthardt-Goerg, 1999).

On the basis of selection, it is possible to find plant species that are appropriately sensitive to ozone and therefore able to indicate O_3 presence in the air. The characteristic symptoms of visible ozone damage may appear in response to long-term treatment with ozone and/or its presence in high concentrations in an environment. They have been given sufficient consideration in section 6.2. However, by using biophysical, biochemical and physiological criteria along with the appropriate equipment (see section 6.2.), it is possible to detect a stress condition caused by O_3 before visible symptoms of the injuries induced by the ozone appear.

6.3.2. Cellular models for studying ozone effects.

Among the possible cellular models for studying ozone effects, autofluorescence and the viability of individual cells should be given some preference (see chapter 5 and sections 6.1 and 6.2.). These parameters are well-seen in generative (pollen) and vegetative microspores, as well as in secretory cells. We will use these three types of cells to discuss some of the models in what follows.

6.3.2.1.Pollen.

Autofluorescence. Sensitive pollen, such as the pollen of *Plantago major* and *Epiphyllum hybridum* (Roshchina and Karnaukhov, 1999), usually has a partial or complete lack of pigment, a factor that contributes to the autofluorescence of the pollen involved. However, pigmented pollen, as shown in Sections 2.1 and 6.2, may be an instrument for observing the fluorescence changes caused by ozone fumigation (Roshchina and Melnikova, 2001). As shown in Table 34, the pollen of *Hemerocallis fulva*, *Passiflora coerulea*, and *Philadelphus grandiflorus,* which is enriched in carotenoids and azulenes (Roshchina et al., 1995), fluoresce in the yellow and orange spectral regions. Fumigation with ozone doses of more than 0.6 ppm for 12 hours causes this fluorescence to change to blue or green-blue, a shift that is easily visible under a luminescent microscope. The principle maximum at 530-560 nm in the fluorescence spectra measured by a microspectrofluorimeter (Karnaukhov, 1978; 2001) is eliminated, and a new peak at 460-500 nm is seen. The fluorescence intensity is also changed and, in the case of *Passiflora coerulea*, increased. Pollen, lacking carotenoids and anthocyanins, such as of *Hippeastrum hybridum*, changes its fluorescent color from green to yellow-orange (maximum shifted from 510 to 550 nm) and the fluorescence intensity is also increased. As for individual pollen components (Table 34), azulene enhances their fluorescence up to 1000 times without changing the maximum, whereas β-carotene and rutin changed their fluorescent color from yellow-orange to green-yellow or blue. Thus, pollen and its individual compounds may use of biosensors for ozone concentrations of more than 0.6 ppm.

In comparing the mode of pollen collection in nature, it should be noted that the easily collected pollen of wind-pollinated plants, belonging to conifers or the Betulaceae family, appears to be unsuitable for use as diagnostics of ozone. Our experiments with pollen of *Pinus sylvestris*, enriched in terpenoids and phenols, and of *Betula verrucosa*, enriched in phenols, showed that both samples fluoresced under normal conditions in the blue region of the spectrum, and exposure to 2.4-5 ppm of ozone only decreased the intensity of the light emission by 50 %. As seen in table 34, pollen enriched in carotenoids and azulenes or in phenols shows the most significant changes in its autofluorescence. Pollen of *Philadelphus grandiflorus* and *Hippeastrum hybridum* could be chosen as cellular models because their cells, as seen below, have high viability in the presence of ozone.

Table 34. The pollen fluorescence after treatment with different doses of ozone (adopted from Roshchina and Melnikova, 2001).

Plant species or individual component	Control		Ozone (ppm)			
			0.6-0.9		2.4-5.0	
	FC	λ (I)	FC	λ (I)	FC	λ (I)
	Pollen enriched in carotenoids and azulenes					
Hemerocallis fulva,	orange	500 (100) 560(100)	blue	475(78)	blue-green	475-500 (78-110)
Passiflora coerulea	orange	450(100), 560(100)	blue	460(100)	blue	460(300)
Philadelphus grandiflorus	yellow	465(100), 535 (100)	green	500(400)	blue	475-480 (160)
	Pollen without carotenoids and anthocyanins					
Hippeastrum hybridum	green-blue	510 (100)	blue-green	495-500 (80)	yellow-orange	550(600)
	Pollen components					
Azulene	Blue	450(100)	blue	450(500)	blue	450 (1000)
β-carotene	Orange	500(100), 585(100)	orange	460(100), 550(100)	green-blue	450(100), 550(75)
Rutin	Green-yellow	460(100), 565	green-yellow	460(78), 565(80)	blue	460(45); shoulder 566(33)

FC- the color of fluorescence excited by ultraviolet light at 360-380 nm, and seen under a luminescent microscope; λ,nm - maximum in the fluorescence spectrum registered by a microspectrofluorimeter, in brackets I – ratio of the fluorescence intensity of the experimental sample to the corresponding intensity of the untreated control (in % of control).

We also compared the effects of ozone on the autofluorescence of these cells with the effects of aging and ultraviolet light. In the case of aging, only more than 1 year of pollen storage, induced changes similar to those resulting from ozone (in doses of 5 ppm for 100 hours). UV-light (300-400 nm, 100 W m^{-2} for 1-2 hours) did not induce sharp changes in the autofluorescence of the cells. Thus, the autofluorescence of the pollen of *Philadelphus grandiflorus* and *Hippeastrum hybridum* could be recommended as cellular biosensors for ozone.

Viability. The viability of pollen grains is evaluated in terms of their germination both in nature and on an artificial medium (*in vitro*). The development of pollen *in vitro*, assayed in terms of its pollen tube growth, may be a useful biosensor for ozone. The information discussed in section 6.2 indicates that the pollen of *Plantago major* is especially sensitive to ozone concentrations < 0.05 ppm, but it easily loses its viability (Section 6.2). The pollen of *Philadelphus grandiflorus* is more tolerant than that of *P. major* (section 6.2.), although its activity may only be seen for a few months after a summer collection. More significant is the pollen of *Hippeastrum* *hybridum* due to the flowering of the species throughout the whole year, the large size of the individual pollen grains, easy germination in a nutrient medium of 10 % sucrose, and its capacity to germinate for up to 6 months (Roshchina et al., 1997). Under a microscope, the nucleus of the developing pollen is easily seen with normal light and either without staining or with staining by rutacridone or acridine orange (Roshchina, 2001d; 2002).

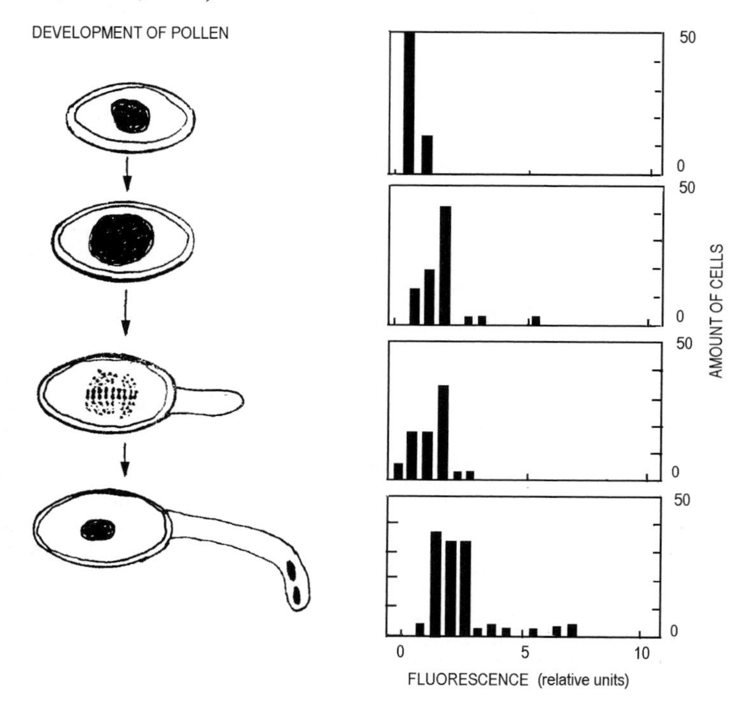

Figure 48. Scheme of pollen tube development. View of pollen from Hippeastrum hybridum in vitro and summed histograms of the green fluorescence intensity at 530 nm of pollen, staining by rutacridone, during its germination. 0 min - dry pollen, 15, 90, 120 min - time after moistening.

The intensity of the green fluorescence after the staining of the nucleus (mainly DNA) and excitation by UV-light (360-380 nm) could be observed under a luminescent microscope and be measured by Karnaukhov-constructed microspectrofluorimeter (Fig.48). The pollen staining by rutacridone was also used in the study of the developing pollen (pollen tube formation) at 0, 15, 90 and 120 min after moistening. An intense the green fluorescence at 530 nm was recorded in the histograms (Fig.48). Dry pollen has one nucleus, and the histogram had a unimodal character. The pollen cell swelled at 15 minutes after moistening, and the green intensity of the nucleus also increased. It is at this stage that mitosis started and new DNA was accumulated. At 90 min after moistening, the pollen tube started to develop, and the divergence of chromosomes was easily seen as the green fluorescent strands of a spindle. The division of the green-fluorescent chromosomes was observed, a fact that confirmed that the DNA was binding with rutacridone. After 120 min of development, the primary nucleus forms two new nucleuses for the generative cells, named spermia. Thus formed, the spermia moved to the tip of the growing pollen tube (if it grows within the pistil, a single spermium interacts with the egg cell, so that fertilization takes place). Visible under a luminescent microscope and on the histogrames were 3 green-fluorescing nuclei - one in the vegetative cell with the pollen tube and the two spermia.

The pollen of *Hippeastrum hybridum,* as seen in Section 6.2, retarded its development when exposed to ozone concentrations > 0.6 ppm for 12 hours. The viability of this pollen may also be estimated by referring both to its autofluorescence (Roshchina et al., 1997) and its fluorescence after staining with the fluorescent dye ethidium bromide (Roshchina et al., 1998 c). This stain can penetrate only into non-viable, non-germinated cells (Table 35). We also compared the effects of ozone on pollen viability with the effects of other factors, such as the duration of pollen storage (aging for 6 months) and UV-light (300-400 nm, 100 W m^{-2} for 2 hours). The development of the pollen grains of *H .hybridum* was only completely blocked after 6 months of storage, and after fumigation with high doses of ozone. Usually non-viable pollen fluoresced 3 times more brightly than viable microspores (Roshchina et al., 1997). In samples on which the effects of aging and were completely blocked after 6 months of storage, and after fumigation with high doses of ozone. Usually non-viable pollen fluoresced 3 times more brightly than viable microspores (Roshchina et al., 1997). In samples on which the effects of aging and ozone were measured (Table 35), the autofluorescence of non-viable

Table 35. Viability of pollen estimated in terms of its germination, autofluorescence and staining with fluorescent dye ethidium bromide (EB).

Factor	*Pollen germination (% of control)*	*Autofluorescence intensity (% of control)*	*Amount of red-fluorescing cells, after staining with EB (% of control)*
Ozone 2.4 ppm for 50 h	0	320±6	94±3
Ageing (6 months)	0	350±10	100±5
UV-light	54±2	200±10	53±3

pollen and its permeability to ethidium bromide was maximal. Unlike aging, short exposures to ultraviolet light had no similar effect, although the number of viable pollen grains decreased by approximately 50 %. The ozone effects do not, therefore, correlate directly with those induced by UV-light. . Consequently, the pollen of *Hippeastrum hybridum* may be used as biosensor for the presence of ozone in a range of doses > 0.6 ppm.

6.3.2.2. Vegetative microspores of spore-breeding plants

The development of the vegetative microspores of spore-breeding plants belonging to the Equisetaceae family, such as the common (field) horsetail *Equisetum arvense*, is ozone sensitive (section 6.2). At small concentrations of ozone, generative organs - the antheridia and archegonia - are not formed. However, first signs of the response to the gas are seen , in particular as changes in the autofluorescence of dry vegetative microsopores (which put on the object glass or on the wet filter paper in a Petri dish) during 24 hours after moistening. The cells alter their luminescence from the blue-green light, characteristic of the dry or non-viable microspores, to the red light, which is peculiar to developed cells (Roshchina et al., 2002a). This shift in the color of fluorescence is a marker for the beginning of development. The wavelength of the light emission mainly reflects the contribution of chlorophyll but, perhaps, additionally that of azulenes. This fact that the blue-red transition indicates the commencement of development was also registered by the dual-wavelength microfluorimeter Radical DMF-2" (Radical, Ltd) interfaced with a PC/AT compatible computer to measure the fluorescence intensities at two separate spectral regions (Roshchina et al., 2002a). A special program "Microfluor" makes it possible to obtain the distribution histograms of the fluorescence intensities in the red 630-700 nm (I_{red}) and blue 430-480 nm (I_{blue}) regions or parameter $\mu = I_{red}/I_{blu}$ ratio (used as a characteristic of synthetic activity) for 50- 200 cells within 15-20 min. The latter allowed the data to be statistically using a student t-test.

As seen in fig. 49, a dry microspore is only one cell, fluorescing mainly in the blue region. In a control sample for 30 min after moistening, the numbers of cells on histograms with the higher parameter μ (I_{red}/I_{blu}) increased due to the stimulation of the red fluorescence. At this stage, chloroplasts are red globules easily seen through the transparent cell cover. They are concentrated around the blue-light emitting nucleus. After 60 min of development, the chloroplasts concentrated on one side of the cell, and the cell started to divide. After 120 min of development, the cover of the microspores was, in most cases, shed and a new second cell could be seen. This new cell later became a rhizoid cell, where chloroplasts are destroyed. Another cell (prothallial) started to divide, representing the beginning of the multicellular thallus (gametophyte). In untreated conditions, this tissue arose after 24 hours of microspore development. Preliminary chronic fumigation of dry microspores with various doses of ozone caused alterations in their development, as can be seen in the histograms of fig. 49.

After accumulated low doses of O_3 reached 0.007-0.05 ppm (0.0015 ppm for 3 h per day), the dry microspores began to emit rose-colored light when subject to excitation by ultraviolet light at wavelengths of 360-380 nm. Subsequent development of microspores was, however, retarded by the ozone treatment. At ozone doses of 0.05 ppm, both the value of the parameter I_{red}/I_{blu} and the number of red-light emitting cells decreased after 120 minutes of development. The decrease

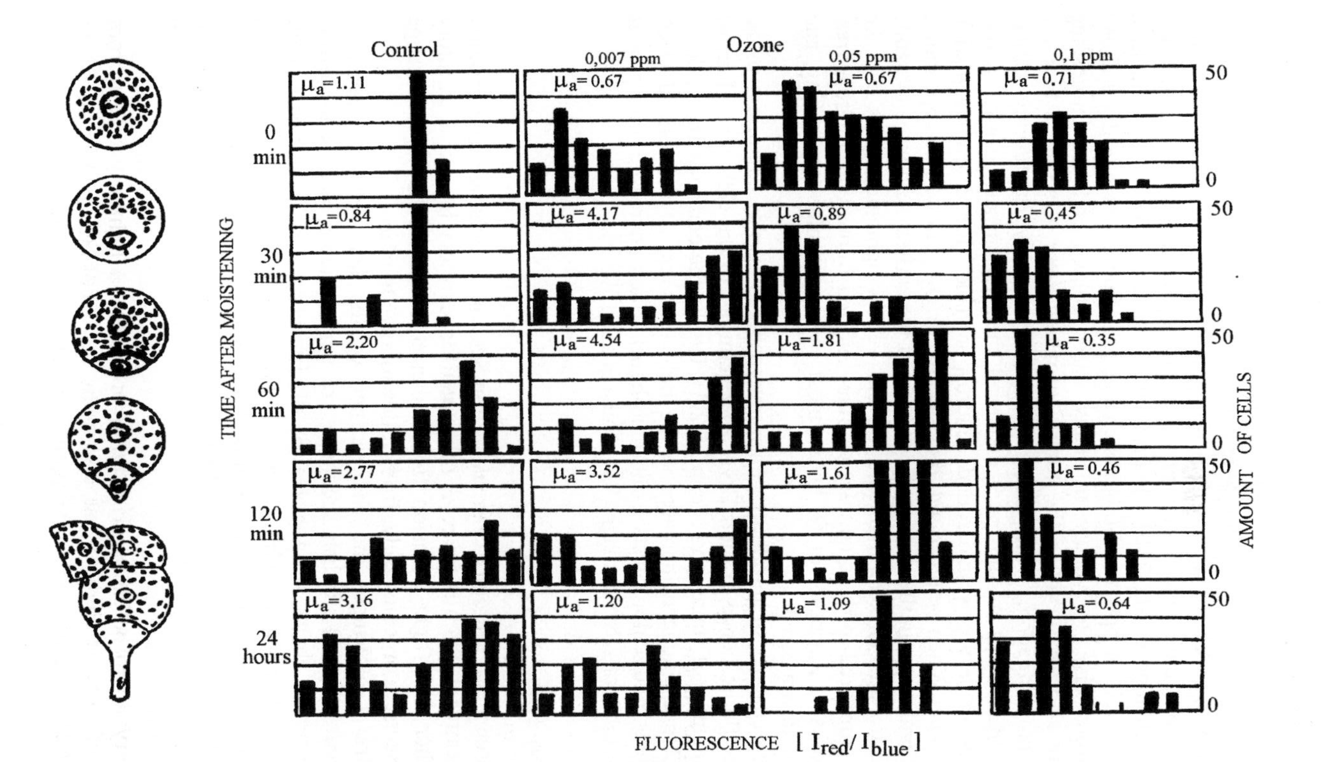

Figure 49. The blue-red transition fluorescence intensity of microspores from Equisetum arvense beginning in a dry state (0 min) and at 30, 60, 120 min, 24 hours after moistening without (Control) and after preliminary treatment with ozone (Ozone). Left - scheme of the microspore development seen under a luminescent microscope. Right - distribution histograms of parameter $\mu = I_{red}/I_{blue}$. μ_a = average value for 50 cells

was even stronger when the dose was 0.1 ppm. As indicated in section 6.2, the development of the microspores undergoing this treatment is significantly retarded.

Estimations of the state of microspores fumigated with higher doses of O_3 may also be based on their autofluorescence, as observed under a luminescent microscope (Roshchina et al., 2001 c). Preliminary treatment with ozone in total concentrations of 0.2-0.4 ppm for 4-6 hours changed the fluorescence of dry and wet microspores from blue to yellow in many cases (blue- and red-fluorescing microspores are also seen), unlike the control in which there was only a blue-red shift in the wet microspore fluorescence. Moreover, weakly fluorescing elaters (structures for attachment to substrate) enhanced their luminescence after ozone treatment. When these are wound round microspore (like thread around wistle), they are bright blue. When be unwound, they became bright yellow. The appearance of yellow-light emissions in the microspores is one of the earliest responses to ozone (see section 2.1., fig.11). The numbers of the yellow-lightening cells can be estimated and compared to the quantities of blue- and red-fluorescing microspores. After ozone doses of 2.5 ppm for 50 hours, most cells were blue-fluorescing, and black necrotic cells appeared.

Besides such changes in autofluorescence, microspores display alterations in the intensity of the fluorescence of their DNA after staining of the cells with Hoechst 33342 (see principles of staining in section 6.2.3). Unlike the control sample, doses of ozone > 0.2 ppm induced increases in the light emission of the stained DNA complex (nucleus and chloroplasts emit blue-light). Such increased intensity is usually characteristic of DNA damage (Roshchina et al., 1998 c).

6.3.2.3.Secretory cells

The autofluorescence of secretory cells, as seen in Section 2.1, changes significantly under ozone treatment. Such is especially the case for secretory hairs and secretory cells, which alter their light emission from either the blue to the yellow-orange region of the spectrum or, conversely, from yellow-orange to blue.

Secretory hairs. O_3 changes the fluorescence of the leaf secretory hairs of *Raphanus sativus* (Fig.12). Radish seedlings are very suitable models for ozone monitoring. Secretory hairs on the leaves of *Fragaria viridis* decrease their fluorescence after fumigation with ozone in doses > 0.1 ppm (Fig. 50). The autofluorescence of secretory hairs differs from the red fluorescence of non-secretory cells, in which only chlorophyll has similar fluorescence. The fact that the autofluorescence of secretory cells is quenched by O_3 makes it possible to use them for ozone monitoring.

Secretory cells in tissue culture. The autofluorescence of secretory cells in root tissue cultures of *Ruta graveolens* has also been studied (Fig.51). After 1-6 h of exposure to ozone (total dose = 0.3-0.6 ppm), the maximum in their fluorescence

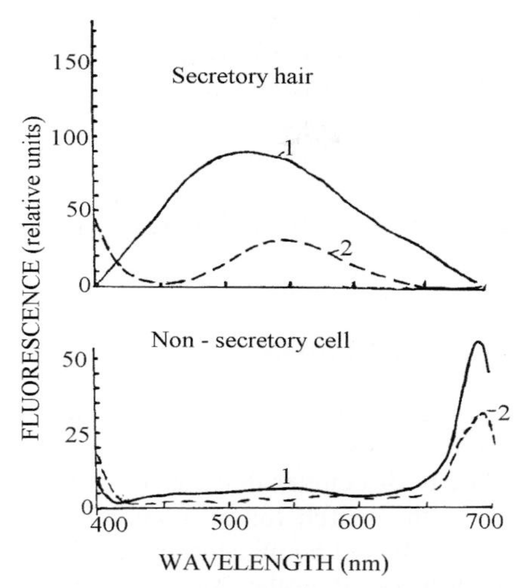

Figure 50. The fluorescence spectra of intact leaf secretory hair and non-secretory cells of Fragaria viridis without (1) and after ozone 0.15 ppm treatment for 3 hours (2).

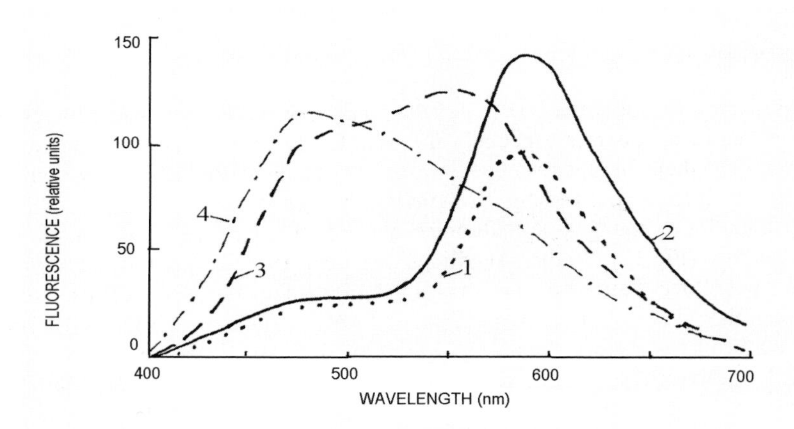

Figure 51. The fluorescence spectra of intact secretory cells of root tissue culture of ruta Ruta graveolens 1. control - without treatment; 2 and 3 - after the ozone treatment 0.3 and 0.6 ppm, respectively; 4. non-secretory cells

spectra was shifted from 590-600 nm to the short-wavelength region. This maximum may belong to the acridine alkaloids of *Ruta graveolens*, especially rutacridone, and/or other acridone alkaloids (Roshchina, 2001d), while flavonoids rutin and quercetin have much lower fluorescence in this region of the spectra. At higher doses, the autofluorescence was seen only in blue region of the spectrum, as was the case for non-secretory cells.

The secretory cells are, consequently, additionally candidates in the search for cellular ozone monitors.

CONCLUSION

Cellular monitoring of ozone consists in: (1) the observation of plant cell sensitivity indoors and outdoors in order to determine sensitive plant cells, species and cultivars; (2) the establishment of cellular diagnostics in terms of their reactions to various doses of ozone; (3) the search for cellular indicators of ozone presence. Among the cellular reactions affected by ozone (both positively and negatively), the germination and autofluorescence of pollen and vegetative microspores, ethylene release, autofluorescence of secretory cells, changes in peroxidase activity are the most sensitive to ozone activity. Sensitive plant cells, such as the pollen of *Hippeastrum hybridum*, *Plantago major* and *Philadelphus grandiflorus*; the vegetative microspores of *Equisetum arvense*; and various secretory cells may serve as biosensors and bioindicators for ozone. Ozone should be considered not only as an air pollutant but, in low concentrations, also a necessary component for the initiation of growth processes.

CONCLUSION

Ozone, a strong oxidizer, is produced in the stratosphere as a result of the photolysis of oxygen by ultraviolet radiation from the Sun. It forms the ozone layer in the atmosphere of our planet, protecting all life from the same UV radiation. Other sources of ozone on the Earth include the oxidation of monoterpenes in woody plants, mainly coniferous trees, by the oxygen in the air and, nowadays, the reaction of O_2 with automobile and industrial wastes containing oxides of nitrogen and CO. Further contributions to the formation of ozone, although not so significant, are made by the electrical discharges in the atmosphere, computers and other techniques emitting ultraviolet radiation. The change in concentration of ozone in the troposphere has a significant influence on living systems.

Among all living systems, the most convenient for research on ozone and for diagnostic uses are plant cells, which display readily discernible pigment changes, structural modifications and other visible symptoms. The effects of ozone on these cells were found to depend on the concentration of the gas in the atmosphere and the duration of the plant's exposure to such conditions. The short-term (1-3 hour) effects of small doses of ozone (0.05-0.1 ppm) only appeared directly on the surface of plant cells; they included changes in cell permeability to ions, in fluorescence of the components of cell wall or cellular excretions and others. Longer (days, months) chronic influences of small doses of O_3 or, conversely, sharp, short-term influences of high concentrations of the gas > 0.1 ppm (for a few hours) cause changes in the primary and secondary metabolism and, ultimately, the appearance and growth of necrotic spots. Sensitivity and resistance to ozone of vegetative species and the various strains (varieties) within species strongly differ and are determined by the set and activity of the antioxidant systems that they contain.

The primary way that ozone penetrates plant tissue is through open stomata and pores on the above-ground parts of plants, or through apertures and pores of microspores. The primary reactions of O_3 after penetration into a living cell occur on the surface of epidermal tissues covered with cuticle or on the surface of microspores (pollen or vegetative microspores) covered with sporopollenin (Fig. 52).

Usually small concentrations of ozone administered in short-term exposures do not penetrate inside a cell but react with superficial components, forming free radicals, ozonides and peroxides, which then effect further influences on a cell.

Antioxidants, which are collected in the cell wall, apoplast and intercellular spaces and which neutralize these reactive oxygen species, serve as a protective barrier against the penetration of ozone and the products of ozonolysis into cells and tissues. Certain amounts of the antioxidants are also concentrated in the plasmic membrane, which also prevents O_3 from penetrating into a cell. The products of ozonolysis that are formed on the cellular surface, the free radicals and peroxides, cause only few and frequently convertible functional changes if their concentrations are small. These include, in particular, the oxidative modification of proteins and lipids, increases or decreases in the activity of membrane-bonded enzymes.

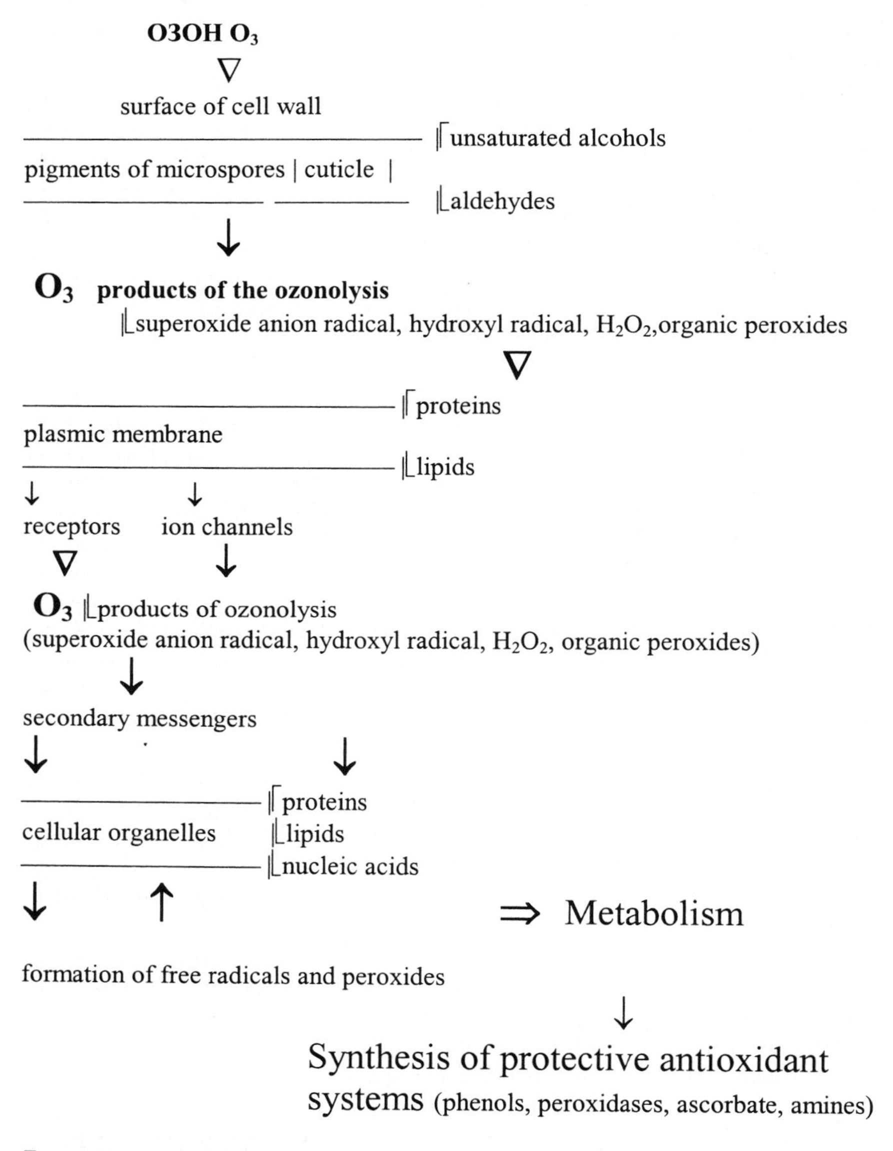

Figure 52. Main pathway of the ozone penetration into plant cell and its cellular targets.

Moreover, ozone and the products of ozonolysis can also oxidize the hormone or mediator receptors in the membranes. This reaction has undoubtedly an influence on the receptor's ability to recognize hormones or mediators, and on the transfer of the chemosignals inside the cell that regulate the activation of systems of such secondary messengers as cyclic nucleotides, inositol-3-phosphate and Ca^{2+} ions. Finally, small doses of ozone and the products of ozonolysis regulate the growth and development of cells without producing any visible symptoms of damage. They can be considered as active modifiers of the hormonal sensitivity of cells.

The reactions of plant cells to small doses of ozone are quite distinct from their response to the gas, when it becomes the air pollutant. If the concentration of O_3 or the duration of its influence are great enough to overcome the cell's protective systems, then ozone itself can injure the cellular plasmalemma and, transgressing this membranous barrier, have a direct effect on the cellular organelles. However, small doses of ozone or the products of ozonolysis do not appear to affect such contact with the internal contents of a cell. The first symptom of damage occurring in a cell is the rapid increase in the permeability of membranes to ions. Later a process of lipid peroxidation occurs. After that, a decrease in photoassimilation and the activation of respiration can also be observed. Finally, the genetic apparatus is affected, a phenomenon that indicates the presence of even deeper damage. The convertibility of the damage depends on the extent of the cell and tissue injuries, the mechanisms of resistance to oxidizing stress, and the ability to undertake reparation.

The toxic effect of ozone can also involve the products of O_3 reactions with water— superoxide anion radical, hydroxyl radical, and hydrogen peroxide—as well as the products of free radical reactions of lipid peroxidation—ozonides and the peroxides of proteins, amines, nucleic acids, the products of a secondary metabolism and others. The effects of the products of ozonolysis depend on their concentration and the functionality of a cell's antioxidant systems regulating formation and utilization of these products. For instance, low concentrations of the products resulting from the interaction of ozone with the surface of pollen (such as peroxides) do not have a negative influence to cell viability and, conversely, even stimulate pollen germination (Roshchina and Melnikova, 2001). The peroxides formed by ozonolysis do not only oxidize the components of the cellular cover, but also act as chemosignals. For instance, when molecules of peroxide are linked to sensors on the cellular surface, and the information transferred into the cytoplasm through special mechanisms, activating systems of secondary messengers, which furthermore transmit it to the organelles (Gamaley and Klyubin, 1996). Visible signs of toxicity are connected with the water content of vegetative cells. Our own analysis, supplemented by a number of literary sources comparing the effects of ozone itself, peroxides, antioxidant-scavengers(traps) of free radicals or decomposers of peroxides, has shown that the toxic effects of ozone in dry pollen do not involve the formation of hydrogen peroxide, any other peroxides, or the superoxide anion radical.

The products of ozonolysis, such as the fluorescing components on the surface of leaves or the products of lipid peroxidation (in particular malondialdehyde), are found after exposures to low concentrations of ozone (0.05 ppm) for periods ranging from 20 min to 1 hour. However, hydroperoxides can be detected after treatment with higher concentrations of O_3 (0.120 ppm) and free radicals when the concentration ranges from 0.07-0.3 ppm (at this concentration, secondary alcohols, diols and fatty acids in the cuticle are lost).

If the influence of ozone is short-term or its concentration is small, visible symptoms of damage (more often necrosis) are not observed, and normal plant development proceeds. Reparation processes or the activation of antioxidant systems—including the enzymes superoxide dismutase, peroxidase, catalase and glutathione reductase—as well as the synthesis of low-molecular substances such as ascorbate, phenols, ubiquinone, uric acid and others, probably ensure that such normal development may occur. We do not precisely know if there is a process of hormesis for ozone—stimulation of cellular viability by small doses of the acting factor and, controversially, its depression by high doses of the same factor (Kuzin, 1995; 1997). The basic experiments on the influence of ozone on plant cells have, till now, been concerned with discovering the mechanisms of damage instigated by this gas, when it is in concentrations great enough to be considered an air pollutant. However, individual instances of hormesis have been noted; these involve the stimulation of respiration and the optimal development of the leaf plate, likely also caused by small amounts of peroxides produced by ozone's interaction with a water environment. It is necessary to consider the possibility that rather low concentrations of ozone, such as those constantly existing under natural conditions, can have a positive effect on plant cellular functions. For instance, preliminary reports have suggested that low doses of O_3 can accelerate growth processes in some plants (chapter 6). First of all, ozone performs a sanitary function, as a strong oxidizer blocking development of parasites and stimulating protective reactions such as the enhanced formation of free radicals and peroxides. These substances form a defensive barrier on the surface of plants and, in turn, promote the synthesis and deposition of phytoalexins, peroxidases and superoxide dismutases on the cellular surface and in the extracellular space. Moreover, it is likely that the superoxide anion radical and hydrogen peroxide, both of which are constantly produced in a water environment as a result of the interaction with ozone, are necessary stimulators of plant cell viability and, simultaneously, of the growth and development of the plant as a whole.

APPENDIX 1

ADDRESSES OF INTERNET RELATED TO PROBLEM OF OZONE

Atmosperic Ozone and Biosphere.
http://bigmac.civil.mtu.edu
http://bigmac.civil.mtu.edu/home/classes/ce459/projects.html
http://bigmac.civil.mtu.edu/home/classes/ce459/public/p16/ozone.html

Ozone Effects on Human Health and Vegetation
http: //Project CE459/public/p16/ce459.html
http://bigmac.civil.mtu.edu/home/classes/ce459/public/p16/ozone.html

Ozone, Plant Reactions and Health. Vegetation Effects
http://bigmac.civil.mtu.edu/home/classes/ce459/public/p16/ozone.html

Indicators for ozone: lobloly pines, ponderosa pines, tobacco, petunia,spinach, tomato, beans, parsley. Photoes of leaf damages of sensitive species are given.

http://www.icb.psn.ru
Autofluorescence of sensitive species

Main Centers of the studies.

EMMB National Exposure Research Laboratory of USA, US EPA Annex, Md-44, Res Triange Park, North Carolina 277111, USA

Air Quality Analysis Section , USEAA Region 6, Ross Avenue 1445, Dallas, Texas, 75200, USA
Man Tech,Environment Technology, Inc 2,m Triangle Drive, PO Box 12313, Res Triangle Park, North Carolina 27709, USA

Department of Biological Scviences, Institute of Environmental and Biological Sciences, University of Lancaster LA14YQ, UK and

GSF-Forschungszentrum fur Umwelt und Gesundheit GmbH, Institute fur Biochemische Pflanzenpathologie, D-85764 Oberschleissheim, Germany

Institute of Environmental and Biological Sciences, Lancaster University, Lancaster LA1 4YQ, UK
Department of Pure and Applied Chemistry, University of Natal, Durban 4001 , South Africa

Agric Plant Physiology and Ecology Research Institute, University of Buenoz Aires, 1417, Argentina

Department of Environmental and Industrial Health, School of Public Health, University of Michigan, Ann Arbor, MI, USA

Michigan Technological University, Houghton, Mi 49931, Spring Quarter, 1997, USA

Russian Academy of Sciences Institute of Cell Biophysics, Pushchino, Moscow Region, 142290, Russia

APPENDIX 2

LIST OF PLANT SPECIES SENSITIVE TO OZONE

List 1. The estimation of plant sensitivity on the leaf damage of grassy species

Very sensitive species (leaf damage is more, than 70 %)	Sensitive species (leaf damage is more, than 45-70 %)
*Anagallis arvensis (*Primulaceae)	*Avena fatua* (Graminae)
*Geum rivale (*Rosaceae*)*	*Conium maculatum* L. (Umbelliferae)
Humulus lupulus (Cannabiaceae)	*Coriandrum sativum* (Umbelliferae)
Lathyrus aphaca (Fabaceae)	*Hypericum perforatum* (Hypericaceae)
Lotus pedunculatus (Fabaceae)	*Hyoscianus niger* (Solanaceae)
Lotus uliginosus (Fabaceae)	*Lathyrus pratense* (Fabaceae)
Medicago lupulina (Fabaceae)	*Papaver rhoeas* L. (Papaveraceae)
Plantago major (Plantaginaceae)	*Phleum pratense* (Gramineae)
Reseda luteola (Scrophulariaceae)	*Polemonium caeruleum* (Polemoniaceae)
*Sanquisorba minor (*Rosaceae)	*Trifolium pratense* (Fabaceae)
Trifolium repens (Fabaceae)	*Ulex europeus* (Fabaceae)
	Vicia hirsuta L.(Fabaceae)

List 2. Sensitive woody species

Ailanthus altissima (Scrophulariaceae) Ailanthus, tree of heaven
Cotoneaster divaricata (Rosaceae) Cotoneaster, spreading
Cotoneaster horizontalis (Rosaceae) Cotoneaster, rock
Fraxinus americana (Oleaceae) Ash, white
Fraxinus pennsylvanica (Oleaceae) Ash, green
Gleditsia triacanlhos inermis (Fabaceae) Honey locust (thornless)
Larix (Pinaceae) Larch, European
Prunus avium var. (Rosaceae) Bing Cherry (Rosaceae) Bing
Rhododendron kaempferi (Ericaceae) Campfire Azalea, campfire
Rhododendron kurume (Ericaceae) Azalea, snow
Rhododendron ohtusum (Ericaceae) Hinodegrii Azalea, Hinodegiri
Rhododendron poukhanensis (Ericaceae) Azalea, Korean
Sorbus aucuparia (Rosaceae) Ash, European mountain
Spiraea vanhouttei (Rosaceae) Bridalwreath
Syringa chinensis (Oleaceae) Lilac, Chinese
Vitis vinifera var. (Vitaceae) Concord Grape, Concord

See also Internet sites: http://www.wsl.ch/forest/infoblatt or http://www.wsl.ch/forest/wus/publi

APPENDIX 3

AUTOFLUORESCENCE OF PLANT CELS AS BIOINDICATORS FOR OZONE

The term "autofluorescence" is used to designate the luminescence of naturally occurring molecules in intact cells induced by ultraviolet light. The phenomenon may be observed both under a luminescent microscope and by employing special microspectroflurimeters. Confocal microscopy could be also recommended for these purposes. Autofluorescence may be a sensitive sensor of ozone in the environment (sections 6.2. and 6.3).

Recommended technique. Microspectrofluorimeters (detectors with optical probes of various diameters up to 2 μm) have been constructed in the Institute of Cell Biophysics of Russian Academy of Sciences (Karnaukhov et al., 1981, 1982; 1983; 1985; 1987). They can detect fluorescence from individual cells and even from a cell wall, large organelles and secretions in periplasmic space (space between plasmalemma and cell wall), as well as from the drops secreted by secretory cells and remaining on the cellular surface (see sections 2.1., 6.2. and 6.3). Microspectrofluorimeters may record the fluorescent spectra or measure the intensities of the fluorescence at two separate wavelengths. A special program "Microfluor" makes it possible to obtain the distribution histograms of the fluorescent intensities and to perform statistical analysis of the data using a student t-test.

Karnaukhov VN, Yashin VA, Kulakov VI, Vershinin VM and Dudarev VV (1981) Einrichtung zur Untersuchung von Lumineszenzeigenschaften der Microobjekte. DDR Patent N147002, 1 - 32.

Karnaukhov VN, Yashin VA., Kulakov VI, Vershinin VM and Dudarev VV (1982) Apparatus for investigation of fluorescence characteristics of microscopic objects. US Patent, N4, 354, 114, 1 - 14.

Karnaukhov VN, Yashin VA, Kulakov VI, Vershinin VM and Dudarev, VV (1983) Apparatus for investigation of fluorescence characteristics of microscopic objects. Patent of England 2.039.03 R5R.CHI.

Karnaukhov VN, Yashin VA. and Krivenko VG (1985) Microspectrofluorimeters.*Proc. of 1-st Sov.-Germany Intern. Symp. Microscop. Fluorimetry and Acoustic Microscopy*,Moscow, p.160-164.

Recommended objects. The secretory cells of vegetative tissues or microspores are suitable objects for O_3 monitoring due to the various fluorescent compounds found in the secretions (see sections 6.2. and 6.3). Low doses of ozone < 0.15 μl/l induce green-yellow fluorescence in the leaves of *Plantago major, Zea mays, Raphanus* *sativus* and *Hippeastrum hybridum* as well as in the vegetative microspores of *Equisetum arvense*. The light emission intensified with the increase of ozone concentration, which also inflicted damage on the light-emitting cells, ultimately resulting in their necrosis. At higher doses of ozone, blue fluorescence at wavelengths of 420-470 nm was observed in pollen containing carotenoids, such as *Passiflora coerulea, Philadelphus grandiflorus* and *Hemerocallis fulva*. Under normal conditions, these cells emit light in the yellow-orange region at wavelengths of 530-560 nm. These pollen grains also lost their fertility, and lipofuscin was formed in their covers (exine). References related to measurements of autofluorescence in secreting plant cells are listed below:

Roshchina VV and Karnaukhov VN (1999).Changes in pollen autofluorescence induced by ozone. Biologia Plantarum 42 (2) : 273-278

Roshchina VV and Melnikova EV (1995) Spectral analysis of intact secretory cells and excretions of plants // Allelopathy J. 2(2): 179-188.

Roshchina VV and Melnikova EV (1996) Microspectrofluorometry: A new technique to study pollen allelopathy // Allelopathy J. 3 (1): 51-58.

Roshchina VV and Melnikova EV (1999) Microspectrofluorimetry of intact secreting cells, with applications to the study of allelopathy. In : Principles and Practices in Plant Ecology. Allelochemical Interactions. Ed. Inderjit, Dakshini KMM. and Foy CL..Boca Raton, USA :CRC Press. pp.99-126.

Roshchina VV and Melnikova EV (2001) Chemosensitivity of pollen to ozone and peroxides. Russian Plant Physiology (Fiziologia Rastenii) 48 (1): 89-99.

Roshchina VV, Melnikova EV and Kovaleva LV (1996) Autofluorescence at the interaction pollen-pistil of *Hippeastrum hybridum*. Doklady of. Russian Acad. Sci 349: 118-120.

Roshchina V.V., Melnikova E.V. and Kovaleva L.V. (1997a) The changes in the fluorescence during the development of male gametophyte Russian Plant Physiol. 47: 45-53.

Roshchina VV, Melnikova EV, Karnaukhov VN and Golovkin BN (1997b) Application of microspectrofluorimetry in spectral analysis of plant secretory cells // Izv. RAN. Ser. Biol. (Biology Bulletin) 2: 167-171.

Roshchina VV, Mel'nikova EV, Mit'kovskaya LI, Karnaukhov VN (1998) Microspectrofluorimetry for the study of intact plant secretory cells. J of General Biology. (Russia) 59: 531-554.

Roshchina VV, Mel'nikova EV, Yashin VA and Karnaukhov VN (2002) Autofluorescence of intact spores of horsetail *Equisetum arvense* L. during their development. Biophysics (Russia) 47 (2): 318-324.

REFERENCES

Adams DO and Yang SF (1981) Ethylene the gaseous plant hormone: mechanism and regulation of biosynthesis. Trends Biochem Sci. 6: 161-164.

Adekenov SM and Kagarlitskii AD (1990) Chemistry of Sesquiterpene Lactones. Gylim, Alma-Ata, 188pp.

Adepipe NO and Ormroid DP (1972) Hormonal regulation of ozone phytotoxicity in *Raphanus sativus*. Z. Pflanzenphysiol. 68 (3): 254-258.

Afanas'ev IV and Kuprianova NS (1993) Kinetics and mechanisms of the reactions of the superoxide ion in solution. III. Kinetics and mechanisms of the reactions of the superoxide ion with ethyl acetate in dimethylformamide, acetonitrile, and their mixtures. Int. J. Chem. Kinet. 15: 1063-1068.

Afanas'ev IV (1989) Superoxide ion: chemistry and biological implications. CRC Press. Florida. vol.1. p.1-300.

Agrawal SB and Agrawal M. (eds.) (2000) Environmental Pollution and Plant Responses. Lewis, Boca Raton, USA.

Airansinen K and Paaso P (1990) Cutin monomers from the needles of two clones of scots pine differ in resistance to *Lophodermelia* needle cast. Physiol. Plant. 79 (2): part 2. p. Ai 102.

Alberts B, Bray D, Lewis J, Raff M, Roberts K and Watson JD (1994) Molecular Biology of the Cell. 3 edition. New York and London, Garland Publishing Inc. 1294 pp.

Alexandrov EL and Sedunov YuS (1979) Human and Stratospheric Ozone. Hydrometeolzdat, Leningrad. 104 pp.

Alexandrov EL, Karol' IL, Rakipova LR, Sedunov YuS and Khrgian AKh (1982) Atmospheric Ozone and Changes of Global Climate. Hydrometeolzdat, Leningrad. 168 pp

Alscher RG and Amthor JS (1988) The physiology of free-radical scavenging: maintenance and repair processes. In: Air Pollution and Plant Metabolism. (Eds. Schulte-Hostede S, Darrall NM, Blank LW and Wellburn AR) pp. 94-115. London, New York. Elsevier Applied Science Publ.

Altshuller AP (1983) Review: natural volatile organic substances and their effect on air quality in the United States. Atmospheric Environment. 17 (11): 2131-2165.

Alvarez ME, Pennell RL, Meijer PJ, Ishikawa A, Dixon RA and Lamb C (1998) Reactive oxygen intermediates mediate a systemic signal network in the establishment of plant immunity. Cell 92: 773-784.

Ames BN, Shigenaga MK and Hagen TM (1993) Oxidants, antioxidants and the degenerative diseases of ageing. Proc Nat. Acad. Sci. USA. 90 (17): 7915-7922.

Amthor JS and Cumming JR (1988) Low levels of ozone increase beam leaf maintenance respiration. Can. J. Bot. 66 (4): 724-726.

Anderson WC and Taylor OC (1973) Ozone induced carbon dioxide evolution in tobacco callus cultures. Physiol Plant. 28 (3): 419-423.

Apasheva LM and Komissarov GG (1996) The effect of hydrogen peroxide on plant development. Biological Bulleten 5 : 621-623.

Apostol I, Heinstein PF and Low PS (1989) Rapid stimulation of an oxidative burst during elicitation of cultured plant cells. Plant Physiol. 90 (1): 109-116.

Archakov AI and Mokhosoev IM (1989) Modification of proteins by active oxygen and their decomposing. Biokhimiya (USSR) 54: 179-186.

Arndt von U and Kaufmann M (1985) Wirkungen von Ozon auf die apparente Photosynthese von Tanne und Buche. Allgemeine Forst Zeitschrift. 40 (1/2): 19-20.

Asada K (1980) Formation and scavenging of superoxode in chloroplasts, with relation to injury by sulphur dioxide. Studies on the Effects of Air Pollutants on Plants and Mechanism of Phytotoxicity. Res. Rep. Natl. Inst. Environ. Stud. Japan. 11: 165-179.

Asada K (1994) Production and action of active oxygen species in photosynthetic tissues. Causes of Photooxidative Stress and Amelioration of Defense Systems in Plants. (Eds. CH Foyer and Mullineaux PM) pp.77-104. Boca Raton, Ann Arbor, London, Tokyo: CRC Press.

Asada K, Takahashi M, Tanaka K and Nacano Y (1977) Formation of active oxygen and its fate in chloroplasts. In: Biochemical and Medical Aspects of Active Oxygen (Ed. O Hayaishi and Asada). pp.45-63.Tokyo. University of Tokyo Press. Japan.

Ashmore MR, Dalpra C and Tickle AK (1988) Effects of ozone and and calcium nutrition on native plant species. In: Air Pollution and Ecosystems (Proceedings of International Symposium Held in

Grenoble, France, 18-22 May, 1987. (Ed. by Mathy P) D. Reidel Publishing Company) Member of the Kluwer Acad. Publ. Group. pp.647-652. Dordrecht. Boston. (Lancaster) Tokyo.

Atkinson R, Hasegawa D and Aschmann SM (1990) Rate constants for the gas phase reactions of O_3 with a series of monoterpenes and related compounds at 296°K. Int. J. Chem. Kinetic. 22: 871-875.

Aucoin RR, Scneider E and Arnason JT (1992) Evaluating the phytotoxicity and photogenotoxicity of plant secondary compounds. In: Plant Toxin Analysis (Eds. Linskens HF and Jackson JF).pp.75-86. Berlin Heidelberg. Springer-Verlag.

Aver'yanov AA (1984) Generation of superoxide anionradicals by some phenols. Biologicheskie Nauki (Russia) 7: 39-44.

Aver'yanov AA and Lapikova VP (1988) Fungitoxicity of the excretions of rice, induced by active oxygen species. Soviet Plant Physiology 35 (6): 1142-1151.

Aver'yanov AA and Lapikova VP (1988b) Generation of free radicals by phenolic compounds related with plant immunity. In: Oxygen Radicals in Chemistry, Biology and Medicine. (Ed. Afanas'ev IB), pp.203-222 Medical Institute, Riga

Aver'yanov AA and Lapikova VP (1995) Peroxidase activity in excretions of healthy and blast-infected rice leaves. Doklady Biological Sciences 340: 702-704.

Backer KH, Fricke W, Löbel J and Schurath U (1985) Formation, transport and control of photochemical oxidants. In: Air Pollution by Photochemical, Oxidants Formation, Transport, Control and Effects on Plants. (Ed. Guderian R) Berlin, Heidelberg, Springer-Verlag. (ser. Ecological Studies). 52: 1-125.

Bae GY, Nakajima N, Ishizuka K and Kondo N (1996) The role in ozone phytotoxicity of the evolution of ethylene upon induction of 1-amino-cyclopropane-1-carboxylic acid synthase by ozone fumigation in tomato plants. Plant Cell Physiology. 37 (2): 129-134.

Bachmann R, Burda C.,Gerson F.,Scholz M and Hansen HJ (1994) Radical anions of polyalkylazulenes: an ESR and EDOR study. Helv Chem Acta 77: 1458-1465.

Bailey PS (1958) The reactions of ozone with organic compounds. Chem. Rev. 58: 926-1110.

Bailey PS (1973) Ozonation in organic chemistry. Vol.1 Olefinic compounds. 493pp.Acad. Press. New York.

Bailey PS (1978) Ozonation in organic chemistry. Vol.1: Olefinic compounds. 272pp Acad. Press. New York, San Francisco, London.

Bailey PS, Mitchard DA and Khashab AIY (1968) Ozonation of amines. The completion between amine oxide formation and side-chain oxidation. J. Org. Chem. 33 (7): 2675-2680.

Baird C (1995) Environmental Chemistry. New York, Freeman and Co, 484 pp

Ball JC, Young WC, Wallington GJ and Japar SM (1992) Mutagenic proper ties of a series of alkyl hydroperoxides. Environ. Sci. Technol. 26 (2): 397-399.

Baraboi VA, Brekhman II, Golotin VG and Kugryashov YuB (1992) Lipid Peroxidation and Stress. Nauka, Sankt-Peterburg.148 pp.

Barber DJW and Thomas JR (1978) Reaction of radicals with lecthin bilayer. Radiation Research. 74 (1): 51-65.

Barbosa LCD, Cutler D, Mann J, Kirby GC and Worhurst DC (1993) Synthesis of some stable ozonides with anti-malarial activity. J. Chem. Soc. Perkin. Trans. 1 (24): N Dec. pp.3251-3252.

Barnes BL (1972) Effects of chronic exposure to ozone on photosynthesis and respiration of Pines. Environ. Pollut. 3: 133-138.

Bauer K, Seiler W and Giehl H (1979) CO-Production höherer Pflanzen an naturlich Standarten CO. Ztschr. Pflanzenphysiol. 94 (3): 219-230.

Baum JA and Scandalios JG (1981) Isolation and characterization of the cytosolic and mitochondrial superoxide dismutases of maize. Archives of Biochem and Biophys. 206 (2): 249-264.

Becker K.H, Fricke W, Löbel J and Schurath U (1985) Formation, transport, and control of photochemical oxidants. In: Air Pollution by Photochemical Oxidants. (Ed.Guderian R.). Berlin Heidelberg : Springer-Verlag, pp.3-125.

Beckerson DW and Hofstra G (1979) Effect of sulphur dioxide and ozone, singly or in combination, on leaf chlorophyll, RNA and protein in white bean. Can. J. Bot. 57: 1940-1945.

Bennett JH and Hill AC (1974) Acute inhibition of apparent photosynthesis by phytotoxic air pollutants. In: Air Pollution Effects on Plant Growth. (Ed. Dugger M) (Symp of DiV. Agrcult and Food Chem at the 167 Meeting of Am Chem Soc, Los Angeles, California, April 1, 1974. ACS Symposium Ser., 3) pp.115-127. Washington. D.C. Am Chem. Soc.

Bennett JH, Lee EH and Heggestad HE (1979 a) Physiological effects of EDU on ozone sensitive bean leaves. Plant. Physiol. 63 (5): (Supplement). p.152.

Bennett JH, Lee EH and Heggestad HE (1984) Biochemical aspects of plant tolerance to ozone and oxyradicals: superoxide dismutase. Gaseous Air Pollutants and Plant Metabolism./Eds. Koziol MJ, Whatley FR pp. 413-424. London, Boston: Butterworths.

Bennett JP, Oshima RJ and Lippert LF (1979 b) Effects of ozone on injury and dry matter partitioning in pepper plants. Environmental and Experimental Botany. 19 (1): 33-39.

Benoit LF, Skelly JM, Moore LD and Dochinger LS (1983) The influence of ozone on Pinus Strobus L. pollen germination. Can J Forest Res 13: 184-186.

Berger RG (1995) Aroma biotechnology. Springer. Berlin. 240 pp.

Bernanose AG and Rene MG (1959) Oxyluminescence of a few fluorescent compounds of ozone. Ozone Chemistry and Technology. Ser. Advances in Chemistry (no editor) Amer Chem. Soc., Washington. (Proc. Int. Ozone Conference, Chicago, November 1956.) pp.7-12.

Bersis D and Vassilliou E (1966) A chemiluminescence method for determining ozone. Analyst. 91 (1085): 499-505.

Bielski BHJ, Cabell DE and Arudi RL (1985) Reactivity of HO_2/O_2 radicals in aqueous solution. J. Phys. Chem. Ref. Data. 14 (4): 1041-1095.

Blum HF (1969) Is sunlight a factor in the geographical distribution of human skin color? Geogr. Rew. 59 (4): 557-581.

Bolsinger M, Lier ME, Lansky DM and Hughes PR (1992) Influence of ozone air pollution on plant-herbivore interactions. Part 1. Biochemical changes in ornamental milkweed (*Asclepias curassavica* L.; Asclepiadaceae) induced by ozone. Environmental Pollution. 72 (1): 69-84.

Bonte I (1982) Effects of air pollutants on flowering and fruiting. Effects of Gaseous Air Pollution in Agriculture and Horticulture. (Eds. Unsworth M.N., Ormrod D.R.) Batterworth Scientific. pp.207-223.London, Boston, Sydney, Wellington, Durban, Toronto.

Bors W, Michel C, Saran M and Lengfelder E (1978) The invovement of oxygen radicals during the autooxidation of adrenaline. Biochim Biophys Acta 540(1): 162-172.

Bors W, Langebartels A, Michel C and Sandermann HJr (1989) Polyamines as radical scavengers and protectants against ozone damage. Phytochemistry. 28: 1589-1595.

Bowler C, Van Montagu M and Inze D (1992) Superoxide dismutase and stress tolerance. Ann Rev. Plant. Physiol. Plant. Mol. Biol. 43: 83-116.

Boyd AW, Willis C and Cyr R (1970) New determination of stoichiometry of the iodometric method for ozone analysis at pH 7.0. Analytical Chemistry. 42 (6): 670-672.

Bridges O and Bridges JW (1996) Loosing Hope: The Environment and Health in Russia. Aldershot, Brookfield USA, Hong Kong, Singapore, Sydney. 266 pp.

Bringer E (1959) Photochemical production of ozone. In: Ozone Chemistry and Technology. Ser. Advances in Chemistry (no editor). Amer Chem. Soc. Washington (Proc. Int. Ozone Conference. Chicago. November 1956). pp.1-6.

Britz SJ and Robinson JM. (2001) Chronic ozone exposure and photosynthate partioning into starch in soybean leaves. Int J Plant Sci 162 (1) : 111-117.

Broad P, Creissen GP, Kular B, Wellburn AR and Mullineaux PM (1995) Oxidative stress responses in transgenic tobacco containing altered levels of glutathione reductase activity. The Plant Journal. 8 (2): 247-256.

Broadbent P., Creissen GP, Kular B, Wellburn AR and Mullineaux PM (1995) Oxidative stress responses in transgenic tobacco containing altered levels of glutathione reductase activity. Plant J * (2) : 247-255.

Brooks RI and Csallany AS (1978) Effects of air, ozone, and nitrogen dioxide exposure on the oxidation of corn and soybean lipids. J. Agricult. and Food Chem. 28 (5): 1203-1209.

Brown KA and Roberts TM (1988) Effects of ozone on foliar leaching in Norway spruce (*Picea abies* L. Karst): confounding factors due to NOx production during ozone generation. Environ. Pollut. 55: 55-73.

Brown DH and Smirnoff N (1978) Observations on the effect of ozone in *Cladonia rangiformis*. Lichenologist 10: 91-94.

Burlakova EB (1981) Role of lipids in cellular process of the information transfer. In: Biochemistry of Lipids and Their Role in Metabolism (Ed. Burlakova EB). pp.23-33.Nauka, Moscow

Burlakova EB and Khrapova NG (1981) Lipid peroxidation in membranes and natural antioxidants. Uspekhi Khimii (Advantages in Chemistry, USSR) 54: 1540-1558.

Butcher SS and Charlson RJ (1972) An Introduction to Air Chemistry. New York, London. Academic Press. 270 pp.

Butt VS (1985) Oxygenation and oxidation in the metabolism of aromatic compounds. Ann. Phytochem. Soc. Europe. 25: 349-365.

Byczkowski JZ and Gessner T (1988) Biological role of superoxide ion-radical. Int. J. Biochem. 20 (6): 569-580.

Byers DH and Salzman BE (1959) Determination of ozone in air by neutral and alkaline iodide procedures. Ozone Chemistry and Technology. (Advances. Chem. Soc.). 21: 93-101.

Bytnerowicz A and Turunen M (1994) Effects of ozone exposure on epicuticular wax of ponderosa pine needles. Air Pollutants and the Leaf Cuticle. (Eds. Percy K.E., Cape J.N., Jagels R., Simpson C.J.). NATO ASI Ser., Ser.G: Ecological Sciences. Berlin Heidelberg. Springer. 36: 305-314.

Caldwell MM, Björn LO, Bornman JF, Flint SD, Kulandaivelu G, Teramura AH and Tevini M (1998) Effects of increase in Ultraviolet radiation on terrestrial ecosystems J. Photochemistry and Photobiology. 46 (1-3): 40-52.

Camp van W, van Montagu M and Inze D (1994 a) Superoxide dismutases. Causes of Photooxidative Stress and Amelioration of Defense Systems in Plants. (Eds. CH Foyer and Mullineaux PM). Boca Raton, Ann Arbor, London, Tokyo: CRC Press. pp.317-341.

Camp van W, Willekens H, Bowler C, Van Montagu M, Inze DM, Reupold-Popp P, Sandermann HJr, Langebartels C (1994 b) Elevated levels of superoxide dismutase protect transgenic plants against ozone damage. Biotechnology. 12 (2): 165-168.

Carlson RW (1979) Reduction in the photosynthetic rate of Acer,Quercus and Fraxinus species caused by sulfur dioxide and ozone. Environment Pollut. 18 (2): 159-170.

Carter GA, Mitchell RJ, Chappelka AH and Brewer CH (1992) Response of leaf spectral reflectance in lobly pine to increase atmosphere ozone and precipitation acidity. J. Experimental Bot. 43 (249): 577-584.

Castillo FJ and Heath RL (1990) Ca^{2+} transport in membrane vesicles from pinto bean leaves and its alteration after ozone exposure. Plant. Physiol. 94 (2): 788-795.

Castillo FJ, Miller PR and Greppin HL (1987) Extracellular biochemical markers of photochemical oxidant air pollution damage to Norway spruce. Experientia. 43 (2): 111-115.

Chameides WL (1989) The chemistry of ozone deposition to plant leaves: role of ascorbic acid. Environ. Sci.Technol. 23 (5): 595-600.

Chameides WL, Lindsay RW, Richardson J and Kiang CS (1988) The role of biogenic hydrocarbons in urban photochemical smog: Atlanta as case study. Science. 241 (4872): 1473-1475.

Chamnongpol S, Willekens H, Moeder W, Langebartels C, Sansermann H JR, van Montagu M, Inze D and van Camp W (1998) Defense activation and enhanced pathogen tolerance induced by H_2O_2 on transgenic tobacco. Proc Natl Acad Sci USA 95: 5818-5823.

Chang CW (1971) Effect of ozone on ribosomes in Pinto bean leaves. Phytochemistry. 10: 2863-2868.

Chang CW and Heggestad HE (1974) Effect of ozone on Photosystem II in *Spinacia oleracea* chloroplasts. Phytochemistry. 13: 871-873.

Chanway CP and Runeck VC (1984) The role of superoxide dismutase in the susceptibility of bean leaves to ozone injury. Can. J. Bot. 62 (2): 236.

Charles SA and Halliwell B (1980) Effect of hydrogen peroxide on spinach (*Spinacia oleracea*) chloroplast fructose bisphosphatase. Biochemical Journal. 189 (2): 373-376.

Chasov AV, Gordon LKh, Kolesnokov OP, Minibaeva FV (2002) Cell surface peroxidase is a generator of superoxide anion in wounded wheat root cells. Cytologia (Russia) 44 (7) 691-696.

Chiron H, Drouet A, Claudot A.C, Eckerskorn C, Trost M, Heller W, Ernst D and Sandermann H.Jr. (2000a) Molecular cloning and functional expression of a stress-induced multifunctional *0-*methyltransferase with pinosylvin methyltransferase activity from Scots pine (*Pinus sylvestris* L.). Plant Molecular Biology 44: 733-745

Chiron H, Drouet A, Lieutier F, Payer HD, Ernst D and Sandermann H.Jr. (2000b) Gene induction of stilbene biosynthesis in Scots pine in response to ozone treatment, wounding, and fungal infection. Plant Physiology 124: 865-872.

Conklin PL and Last RL.(1995) Differential accumulation of antioxidant mRNAs in *Arabidopsis thaliana* exposed to ozone.Plant Physiology109(1):203-212,

Chevrier N (1986) Etude des changements physiologiques, biochimiques et ultra-structuraux induits par l'ozone chez *Euglena gracilis* Z., Ph.D.Thesis. University of Montreal. p.147.

Chevrier N, Chung YS and Sarhan F (1990) Oxidative damages and repair in *Euglena gracilis* to ozone. II. Membrane permeability and uptake of metabolites. Plant and Cell Physiology. 31 (7): 987-992.

Chevrier N and Sarhan F (1992) Effect of ozone on energy metabolism and its relation to carbon dioxide. Fixation in *Euglena gracilis*. J. Plant. Physiol. 140 (5): 521-526.

Chevrier N, Sarhan F and Chung YS (1988) Oxidative damages and repair in *Euglena gracilis* to ozone. I. SH groups and lipids. Plant and Cell Physiology. 29: 321-327.

Chimiklis PE and Heath RL (1972) Effluxes of K^+ and H^+ from *Chlorella sorokiana* as affected by ozone. Plant Physiol. (Suppl.). 49 (3).

Chimiklis PE and Heath RL (1975) Ozone-induced loss of intracellular potassium ion from Chlorella sorokiniana. Plant Physiol. 56: 723-727.

Chrestin H, Bangratz J, d'Auzac and Jacob JL (1984) Role of the lutoid tonoplast in the senescence and degeneration of the laticifers of *Hevea brasiliensis*. Z. Pflanzenphysiol. 114 (3): 261-268.

Clark A.J., Landolt W., Bucher J.B., Strasser R.J., 1999: Ozone exposure response of beech quantified with a chlorophyll A fluorescence performance index. In: Fuhrer J., Achermann B. (eds.), Critical Levels for Ozone - Level II. Environmental Documentation 115: 215-218. Swiss Agency for Environment, Forest and Landscape, Bern, Switzerland.

Clark A.J., Landolt W., Bucher J.B., Strasser R.J., 2000: Beech (*Fagus sylvatica*) response to ozone exposure assessed with a chlorophyll *a* fluorescence performance index. Environ. Pollut. 109: 501-507.

Clayton H, Knight MR, Knight H, McAinsh MR and Hetherington AM (1999) Dissection of the ozone-induced calcium signature. The Plant Journal 17 (5): 575-579.

Comeau G and Le Blanc F (1971) Influence de l'ozone et l'anhydride sulfureux sur la regeneration desfeulles de *Funaria hygrometrica*. Hedw Nat Can.98: 347-358

Constantinidou HA and Kozlowski TT (1979) Effects of sulfur dioxide and ozone on *Ulmus americana* seedlings. II. Carbohydrates, protein and lipids. Can J. Bot. 57: 176-184.

Coohill TP (1991) Stratospheric ozone depletion as it affects life on earth - the role of Ultraviolet action spectroscopy. Impact of Global Climatic Changes on Photosynthesis and Plant Productivity. P.

Corkill GA (1988) High Resolution Chromatogr. Comm. 11: 12-215.

Coulson CL and Heath RL (1974) Inhibition of the photosynthetic capacity of isolated chloroplasts by ozone. Plant Physiology. 53 (1): 32-38.

Cowling DW and Koziol MJ (1982) Mineral nutrition and plant response to air pollutants. Effects of Gaseous Air Polution in Agriculture and Horticulture (Ed. by MN Unsworth and DP Ormrod). pp.349-375. London. Butterworth Scientific.

Craker LE (1971) Ethylene production from ozone in injured plants. Environmental Pollution. 1 (4): 299-304.

Craker LE and Feder WA (1972) Development of the inflorescence in petunia, geranium, and poinsettia under ozone stress. Hort. Science. 7 (1): 59-62.

Creissen G, Edwards EA and Mullineaux PM (1994) Glutathione reductase and ascorbate peroxidase. Causes of Photooxidative Stress and Amelioration of Defense Systems in Plants. (Eds. CH Foyer and Mullineaux PM). pp.343-364. Boca Raton, Ann Arbor, London, Tokyo: CRC Press.

Criege R (1959) The products of ozonation of some olefins. Ozone Chemistry and Technology. Washington: Amer. Chem. Soc. pp.133-135

Criege R (1975) Mechanism of ozonolysis. Angewandte Chemie. 14 (11): 745-752

Cross CE and Halliwell B (1994) Evaluation of biomolecular damage by ozone. Methods in Enzymology. Boston, London, Sydney, Tokyo, Toronto, San Diego New York. Academic Press. 234: Part D. pp. 252-256.

Cueto R, Squadrito GL and Pryor WA (1994) Quantifying aldehydes and distinguishing aldehydic product profiles from autooxidation and ozonation of unsaturated fatty acids. Methods in Enzymol 233 (part C): 174-182.

Curtis CR, Howell RK and Kremer DF (1976) Soybean peroxidases from ozone injury. Environ Pollution. 11 (3): 189-194.

Dahse I, Bernstein M, Müller E and Petzold U (1989) On possible functions of electron transport in the plasmalemma of plant cells. Biochem. Physiol. Pflanzen. 185 (¾): 145-180.

Dann MS and Pell EJ (1989) Decline of activity and quantity of ribulose bisphosphate carboxylase/oxygenase and net photosynthesis in ozone-treated potato foliage. Plant Physiol. 91 (1): 427-432.

Darrall NM (1989) The effect of air pollutants on physiological processes in plants. Plant, Cell and Environment. 12 (1): 1-30.

Dass HC and Weaver GM (1972) Enzymatic changes in intact leaves of *Phaseolus vulgaris* following ozone fumigation. Atmos. Environ. 6 (10): 759-763.

Davies KJA (1986) Intracellular proteolytic systems mag function as a secondary antioxidant defences. An hypothesis. J. Free Rad. Biol. Med. 2: 155-173.

Davies KJA (1987) Protein damage and degradation by oxygen radicals. 1. General aspects. J. Biol. Ghem. 262 (20): 9895-9901.

Davies KJA and Delsignore ME (1987) Protein damage and degradation by oxygen radicals. III Modification of secondary and tertiary structure. J. Biol. Chem. 262 (20): 9908-9913.

Davies KJA, Delsignore ME and Lin SW (l987) Protein damage and degradation by oxygen radicals. II Modification of amino acids. J. Biol. Chem. 262 (20): 9902-9907.

Davies KJA and Goldberg AL (1987) Oxygen radicals stimulate intracellular proteolysis and lipid peroxidation by independent mechanisms in erithrocytes. J. Biol. Ghem. 262 (17): 8220-8226.

Davies KJA, Lin SW and Pacifici RE (1987 c) Protein damage degradation by oxygen radicals. IV. Degradation of denatured protein. J. Biol. Chemistry. 262 (20): 9914-9920.

Davis DD (1970) The influence of ozone on conifers. Ph.D.Thesis. University Park: Pennsylvania State Univ. 93 pp.

Davis DD and Copploino IB (1974) Relationship between age and ozone sensitivity of current needles of ponderosa pine. Plant Dis.Rep. 58 (3): 660-665.

Davison AW and Barnes JD (1993) Predisposition to stress following exposure to air pollution. Interacting Stresses on Plants in a Chenging Climate. Eds. Jacson MB and Black CR, (NATO ASI SER. ser.I. Global Environmental Change). (16): pp.11-123. Berlin. Springer.

Decleire M, de Cat W, de Temmerman and Baeten H (1984) Modifications de l'activite' des peroxidase, catalase et superoxide dismutase dans de feuilles d'epinard traite' a l'ozone. J. Plant Physiol. 116 (2): 147-152.

Degousee N, Triantaphylides C and Montillet JL (1994) Involvement of oxidative processes in the signaling mechanisms leading to the activation of glyceollin synthesis in soybean (Glycine max). Plant Physiol. 104 (3): 945-952.

Delwiche CC (1970) Carbon monoxide production and utilization by higher plants. Ann. N.Y. Acad. Sci. 174 (1): 116-121.

Dhindsa RS (1982) Inhibition of protein synthesis by products of lipid peroxidation. Phytochemistry. 21 (2): 309-313.

Dhindsa RS, Amaral AC and Cleland RE (1984) Rapid reduction of IAA of malondialdehyde leaves in *Avena* coleoptiles, a possible effects on lipid peroxidation. Biochim. Biophys. Res. Commun. 125 (1): 76-81.

Dimitriades B (1981) The role of natural organics on photochemical air pollution. Issues and research needs. J. Air Pollut Control Association. 31 (3): 229-235.

Dittrich H, Kutchan TM and Zenk MH (1992) The jasmonate precursor, 12-oxo-phytodiennoic acid, induces phytoalexin synthesis in (*Petroselinum crispum* cell cultures). FEBS Letters. 309 (1): 33-36.

Dizengremel P and Citerne A (1988) Air pollutant effects on mitochondria and respiration. In: Air Pollution and Plant Metabolism. (Eds. Schulte-Hostede S, Darrall NM, Blank LW, Wellburn AR) pp.169-188. London. New York. Elsevier Applied Science.

Dizengremel P, Sasek TW, Brown KJ and Richardson CJ (1994) Ozone-induced changes in primary carbon metabolism enzymes of loblolly pine needles. J. Plant Physiol. 144 (3): 300-306.

Dodd NJF and Ebert M (1971) Demonstration of surface free radicals on spore coats by ESR techniques. Sporopollenin: Proc. Symp. at Geology Department, Imperial College, London, 1970, 23-25 September. (Eds. Brooks J, Grant PR, Muir M, Gijzel PR) pp. 408-421. London, New York. Acad. Press.

Doke N, Miura Y, Sanchez LM and Kawakita K (1994) Involvement of superoxide in signal transduction: responses to attack by pathogens, physical and chemical shocks, and UV-irradiation. In: Causes of Photooxidative Stress and Amelioration of Defense Systems in Plants. (Eds. CH Foyer and Mullineaux PM) pp. 177-197. Boca Raton, Ann Arbor, London, Tokyo: CRC Press.

Dominy PJ and Heath RL (1985) Inhibition of the K^+- stimulated ATPase of the plasmalemma of pinto bean leaves by ozone. Plant Physiol. 77 (1): 43-45.

Dubinina EE and Shugalei IV (1993) Oxidative modification of proteins. Uspekhi Sovremennoi Biologii (Advantages of Modern Biology, Russia) 113: 71-81.

Duce RA, Mohnen VA, Zimmerman PR, Grosjean D, Caustreels W, Chatfield R, Doskotch RW, El-Feraly FS, Fairchaild EN and Huang CT (1976) Peroxyferolide: a cytotoxic germacranolide from *Liriodendron tulipifera*. J.C.S. Chem. Commun. 11: 402-403.

Duce RA, Mohnen VA, Zimmerman PR, Grosjean D, Caustreels W, Chatfield R, Jacnicke R, Orgen JA, Pellizzari ED and Wallace GT (1983) Organic material in the global troposphere. Rev. Geophys and Space Physics. 20 (4): 921-952.

Dugger WM and Palmer RL (1969) Carbohydrate metabolism in leaves of rough lemon as influenced by ozone Proc. First Intern.Citrus Symp. 2: 711-715.

Dugger WM, Taylor OC, Cardiff EA and Thomson CR (1962 a) Relationship between carbohydrate content and susceptibility of pinto bean plants to ozone damage. Proc. Am Soc. Hortic. Sci. 81: 304-315.

Dugger WM, Taylor OC, Cardiff EA and Thomson CR (1962 b) Stomatal action in plants as related to damage from photochemical oxidants. Plant Physiol. 37 (3): 487-491.

Dugger WM and Ting IP (1970) Air pollution oxidants-their effects on metabolic processes in plants. Annu Rev Plant Physiol. 21: 215-234.

Dugger WMJr, Taylor OC, Klein WH and Shropshire W (1963) Action spectrum of peroxyacetylnitrate damage to bean plants. Nature. London. 198 (4875): 75-76.

Dunning JA and Heck WW (1977) Respone of bean and tobacco to ozone: effect of light intensity, temperature and relative humidity. J. Air Pollut Control Assoc. 27: 882-886.

Dunning JA, Heck WW and Tingey DT (1974) Foliar sensitivity of pino bean and soybean to ozone as affected by temperature, potassium nutrition, and ozone dose. Water Air Soil Pollut. 3: 305-313.

Eckey-Kaltenbach H, Ernst D, Heller W and Sandermann HJr (1994 a) Cross-induction of defensive phenylpropanoid pathways in parsley plants by ozone. Acta Horticulturae 381: 192-198

Eckey-Kaltenbach H, Ernst D, Heller W and Sandermann HJr (1994 b) Biochemical plant responses to ozone. IV. Cross-induction of defensive pathways in parsley (*Petroselinum crispum* L.) plants by ozone. Plant Physiol. 104: 67-74.

Eckey-Kaltenbach H, Groβkopf E, Sandermann HJr and Ernst D (1994 c) Induction of pathogen defence genes in in parsley (*Petroselinum crispum* L.) plants by ozone. Proceedings of the Royal Society of Edinburgh. 102 B: 63-74.

Eckey-Kaltenbach H, Kiefer E, Groβkopf E, Ernst D and Sandermann HJr (1997) Differential transcript induction of parley pathogenesis-related proteins and of a small heat shock protein by ozone and heat shock. Plant Molecular Biology. 33: 343-350

Eerden van der L, Tonneijck A, Jarosz W, Bestebroer S and Dueck T (1993) Influence of nitrogenous air pollutants on carbon dioxide and ozone effects on vegetation. In: Interacting Stresses on Plants in a Changing Climate. (Eds. Jacson MB and Black CR) (NATO ASI SER. ser. I. Global Environmental Change. Vol.16) pp. 125-137. Berlin. Springer.

Ehmert A (1959) Chemical ozone measurement. In: Ozone Chemistry and Technology (Advances Chem. Soc.). 21: 128-132.

Elkiey T and Ormrod DP (1979) Ozone and/or sulfur dioxide effects on tissue permeability of petunia leaves. Atmos Environ. 13: 1165-1168.

Elkiey T and Ormrod DP (1980) Sorption of ozone and sulfur dioxide by Petunia leaves. J. Environ Qual. 9 (1): 93-95.

Elstner EF, Fischer HP, Oswald W and Kwiatkowski G (1980) Superoxide and ethane formation in subchloroplast particles catalysis by pyridium derivatives. Z. Naturforsch. Teil C. Bd.35: 220-775.

Elstner EF (1982) Oxygen activation and oxygen toxicity. Ann Rev. Plant. Physiol. 33: 73-96.

Elstner EF, Oswald W and Youngman RJ (1985) Basic mechanisms of pigment bleaching and loss of structural resistance in spruce (*Picea abies*) needles: advances in phytomedical diagnostics. Experientia. 41: 591-597.

Emik LQ, Plata RL, Cambell KI and Clarke GL (1971) Biological effects of urban air pollution. Arch Environ Health. 23: 335-342.

Enders G (1985) Deposition of ozone to a mature spruce forest: measurements and comparison to models. Environmental Pollution. 75 (1): 61-67.

Endress AG and Grünwald C (1985) Impact of chronic ozone on soybean growth and biomass partitioning. Agri. Ecosyst. Environ. 13 (1): 9-23.

Enge RL and Gabbeman WR (1966) The effect of low levels of ozone on Pinto beans, *Phaseolus vulgaris* L. Proc. Am. Soc. Hort. Sci. 91: 304-309.

EPA (US Environmental Protection Agency) (1976) The photochemical oxidants. In: Diagnosing vegetation injury caused by air pollution. Appl Sci Assoc Inc, EPA Contr 68-02-1344.

EPA (US Environmental Protection Agency) (1978 a) Effects of photochemical oxidants. On vegetation and certain microorganisms. In: Air Quality Criteria for Ozone and other Photochemical Oxidants, EPA600/8-78-004, pp. 253-293

EPA (US Environmental Protection Agency) (1978 b) Ecosystems. In:Air Quality Criteria for Ozone and other Photochemical Oxidants The photochemical oxidants, EPA600/8-78-004, pp. 294-328.

EPA (US Environmental Protection Agency) (1976) The photochemical oxidants. In:Diagnosing vegetation injury caused by air pollution. Appl Sci Assoc Inc, EPA Contr 68-02-1344.

Erickson LC and Wedding RT (1956) Effects of ozonated hexene on photosynthesis and respiration of *Lemna minor*. Am J. Bot. 43 (1): 32-36.

Ernst D, Eckey H and Sandermann H (1992 b) Biochemical and moleculars responses of ozone in *Petroselinum crispum* L. Physiol Plantarum. 85: (Suppl.) A 80.

Ernst D, Schraudner M, Langebartels C and Sandermann HJr (1992) Ozone-induced changes of mRNA levels of β-1,3-glucanase, chitinase and 'pathogenesis-related' protein 1b in tobacco plants. Plant Molecular Biology. 20: 673-682.

Ernst D, Bodemann A, Schmeizer E, Langebartels C and Sandermann HJr (1996) β-1,3-Glucanase mRNA is locally, but not systemically induced in *Nicotiana tabacum* L. cv.BelW3 after ozone fumigation. J. Plant Physiol. 148: 215-221.

Evans LS and Ting JP (1974) Effect of ozone 86 Rb- Ca-labeled potassium transport in leaves of *Phaseolus vulgaris* L. Atmos. Environ. 8: 855-861.

Evans LS and Ting JP (1973) Ozone - induced membrane permeability changes. Amer. J. Bot. 60: 155-162.

Evans LS and Ting JP (1974) Ozone - sensitivity of leaves: Relationship to leafwater content, gas transfer resistance and anatomical characteristic. Amer. J. Bot. 61: 592-597.

Executive Summary. Special of Atmospheric Science Panel J. Photochemistry and Photobiology. (1998) 46 (1-3): 1-4.

Farage RK, Long SP, Lechner EG and Baker NR (1991) The sequence of change within the photosynthetic apparatus of wheat following short-term exposure to ozone. Plant Physiol. 95 (2): 529-535.

Farage RK and Long SP (1995) The effects of acute ozone exposure on photosynthesis in contrasting species: An in vivo analysis. Photosynthesis Research 43: 11-18.

Feder WA (1968) Reduction in tobacco pollen germination and tube elongation, induced by low levels of ozone. Science. 160 (3832): 1122.

Feder WA (1970) Plant response to chronic exposure of low levels of oxidant type air pollution. Environ. Pollut. 1 (1): 73-79.

Feder WA (1973) Cumulative effects of chronic exposure of plants to low levels of air pollutants. Air Pollution Damage to Vegetation (Ed. JA Naegele). (Advances in Chemistry series 122). American Chemical society. pp. 21-31.Washington. D.C. USA.

Feder WA (1978) Plants as bioassay systems for monitoring atmosphere pollutants. Environ. Health Perspectives. 27 (1): 139-147.

Feder WA and Campbell FJ (1968) Influence of low levels of ozone on flowering of carnations. Phytopathology. 58: 1038-1039.

Feder WA, Fox FL, Heck WW and Campbell FJ (1969) Varietal responses of petunia to several air pollutants. Plant Disease Reproduction. 53: 506-510.

Feder WA, Krause JHM, Harrison BH and Riley WD (1982) Ozone effects on pollen-tube growth *in vivo* and *in vitro*. In: Effects of gaseous air pollution in agriculture and horticulture (Eds. Unsworth MN, Ormrod DP, Fujii R, Novales RR) Butterworths Sci. pp. 482. London.

Feder WA and Manning WJ (1979) Living plants as indicators and monitors. Handbook of Methodology for the Assessment of Air Pollution Effects on Vegetation. (Eds. Heck WW, Krupa SV, Linzon SN) pp. 9-1, 9-14. Pittsburgh. Air Pollution Control Association.

Feder WA and Sullivan F (1969) Differential susceptibility of pollen grains to ozone injury. Phytopathology. 59: 322.

Federico R and Angelini R (1986) Occurence of diamine oxidase in the apoplast of pea epicotyls. Planta. 167: 300-302.

Fetner RH (1962) Ozone-induced chromosome breakage human cell cultures. Nature (London). 194: 793-794.

Finlayson-Pitts BJ and Pitts JN, Jr (1986) Atmospheric Chemistry. Fundamental and Experimental Techniques. Wiley, New York, Chichester, Brisbane, Toronto, Singapore. 1098 pp.

Fletcher RA (1969) Retardation of leaf senescence by benzyladenine in intact bean plants. Planta. 89 (1): 1-8.

Fletcher RA, Adedipe NO and Ormond DP (1972) Abscisic acid protects bean leaves from ozone-induced phytotoxcity. Can J. Bot. 50: 2389-2391.

Foyer C, Rowell J and Walker D (1983) Measurement of the ascorbate content of spinach leaf protoplasts and chloroplasts during illumination. Planta. 157 (3): 239-244.

Foyer CH and Harbinson J (1994) Oxygen Metabolism and the Regulation of Photosynthetic Electron Transport. Causes of Photooxidative Stress and Amelioration of Defense Systems in Plants. (Eds. CH Foyer and Mullineaux PM) Boca Raton, Ann Arbor. pp. 1-42. London, Tokyo: CRC Press.

Foyer CH (1993) Ascorbic acid. Antioxidants in Higher Plants. (Ed. Alscher RG and Hess JL) Boca Raton, Ann Arbor. pp. 31-58. London, Tokyo: CRC Press.

Frage PK, Long SP, Lechner EG and Baker NR (1991) The sequence of change within the photosynthetic apparatus of wheat following short-term exposure to ozone. Plant Physiol. 95 (2): 529-535.

Frankel EN (1983) Volatile lipid oxidation products. Prog. Lipid Research. 22 (1): 1-33.

Frederick PE and Heath RL (1975) Ozone-induced fatty acid and viability changes in Chlorella. Plant Physiol. Academic Press. NY. 55 (1): 15-19.

Fridovich I (1984) Overview:biological sources of $O_2^{\cdot -}$.Methods Enzymol 105: 59-61.

Fridovich I (1986) Biological effects of superoxide radical. Arch. Biochem. Biophys. 247 (1): 1-11.

Fridovich I (1976) Oxygen radicals, hydrogen peroxide and oxygen toxicity. In: Free Radicals in Biology (Ed. WA Prior). pp. 239-277.

Fridovich I (1981) Superoxide radical and superoxide dismutases. In: Oxygen and Living Processes. An Interdisciplinary Approach. (Ed. DL Gilbert) pp. 250-272. New York, Heidelberg, Berlin. Springer.

Fridovich I (1983) Superoxide radical: An Endogenous toxicant. Ann. Review of Pharmacology and Toxicology. 23: 239-257.

Friedrich R and Reis S (Eds) (2000) Tropospheric Ozone Abatement. Berlin, Heidelberg: Springer - Verlag. 237 pp.

Fuentes JD, Gillespie TJ and Bunce NJ (1994 a) Effects of foliage wetness on the dry deposition of ozone onto red maple and poplar leaves. Water, Air, Soil Pollut. 74 (1): 189-210.

Fuentes JD, den Hartog G, Neumann HH, Gillespie TJ (1994 b) Measurements and modelling of ozone deposition to wet folliage. Air Pollutants and the Leaf Cuticle. (Eds. Percy KE, Cape JN, Jagels R, Simpson CJ) NATO ASI Ser., Ser.G.: Ecological Sciences. Berlin, Heidelberg. Springer. 36: 239-253.

Fuhrer J and Achermann B (Eds) (1999) Critical Levels for Ozone – Level II. Environmental Documentation Vol. 115. Swiss Agency for Environment, Forest and Landscape, Bern.

Fujii R and Novales RR (1968) Reduction in tobacco pollen germination and tube elongation induced by low levels of ozone. Science. 160 (3832): 1122.

Fukunagu K, Susuki T, Hara A and Takama K (1992) Effect of ozone on the activities of reactive oxygen scavenging enzymes in RBC of ozone exposed japanese Charr. (Salveninus-Leucomaenis). Free Rad Reseach Comm. 17 (5): 327-334.

Furucawa A and Kadota M (1975) Environment Control in Biology. 13 (1): 1-13.

Furucawa A and Totsuka T (1979) Effects of NO_2, SO_2 and ozone alone end in combinations on net photosynthesis in sunflower. Environ. Control. Biol. 17: 161-166.

Gäb S, Hellpointner E, Turner WV and Korte F (1985) Hydroxymethyl hydroperoxide and bis (hydroxymethyl) peroxide from gas-phase ozonolysis of naturally occurring alkenes. Nature. 316 (6028): 535-536.

Gabbita SP, Robinson KA, Stewart CA, Floyd RA and Hensley K (2000) Redox regulatory mechanisms of cellular signal transduction. Arch Biochem and Biophys 376: 1-13.

Gabrielli F (1983) Roles of turnover and repair of macromolecules and supramolecular structural components. Life Sciences. 33 (9): 805-816.

Galliano H, Cabane M, Eckerskorn C, Lottspeich F, Sandermann H and Ernst D (1993 a) Molecular cloning, sequence analysis and elicitor-induced ozone-induced accumulation of cinnamylalcohol dehydrogenase from Norway Spruce *(Picea abies* L.). Plant Mol. Biol. 23 (1): 145-156.

Galliano H, Heller W and Sandermann H (1993 b) Ozone induction and purification of spruce (*Picea abies* L.Karst.) cinnamyl alcohol dehydrogenase. Phytochemistry. 32: 557-563.

Galliano H, Cabane M, Eckerskorn C, Lottspeich F, Sandermann HJr and Ernst D (1993 c) Molecular cloning, sequence analysis and elicitor-/ozone-induced accumulation of cinnamyl alcohol dehydrogenase from Norway spruce (*Picea abies* L.). Plant Molecular Biology 23: 145-156

Gamaley IA and Klyubin IV (1996) Hydrogen peroxide as signalling molecule. Tsitologia (Cytologia, Russia) 38 (12): 1233-1247.

Gaponenko VI, Baleeva EF, and Zhebrakova IV (1988) Phenomenon of metabolic heterogeny of chlorophyll under ozone action. Vestnik of Belorussian Acad Sci, Ser BiolSciences. 3: 38-41.

Garland JA and Penkett SA (1976) Absorption of peroxyacetyl nitrate and ozone by natural surfaces. Atmos Environ. 10: 1127-1131.

Garreo JP (1994) Cuticular characteristics in the detection of plant stress due to air pollution - new problems in the use of these cuticular characteristics. In: Air Pollutants and the Leaf Cuticle. (Eds. Percy KE, Cape JN, Jagels R, Simpson CJ) NATO ASI Ser., Ser.G: Ecological Sciences. 36: 113-122. Berlin, Heidelbeg. Springer.

Gavrilova AA, Churmasov AV and Rezchikov VG (1999) Mechanism of the ozone action on plant growth processes. In: Materials of 2 nd Meeting of Russian Biophysists. 23-27 August 1999, Moscow. pp.873-874. Biological Center of Russian Academy of Sciences, Pushchino.

Gebicki S and Gebicki JM (1993) Formation of peroxides in amino acids and proteins exposed to oxygen free radicals. Biochemical Journal. 289 (3): 743-749.

Genoud T and Metraux JP (1999) Crosstalk in plant cell signalling: structure and function of the genetic network Trends in Plant Science $ (12): 503-507

Gerini O, Guidi L, Lorenzini G and Soldafinig. J (1990) Environ Qual. 19 (1): 154.

Gibbs M (1966) Carbohydrates: Their role in plant metabolism and nutrition. Plant Physiology. (Ed. FC Steward). New York, London. Academic Press. P. IVB: 3-115.

Gilbert DL (1981) Significance of oxygen on Earth. In: Oxygen and Living Processes. An Interdisciplinary Approach. (Ed. DL Gilbert). pp. 73-101. New York. Heidelberg. Berlin. Springer.

Gillham DJ and Dodge AD (1986) Hydrogen-peroxide-scavenging systems within pea chloroplasts. A quantitative study. Planta. 167 (2): 246-251.

Glater RB, Solberg RA and Scott FM (1962) A developmental study of the leaves of *Nicotiana glutinosa* as related to their smog-sensitivity. Am. J Bot. 49 (2): 954-970.

Gleason JF, Bhartia PK, Herman JR, McPeters R, Neuman P, Stolarski RS, Flynn L, Labow G, Larko D, Seftor C, Wellemeyer C, Komhyr WD, Miller AJ and Planet W (1993) Record low global ozone in 1992. Science. 260 (5107): 523-526

Goldstein BD, Buckley RD, Cardenas R and Balchum OJ (1970) Sensitivity of some humans to lethal doses of ozone (up to 14 ppm) is connected with deficite of vitamine E. It is due unnormal lipid oxidation that induces the earlier aging. Ozone and vitamin E. Science. 169 (3945): 605.

Goldstein BD, Levine MR, Cuzzi-Spada RMS, Cardenas R, Buckley RD and Balchum OJ (1972) p-Aminobenzoic acid as a protective agent in ozone toxicity. Arch Environ Health. 24 (4): 243-247.

Goodwin T W. and Mercer E I (1983) Introduction to Plant Biochemistry. Second Edition. Pergamon Press. Oxford, New York.

Gordon T, Taylor B.F. and Amdur MO (1981) Ozone inhibition of tissue cholinesterase. Arch. Environ.Health 36: 284-288

Gordon LKx, Kolesnikov OP and Minibayeva Ph.V (1999) Formation of superoxide by redox system of plasmalemma of root cells and its participation in detoxification of xanobiotics. Doklady of Russian Academy of Sciences 367 (3): 409-411.

Grushko YaM. Hazzard Inorganic Compounds in Industrial Wastes to Atmosphere. Khimiya, Leningrad, 192 pp.

Goto N and Niki E (1994) Measurement of superoxide reaction by chemiluminescence. Methos in Enzymology 233 (part C): 154-160.

Greitner CS and Winner WE (1989) Effects of O_3 on alder photosynthesis and symbiosis with *Frankia*. New Phytologist. 111 (4): 647-656.

Gressel J and Galun E (1994) Genetic controls of photooxidant tolerance. Causes of Photooxidative Stress and Amelioration of Defense Systems in Plants. (Eds. C.H. Foyer and P.M. Mullineaux) CRC Press: Boca Raton, Ann Arbor. pp. 237-273. London, Tokyo.

Grimmig B, Schubert R, Fischer R, Hain R, Schreier PH, Betz C, Langebartels C, Ernst D and Sandermann HJr (1997) Ozone- and ethylene-induced regulation of a grapevine resveratrol synthase promoter in transgenic tobacco. Acta Physiologiae Plantarum 19: 467-474.

Grimsrud EP, Westberg HH and Rasmussen RA (1974) Atmospheric reactivity of monoterpene hydrocarbons., NO_x photooxidation and ozonolysis. Proc Symp., Chemical. Kinetic data for the upper and lower atmosphere. W. Arrenton, Virginia, September 15-18, 1974. (Eds. Benson SW, Golden DM, Barker JR). Int J. Chem. Kinet., VII symposium 1. pp. 183-195

Groβkopf E, Wegener-Strake A, Sandermann HJr and Ernst D (1994) Ozone-induced metabolic changes in Scots pine: mRNA isolation and analysis of in vitro translated proteins. Can. J.For.Res. 24: 2030-2033.

Grosso JJ, Menser NA, Hodges GA and McKinney HH (1971) Effects of air pollutants on *Nicotiana* cultivars and species used for virus studies. Phytopathology. 61: 945-950.

Grulke NE and Miller PR (1994) Changes in gas exchange characteristics during the life span of giant sequoia: implications for response to current and future concentrations of atmospheric ozone. Tree Physiology. 14 (7-9): 659-668.

Grünwald C and Endress AG (1985) Foliar sterols in a soyabeans exposed to chronic levels of ozone. Plant Physiol. 77 (1): 245-247.

Guderian R. (ed.) Air Pollution by Photochemical oxidants. Formation, Transport, Control and Effects on plants. Berlin, Heidelberg, New York, Tokio. Ecological Studies. Springer-Verlag. 1985. Vol.52. P.129-334.

Guderian R (1999) (Unuversitat Essen) Handbuch der Umweltveranderunden und Okotoxikologie. Bd. 1 B Atmosphere: Multiphasenchemie - Ozonebbau-Deposition-Trebhaus-effect-Konsequenzen. Berlin-Heidelberg. 330 ss.

Guderian R and Reidl K (1982) Höhere Pflanzen als Indikatoren fur Immissiosbelastungen im terrestrischen Bereich. Decheiana Beih. 26: 6-22.

Guderian R, Tingey DT and Rabe R (1985) Effects of photochemical oxidants on plants. In: Air Pollution by Photochemical Oxidants. Formation, Transport, Control and effects on Plants. (Ed. R Guderian). pp .127-333. Berlin, Heidelberg. Springer-Verlag.

Günthardt-Goerg M.S., 1999: Adaptation of plants to anthropogenic and environmental stresses: The effects of air constituents and plant-protective chemicals (Authors: Bader K.P., Abdel-Basset R.). In: Pessarakli M. (ed.), Handbook of Plant and Crop Stress: 991. Marcel Dekker, New York.

Günthardt-Goerg M.S. , 2001: Ozonsymptome an Waldbaumarten. Wald und Holz 10: 30-33.

Günthardt-Goerg M.S., Maurer S., Bolliger J., Clark A.J., Landolt W., Bucher J.B., 1999: Responses of young trees (five species in a chamber exposure) to near-ambient ozone-concentration. Water Air Soil Pollut. 116: 323-332.

Günthardt-Goerg M.S., McQuattie C.J., Scheidegger C., Matyssek R., Rhiner C., 1997: Ozone-induced cytochemical and ultrastructural changes in the leaf mesophyll cell wall. Can. J. For. Res. 27: 453-463.

Günthardt-Goerg M.S., McQuattie C.J., Mathies D., Frey B., 1999: Visible and microscopical injury in leaves of 5 deciduous tree species related to current critical ozone levels. In: Fuhrer J., Achermann B. (eds.), Critical Levels for Ozone - Level II. Environmental Documentation 115: 181-185. Swiss Agency for Environment, Forest and Landscape, Bern, Switzerland.

Günthardt-Goerg M.S., McQuattie C.J., Maurer S., Frey B., 2000: Visible and microscopical injury in leaves of five deciduous tree species related to current critical ozone levels. Environ. Pollut. 109: 489-500.

Günthardt-Goerg M.S., Vollenweider P., Egli P., 2000a: Sichtbare Reaktionen auf Ozon und/oder Schwermetalle bei Pflanzen. Schriftenreihe KRdL 32: 349-354.

Günthardt-Goerg M.S., Vollenweider P., Egli P., 2000b: Sichtbare Reaktionen auf Ozon und/oder Schwermetalle bei Pflanzen. Informationsblatt Forschungsbereich Wald, WSL 3: 3-5.

Gupta AS, Alscher RG and McCune D (1991) Response of photosynthesis and cellular antioxidants to ozone in *Populus* leaves. Plant Physiol. 96 (2): 650-655.

Gurol MD and Singer PC (1982) Kinetics of ozone decomposition: a dynamic approach. Environ. Sci. Technol. 16: 377-383.

Gutteridge JM (1995) Signal messenger and trigger molecules from free radical reactions and their control by antioxidants. In: Signalling Mechanisms - from transcription factors to oxidative stress. (Eds. LP Packer and KWA Wirtz) Springer in cooperation with NATO Sci. Affairs Division. Berlin, Heidelberg. (NATO ASI Ser.) H92: 157-164.

Haber F and Weiss J (1932) Uber die Katalyse des Hydroperoxydes. Naturwissenschaften. 20 (51): 948-950.

Häder DP, Kumar HD, Smith RC and Worrest RC (1998) Effect on aquatic ecosystems. J. Photochemistry and Photobiology. 46 (1-3): 53-68.

Hadjur C and Jardon P (1995) Quantative analysis of superoxide anion radicals photosentized by hypericin in a model membrane using the cytochrome c.reduction method. J. of Photochemistry and Photobiology. 29 B., Biology. (2-3): 147-156.

Hahlbrock K, Knobloch KH, Kreuzaler F, Potts JRM and Wellmann E (1976) Coordinated induction and subsequent activity change of two groups of metabolically interrelated enzymes. Light-induced synthesis of flavonoid glycosides in cell suspension cultures of *Petroselinum hortense*. Eur J. Biochem. 61 (1): 199-206.

Hahlbrock K and Scheel D (1989) Physiology and molecular biology of phenylpropanoid metabolism. Annu Rev. Plant. Physiol. Mol. Biol. 40: 347-369.

Hahlbrock K and Wellmann E (1970) Light induced flavone biosynthesis and activity of phenylalanine ammonia-lyase and UDP-apiose synthetase in cell suspension cultures of *Petroselinum hortense*. Planta. 94: 236-239.

Haraguchi, H., Saito, T., Okamura, N. and Yagi, A. (1995) Inhibition of lipid peroxidation and superoxide generation by diterpenoids from *Rosmarinus officinalis*. Planta Medica **61**: 333-336

Haugland RP (1996) Handbook of Fluorescent Probes and Research Chemicals. Molecular Probes, Eugene.

Heiden AC, Hoffmann T, Kahl J, Kley D, Klockow D, Langebartels C, Melhorn H, Sandermann JrH, Schraudner M, Schuh G and Wildt J (1999) Emission of volatile organic compounds from ozone-exposed plants. Ecol Applic 9: 1160-1167.

Halliwell B (1981) Chloroplast metabolism. Oxford.: Clarendon Press. 257 pp.

Halliwell B (1987) Oxidative damage, lipid peroxidation and antioxidant protection in chloroplasts. Chem. Phys. Lipids. 44: 327-340.

Halliwell B (1984) Oxygen radicals: a common sense look at their nature and medical importance. Medical Biology. 62 (2): 71-77.

Halliwell B (1979) Oxygen free radicals in living systems: dangerous but usefull. Strategies of Microbial Life in Extreme Environments. Life in Sci. Res. Reports 13. pp. 195-221. Berlin. Springer.

Halliwell B and Gutteridge JMC (1985) Free Radicals in Biology and Medicine. Oxford: Clarendon Press. 346 pp.

Halliwell B and Gutteridge JMC (1989) Free radicals. M. Biology and Medicine. Oxford: Clarendon Press. 215 pp.

Halliwell B and Gutteridge JMC (1984) Oxygen toxicity, oxygen radicals, transition metals and disease. Biochem. J. 219 (1): 1-14.

Hamelin C (1985) Production of single- and double-strand breaks in plasmid DNA by ozone. Int J. Radiat. Oncol. Biol. Phys. 11: 253-257.

Hamelin C and Chung YS (1975) Characterization of mucoid mutants of *Escherichia coli* K 12 isolated after exposure to ozone. J. Bacteriol. 122 (1): 19-24.

Hamelin C and Chung YS (1974 a) Lethal and mutagenic effects of ozone on *Escherichia coli*. Can J. Genet. Cytol. 16: 706.

Hamelin C and Chung YS Optimal conditions for mutagenesis is by ozone in *Escherichia* coil K 12. Mutant Res. 1974 b. 24: 271-279.

Hamelin C and Chung YS (1976) Rapid test for assay of ozone sensitivity in *Escherichia coli*. Mol. Genet. 45 (2): 191-194.

Hamelin C, Sarkan F and Chung YS (1977) Ozone - induced DNA degradion in different DNA polymerase 1 mutants of *Escherichia coli* K 12. Biochem. Biophys. Res. Commun. 77 (1): 220-224.

Chanson GP and Steward WS (1970) Photochemical oxidants: effect on starch hydrolysis in leaves. Science. 168 (3936): 1223-1224.

Chanson GP, Thorne L and Addis DH (1975) The ozone sensitivity of *Petunia hybrida* Vilm as related to physiological age. J. Am Soc. Hortic. Sci. 100: 188-190.

Hanst PL and Gau BW (1983) Atmospheric oxidation of hydrocarbons; formation of hydroperoxides and peroxyacids. Atmospheric Environment. 17 (11): 2259-2265.

Harrison BH and Feder MA (1974) Ultrastructural changes in pollen exposed to ozone. Phytopathology. 64 (2): 257-258.

Harrison KA and Murphy RC (1996) Direct mass spectrometric analysis of ozonides. Application to unsaturated glycerophosphocholine lipids. Anal. Chem. 68 (18): 3224-3230.

Hartmann T (1985) Prinzipien des pflanzlichen Sekundastoffwechsels. Plant. Syst. and Evol. 150 (1-2): 15-34.
Hatakeyama S, Ohna M, Weng JM, Takagi H and Akimoto H (1987) Mechanism for the formation of gaseous and particulate products from ozone-cycloalkane reactions in air. Environ. Sci. Technol. 21 (1): 51-52.
Hauffe KD, Hahlbrok K and Scheel D (1986) Elicitor - stimulated furanocoumarin biosynthesis in cultured parsley cells: S-adenosyl-L-methionine: bergaptol and S-adenosyl-L-methionine: xanthotoxol o-methyl-transferases. Z. Naturforsch. 41 (2): 228-239.
Hausladen A and Alscher RG (1993) Glutathione. In Antioxidants in Higher Plants. (Eds. Alcher RG, Ness JL) Boca Raton, Ann Arbor. pp. 1-30. London, Tokyo: CRC Press.
Hazucha MJ, Folinsbee LJ and Seal E (1992) Effects of steady-state and variable ozone concentration profiles on pulmonary function. Am. Rev. Resp. Dis. 146 (6): 1474-1479.
Heagle AS (1973) Interactions between air pollutants and plant parasites. Annu Rev Phytophathol 11; 365-488.
Heagle AS (1982) Interactions between air pollutants and parasitic plant diseases. In: Effects of Gaseous Air Pollution in Agriculture and Horticulture. pp.333-348.
Heagle AS (1989) Ozone and crop yield. Ann. Rev. Phytopathol. 27: 397-423.
Heagle AS and Key LW (1973 a) Effect of ozone on the wheat stem rust fungus. Phytopathology. 63 (3): 397-400.
Heagle AS, Miller JE and Sherrill DE (1994) Atmospheric pollutants and trace gases. A white clover system to estimate effects of tropospheric ozone on plants. J. Environ. Qual. 23 (3): 613-621.
Heagle AS, Philbeck RB, Roges HH and Letchworth MB (1979) Dispensing and monitoring ozone in open-top field chambers for plant effect studies. Phytopathology. 69 (1): 15-20.
Heagle AS and Strikland A (1972) Reaction of *Erysiphe gramines* f. Sp. hordei to low levels of ozone. Phytopathology. 62: 1144-1148.
Heath RL (1980) Initial events in injury to plants by air pollutants. Annu. Rev. Plant Physiol. 31: 395-431.
Heath RL (1988) Biochemical mechanisms of pollutant stress. Assessment of Crop* Loss From Air Pollutants (Eds. Heck WW, Taylor OC, Tingey DT) pp. 259-286. London: Elsevier
Heath RL (1975) Ozone. Responses of Plants to Air Pollution. (Ed. I.B. Nudd and T.T. Kozlowski) New York - San Francisco. pp. 23-55. London.: Academic Press.
Heath RL (1987) The biochemistry of ozone attack on the plasma membrane of plant cells. Adv. Phytochem. 21: 29-54.
Heath RL (1978) The inhibition of respiration and glycolysis in *Chlorella* exposed to ozone. Plant Physiol. 61 (1): 93.
Heath RL and Castillo FJ (1988) Membrane disturbance in response to air pollutants. In: Air Pollution and Plant Metabolism. (Eds. Schulte-Hostede S, Darrall RH, Blank LM, Wellburn AR) pp. 55-75. London and New York: Elsevier. Applied Science.
Heath RL, Frederick PE and Chimikies PE (1982) Ozone inhibition of photosynthesis in *Chlorella sorokiniana*. Plant Physiol. 69: 229-253.
Heath RL and Frederick PN (1979) Ozone alteration of membrane permeability in *Chlorella*. Plant Physiol. 64: 455-459.
Heath RL and Tappet AL (1976) A new sensitive assays for the measurement of hydroperoxides. Analyt. Biochem. 76 (1): 184-191.
Heath RL and Taylor GE Jr(1997a) Physiological processes and plant responses to ozone exposure. Forest Decline and Ozone. A Comparison of Controlled Chamber and Field Experiments /Eds. Sandermann H, Wellburn AR, Heath RL (Ecological Studies, vol. 127). pp. 317-368. Berlin, Heidelberg: Springer-Verlag.
Heath RL and Taylor GE Jr(1997b) Physiological processes and plant responses to ozone exposure. Ecol Stud 127: 317-368.
Heath RL, Chimiklis PE and Frederick PE (1974) The potassium and lipids in ozone injury to plant membranes. Air pollution Effect of on Plant Growth (Ed. WM Dugger) Washington: Am Chem. Soc. Ser. (3): 58-75.
Hebben CR (1966) Sensitivity of fungal spores to sulphur dioxide and ozone. Abstr. Phytopathology. 56: 880-881.
Heber U, Kaiser W, Luwe M, Kindermann G, Veljovic-Javonovic S, Yin Z, Pfanz H and Slovik S (1994) Air pollution, photosynthesis and forest decline: Interactions and consequences. Ecology of

photosynthesis. (Eds. Schulze ED, Caldwell MM) pp. 179-296. Berlin, Heidelberg, New York, London, Paris, Tokyo, Hong Kong, Barcelona, Budapest: Springer.

Heck WW (1966) The use of plants as indicators of air pollution. Internat. J. of Air and Water Poll. 10 (1): 99-111.

Heck WW, Dunning JA and Hindawi IJ (1965) Interactions of environmental factors on the sensitivity of plants to air pollution. J. Air Pollut Control Assoc. 15: 511-515.

Heck WW, Dunning JA and Hindawi IJ (1966) Ozone: Nonlinear relation of doze and injury to plants. Science. 151: 577-578.

Heck WW, Heagle AS and Shriner DS (1986) Effect on vegetations: native crop forest. Air Pollution (Ed. AS Stern). New York: Acad Press. 6: 247--350.

Heck WW, Krupa SV and Linzon SN (1978) Handbook of Methodology for the Assesment of Air Pollution Effects on Vegetation. Pittsburg: Pa. Air Poll. Contr. Assoc. 392 pp.

Heck WW, Taylor OC, Adams R, Bingham G, Miller J, Prenston E and Weinstein L (1982) Assessment of crop loss from ozone. J. Air Pollut Gontrol Assoc. 32: 353-361.

Heck WW, Taylor OC and Tingey DT (Eds.) (1988) Assessment of Crop Loss from Air Pollutants. London. Elsevier.

Heggelstad HE, Menser HA (1962) Leaf spot-sensitive tobacco stain Bel W_3, a biological indicator at the air pollutant ozone. Phytopathology. 52: 735.

Heggelstad HE (1971) Plants as indicators of the air pollutants ozone and PAN. Air Pollution (Proc. of 1st Eur. Congress on the Influence of Air Pollution on Plants and Animals. Wageningen, April 22-27, 1968. Second impression 1971. Wageningen: centre for Agriculture Publishing and Documentation Veenman H and Zonen NV. pp. 329-335.

Heggelstad HE and Heck WW (1971) Nature, Extent and Variations of Plant Response to Air Pollutants. Advances in Agronomy. 23: 111-145.

Heggelstad HE (1969) Plants as indicators of the air pollutants ozone and PAN. Air Pollution (Proc. 1st Eur. Congress on the Influence of Air Pollution on Plants and Animals. Wageningen, April 22 -27, 1968) Second impression.1971. Eds. Veenman H and Zonen NV Wageningen: Centre for Agricultural Publishing and Documentation. pp. 329-335.

Heiden AC, Hoffmann T, Kahl J, Kley D, Klockow D, Langebartels C., Melhorn H, Sandermann H Jr., Schraudner M., Schuh G. and Wildt J (1999) Emission of volatile organic compounds from ozone-exposed plants. Ecol Applic 9: 1160-1167

Heidenreich B, Seidlitz H, Ernst D and Sandermann HJr (1999) Mercuric-ion-induced gene expression in *Arabidopsis thaliana.* International Journal of Phytoremediation 1: 153-167.

Heinecke JW, Rosen H, Suzuki LA and Chait A (1987) The role of sulfur containing amino acids in superoxide production and modification of low density lipoprotein by arterial smooth muscle cells. J. Biol. Chem. 262: 10098-10103.

Hellpointner E and Gab S (1989) Detection of methyl, hydroxymethyl and hydroxyethyl hydroperoxides in air and precipitation. Nature. 337 (6208): 631-634.

Hendricks SB and Taylorson RB (1975) Breaking of seed dormancy by catalase inhibition. Proc Natl Acad Sci USA. 72 (1): 306-309.

Hersch P and Deuringer R (1963) Galvanic monitoring of ozone in air. Analytical Chemistry. 35 (7): 897-899.

Hewitt CN and Kok GL (1991) Formation and occurrence of organic hydroperoxides in the troposphere: laboratory and field observations. J. Atmos Chem. 12 (2): 181-194.

Hewitt CN, Kok GL and Fall R (1990) Hydroperoxides in plants exposed to ozone mediate air pollution damage to alkene emitters. Nature. 344 (6261): 56-58.

Hewitt CN, Monson RK and Fall R . (1990) Isoprene emission from the grass *Arundo donox L.* are not linked to photorespiration. Plant Sci. 66: 139-144.

Hewitt CN and Terry G (1992) Understanding ozone plant chemistry. Environ Sci. Technol. 26 (10): 1890-1891.

Heyden van der D.J., Skelly J.M., Innes J.L., Hug C., Landolt W., Bleuler P., Zhang J., 1999: Ozone exposure thresholds and foliar injury on native plants of Switzerland. In: Fuhrer J., Achermann B. (eds.), Critical Levels for Ozone - Level II. Environmental Documentation 115: 77-81. Swiss Agency for Environment, Forest and Landscape, Bern, Switzerland.

Heyden van der D.J., Skelly J.M., Innes J.L., Hug C., Zhang J., Landolt W., Bleuler P., 2001: Ozone exposure thresholds and foliar injury on native plants of Switzerland. Environ. Pollut. 111: 321-331.

Hibben CR (1969) Ozone toxicity to sugar maple. Phytochemistry. 59 (4): 399.

Hibben CR and Stotzky G (1969) Effects of ozone on the germination of fungus spores. Can J. Microbiol. 15: 1187-1196.

Hicks DR (1978) Diagnosing Vegetion Injury Caused by Air Pollution. Washington, DC, USA. Environmental Protection Agency. 182 p.

Hildebrandt AG, Roots I, Tjoe M and Heinemeyer G (1978) Hydrogen peroxide in hepatic microsomes. Methods Enzymol. 52: 342-350.

Hill AC, Heggestad HE and Linzon SN (1970) Ozone. Recognition of air pollution injury to vegetation: A Pictorial Atlas. (Eds. Jacobson JS, Hill AC) pp. B1-B22. Pittsburg: Pa. Air Pollut Controll Assoc.

Hill AC, Pack MR, Treshow M, Downs RJ and Transtrum LC (1961) Plant injury induced by ozone. Phytopathology. 51: 356-368.

Hill AG and Liittlefield N (1969) Ozone. Effects on apparent photosynthesis rate of transpiration and stomatal closure in plants. Environ. Sci. Technol. 3 (1): 52-56.

Hill CA (1971) Vegetation: a sink for atmospheric pollutants. Air Pollut. Control Assog. J. 21 (6): 341-346.

Hiller W, Rosemann D, Pflanz K and Sandmann H (1990) Ozone-induction of secondary metabolism in Scots pine and Norway spruce. Bull de Liaison. 15: 104.

Hoigne J and Bader H (1983) Rate constants of reactions of ozone with organic and inorganic compounds in water.I. Non-dissociating organic compounds. Water Res. 17 (2): 173-183.

Holloway PJ (1994) Plant cuticles: physicochemical characteristics and biosynthesis. Air Pollutants and the Leaf Cuticle. (Eds. Percy KE, Cape JN, Jagels R, Simpson CJ) NATO ASI Ser., Ser. G: Ecological Sciences. Berlin Heidelbeg: Springer. 36: 1-13.

Horsman DC and Wellburn AR (1982) Appendix II. Guide to the metabolic and biochemical effects of air pollutants on higher plants. In: Effects of Air Pollutants on Plants (Ed. Mansfield TA) pp.186-199. Cambridge University Press, London, New York, Melbourne.

Houston DB (1974) Responses of selected *Pinus strobus L.* clones to fumigations with sulfur dioxide and ozone. Can. J. For Res. 4 (1): 65-68.

Houten JG (1966) Bezwaren van Luchtverontreiniging voor de bandbouw. Landbouwkd Tijdschr. 78: 2-13.

Hov O (1997) Tropospheric Ozone Research. Tropospheric Ozone in the Regional and Sub-regional Context. Berlin, Heidelberg : Springer-Verlag. 460 pp.

Hove LWA van and Bossen ME (1994) Physiological effects of five month exposure at low concentrations of O_3 and NH_3 on Douglas fir (*Pseudotsuga menziesii*). Physiologia Plantarum. 92 (1): 140-148.

Howell RK (1970) Influence of air pollution on quantities of caffeic acid isolated from leaves of *Phaseolus vulgaris*. Phytopathology. 60: 1626-1629.

Howell RK (1974) Phenols, Ozone and their involvement in pigmentation and physiology of plant injury. Air Pollution Effects on Plant Growth. (Symp. of Div. Agrcult. and Food Chem. at the 167 Meeting of Am Chem. Soc. Los Angeles, California, April 1. 1974), (Ed. Dugger M) Washington. ACS Symposium Ser. 3 Am Chem Soc. D.C. pp. 94-105.

Howell RK and Kremer DF (1973) The chemistry and physiology of pigmentation in leaves injured by air pollution. J. Environ Qual. 2: P.

Hull LA (1981) Terpene ozonolysis products. Atmospheric Biogenic Hydrocarbons. (Eds. Bufalini JJ and Ants RR) Ambient Concentrations and Atmospheric Chemistry. Minnesota. Ann Arbor Science. 2: 161-186.

Hunter FE. Scott A, Hoffsten PE, Guerra F, Weinstein J, Scneider A, Schuta B, Fink J, Ford L and Smith E (1964) Studies on the mechanism of ascorbate-induced swelling and lysis of isolated liver mitochondria. J. Biol. Chem. 239:604-613.

Hurwitz B., Pell EJ and Sherwood RT (1979) Status of coumestrol and 4'-7-dihydroxyflavone in alfalfa foliage exposed to ozone. Phytopathology. 69: 810-813.

Huttunen S (1994) Effects of air pollutants on epicuticular wax structure. Air Pollutants and the Leaf Cuticle. (Eds. Percy KE, Cape JN, Jagels R, Simpson CJ) NATO ASI Ser., Ser.G. Ecological Sciences. Berlin. Heidelberg: Springer. 36: 81-96.

Ikemeyer D, Buttner P and Barz W (1993) Seasonal changes in the activities of apoplastic, cytoplasmic, ionically and covalently bound isoperoxidases from Norway spruce (*Picea abies* (L.) Karst) needles: A comparison between three collection sites with different ambient ozone concentrations. Z. Naturforsch. C. 48 (11-12): 903-910.

Ingolds RS, Fetner RH and Eberhardt WH (1959) Determining ozone in solution. Ozone Chemistry and Technology (Advances Chemical. Ser.) 21: 102-107.

Inness MR (1994) Ozonized water as a pesticide for plants. PCT Int. Appl. WO 9400, 014 (Cl. AO1N59/00).(referat 994129c in Chem. Abstracts) 120 (9): 14 p.

Inoue M and Koyama K (1994) In vivo determination of superoxide and vitamin C radicals, using cytochrome c and superoxide dismutase derivatives. Methods in Enzymology, San Diego New York. Boston. London. Sydney. Tokyo. Toronto: Academic Press. 234 (Part D): 338-343.

Inoue M, Suzuki R, Koide T, Sakaguchi N, Ogiharo Y and Yabu Y (1994) Antioxidant, gallic acid, induced apoptosis in HL-60RG cells. Biochim. Biophys Research Communication. 204 (2): 898-904.

Iokouching Ambe G (1985) Aerosole formed from the chemical reaction of monoterpenes. Atmos Environment. 19: 1271-1276.

Ishizaki K, Sawadaishi K, Miura K and Shinriki N (1987) Effect of ozone on plasmic DNA of *Escherichia coli* in situ. Water Res. 21 (7): 823-827.

Ishizaki K, Shinriki N, Ikehata A and Ueda T (1981) Degradation of nucleic acids with ozone I degradation of nucleobases, ribonucleosides and ribonucleosides 5'monophosphates. Chem. Pharm. Bull. 29 (3): 868-872.

Isidorov VA and Zenkevich IBA (1985) Volative organic compounds in the atmosphere of forests. Atmos. Environ. 19 (1): 1-8.

Jabs T (1999) Reactive oxygen intermediates as mediators of programmed cell death in plants and animals. Biochemical Pharmacology 57 (3) : 231-245.

Jacob B and Heber U (1998) Apoplastic ascorbate does not prevent the oxidation of fluorescent amphiphilic dyes by ambient and elevated concentrations of ozone leaves. Plant Cell Physiol. 39 (3): 313-322.

Jacob DJ and Wofsi. SC (1988) Photochemistry of biogenic emission over the Amazon forest. J. Geophys Res. 95: 1477-1486.

Jacobson JS (1982) Ozone and the Growth and Productivity of agricultural crops. Effects of Gaseous air Pollution in Agriculture and Horticulture (Ed. Unsworth MH and Ormrod DP) Butterworth Sci. pp. 295-304. London.

Jacobson JS (1977 a) The effects of photochemical oxidants on vegetation VDL. 270: 163-173.

Jaenicke R, Ogren JA, Pellizzari ED and Wallace GT (1983) Organic material in the global troposphere. Rev Geophys and Space Physics. 21 (4): 921-952.

Jagels R (1994) Leaf wettability as a measure of air pollution effects Air Pollutants and the Leaf Cuticle. (Eds. Percy KE, Cape JN, Jagels R, Simpson CJ) NATO ASI Ser. Ser.G. Ecological Sciences. Berlin. Heidelbeg. Springer. 36: 97-105.

Jakomoko F, Akimoto H and Okuda M (1981) Photochemical reactivity and ozone formation in 1-olefin-nitrogen oxide-oil system. Environ. Sci. Technol. 15: 665-671.

Jeong JH, Nakamura H and Ota J (1980) Physiological studies on photochemical oxidants injury in rice plants l. Varietal difference of abscisic acid content and its relations to the resistance to ozone. Jap. J. Crop. Sci. 49: 456-460.

Jocelyn PC (1972) The Biochemistry of the SH group. New York. Academic Press. pp. 94-136.

Juhren M, Noble W and Went FW (1957) The standardization of the *Poa annua* as an indicator of smog concentrations. 1. Effects of temperature, photoperiod, and light intensity during growth of the test plants. Plant Physiol. 32: 576-586.

Juttner F (1988) Changes of monoterpene concentrations in needles of pollution injured *Picea abies* exhibiting montane yellowing. Physiol Plantarum. 72(1): 48-56.

Kaizong C, Gengsheng X., Xinmin L., Gengmei I and Yafu W (1999) Efect of hydrogen peroxide on somatic embriogenesis *of Lycium barbarum* L. Plant Science. 146 (1): 9-16.

Kanfer S and Turro NJ (1981) Reactive forms of oxygen. In: Gilbert DL (ed) Oxygen and Living Processes. An Interdisciplinary Approach. Springer-Verlag: New York, Heidelberg, Berlin. pp.47-72.

Kangasjarvi J, Talvinen J., Utriainen M., Karjalainen R. Plant defense systems induced by ozone. Plant Cell Environment. 1994. Vol.17. N 7. P.783-794.

Kanofsky JR and Sima PD (1995) Reactive absorption of ozone by aqueous biomolecule solutions implicallions for the role of sulfhydryl compounds as targets for ozone. Arch Biochem Eiophys. 316 (1): 52-62.

Karnosky DF (1976) Threshold levels for foliar injury to *Populus tremuloides* by sulfur dioxide and ozone. Can. J. For Res. 6: 166-169.

Karnaukhov VN (1978) Luminescent Spectral Analysis of Cells. Nauka, Moscow. 204 pp.

Karnaukhov VN (1988) Biological Functions of Carotenoids. Nauka, Moscow. 240 pp
Karnaukhov VN (1990) Carotenoids: Progress, problems and prospects. Copm Biochem Physiol B 95 (1): 1-20.
Karnaukhov VN (2001) Spectral Analysis in Cell-level Monitoring of Environmental State. Moscow, Nauka, 186 pp.
Kangasjärvi, J., Tuomainen, J., Betz, C., Ernst, D., Langebartels, C. and Sandermann, H. Jr. 1997. Ethylene synthesis in tomato plants exposed to ozone - The role of ethylene in ozone damage. In: Agellos Kanellis, Caren Chang, Hans Kende, Don Grierson (Eds.) The Biology and Biotechnology of the Plant Hormone Ethylene. Kluwer Academic Publishers. 259-265 .
Kettunen, R., Overmyer, K. and Kangasjärvi, J. 1999. The role of ethylene in the formation of cell damage during ozone stress. Does ozone induced cell death require concomitant AOS and ethylene production? In.: Kanellis, A., Chang, C., Klee, H., Bleecker, A.B., Pech, J.C. & Grierson, D. (Eds), Biology and Biotechnology of the Plant Hormone Ethylene II. Kluwer, Dortrecht, 209-305.
Kiiskinen, M., Korhonen, M. and Kangasjärvi, J. 1997. Isolation of and characterization of cDNA for a plant mitochondrial phosphate translocator (*Mpt1*). Ozone stress induces Mpt1 mRNA accumulation in birch (*Betula pendula* Roth). *Plant Mol Biol*, **35**:271-279.
Korhonen, M. S., R. Pellinen, M. Kiiskinen and J.S. Kangasjärvi. 1998. Cellular-molecular diagnostics of ozone injury in birch (*Betula pendula* Roth). In: De Kok, L.J. (Ed.), Responses of Plant Metabolism to Air Pollution and Global Change, pp..351-354. SPB Academic Publishing, The Hague, The Netherlands.
Kawano T and Muto S (2000) Mechanism of peroxidase actions for salicylic acid-induced generation of active oxygen species and an increase in cytosolic calcium in tobacco cell suspension culture. J Expt Bot 51: 685-693.
Karol' IL (1992) Ozonic Shield of Earth and Human. Hydrometizdat, Sankt-Peterburg.
Kayushin LP, Pulatova MK and Krivenko VG (1976) Free Radicals and Their Transformations in Irradiated Proteins. Moscow: Atomizdat.
Keen NT and Taylor OC (1975) Ozone injury in soybeans. Isoflavonoid accumulation is related to necrosis. Plant Physiol. 55 (4): 731-733.
Keller Th(1974) The use of peroxidase activity for monitoring and mapping air pollution areas. Eur J. For Pathol. 4 (1): 11-19.
Kerner K, Langebartels C, Meyer K and Sandermann HJr (1987) Ozone - induced changes in polyamines and enzymes in plants and callus cultures of tobacco. Abstracts of 14th lnt Bot. Congress, 1987. Berlin West, Dahlem, 24-July- I August. P. 81.
Kerr RA (1993) Ozone take a nose after the eruption of Mt. Pinatubo. Science. 260 (5107): 490-491.
Kerstiens G (1994) Air pollutants and plant cuticles: mechanisms of gas and water transport, effects on water permeability. Air Pollutants and the Leaf Cuticle. (Eds. Percy KE, Cape JN, Jagels R, Simpson CJ) NATO ASI Ser.G: Ecological Sciences. Berlin. Heidelberg. Springer. 36: 39-53.
Kerstiens G and Lendzian KJ (1989) Interactions between ozone and plant cuticles. 1. Ozone deposition and permeability. New Phytol. 112 (1): 13-19.
Keutsch F (1971) Chromosomen mutationen durch ein organisches Peroxyd. Ber. Schweiz Bot. Ges. 81: 180-185.
Khalil MAK (Ed) (1999) Atmospheric Methane. Berlin, Heidelberg. 335 pp.
Kikugawa K and Beppu M (1987) Involvement of lipid oxidation products in the formation of fluoresent cross-liked proteins. Chem. Ph. Phys. Lipids. 44: 277-296.
Kirtikara K and Talbot D (1990) Biochemical antioxidants are induced in plant tissue responding to oxidative stress. J. Cell Biol. 111 (5 Part. 2): 105.
Kisaki T (1973) Mixed function of oxidase inhibitors protect plants from ozone injury. Arris. Biol. Chem. 37: 2449-2450.
Kley D, Kleinman M, Sandermann H Jr, and Krupa S (1999) Photochemical oxidants: State of the science. Environ Pollut 100: 19-24.
Knogqe W, Kombrink E, Schmelzer E and Hahlrock (1987) Occurrence of phytoalexins and other putative defense-related substances in uninfected parsley plants. Planta. 171: 279-287.
Knox JP and Dodge AD (1985 b) Isolation and activity of photodynamic pigment hypericin. Plant Cell Environ. 8 (1): 19-25.
Knox JP and Dodge AD (1985 a) Singlet oxygen and plants. Phytochemistry. 24 (5): 889-896.
Knudson LL, Tibbits TW and Edwards GE (1977) Measurement of ozone injury by determination of leaf chlorophyll concentration. Plant Physiol. 60: 606-608.

Koch JR, Creelman RA, Eshita SM, Seskar M, Mullet JE and Davis KR (2000) Ozone sensitivity in hybrid poplar correlates with insensitivity to both salicylic and jasmonic acid. The role of programmed cell death in lesion formation. Plant Physiol 123: 487-496.

Kochhar M, Blum U and Reinert RA (1980) Effect of O_3 and (or) fescue on on ladino clover: interactions. Can J. Bot. 58: 241-249.

Kogan FKh, Grachev SV and Eliseeva SV (1998) Carbon dioxide - natural regulator of the free radical homeostasis at the different steps of evolution. Doklady of Russian Academy of Sciences 362 (5): 705-708.

Kohler A (1978) Wasserpflanzen als Bioindikatoren. Bei Veroff Naturschutz Landschaftspfl Baden-Wurttemberg. 11: 259-281.

Kohut RJ, Laurenoe JA and Amundson RG (1980) Effect of ozone and sulfur dioxide on yield of red clover and timothy. J. Environ Qual. 17: 580-585.

Koiwai A and Kisaki T (1973) Mixed function of oxidase inhibitors protect plants from ozone injury. Agr. Biol. Chem. 37 (10): 2449-2450.

Kolattukudy PE (1987) Lipid-derived defensive polymers and waxes and the role in plant-microbe interaction. Biochemistry of Plants. Lipids: Structure and function. Orlando etc: Acad. Press. 9: 291-347.

Kolattukudy P and Soliday CL (1986) Effects of stress on the defensive barriers of plants. Cellular and Molecular Biology of Plant Stress. (Ed. JL Key, T Kosuge) pp. 381-400. New York: Liss.

Koning HWDe and Jegier Z (1968 a) A study of the effects of ozone and sulfur dioxide on the photosynthesis and respiration of *Euglena gracilis*. Atmos Environ. 2: 321-326.

Koning HWDe and Jegier Z(1968 b) Quantitative relation between ozone concentration and reduction of photosynthesis of *Euglena gracilis*. Atmos Environ. 2: 615-616.

Konkol J and Dugger WM Jr (1967) Anthocyanin formation as a response to ozone and Smog treatment in *Rumex crispus* L. Plant Physiol. 42: 1023-1024.

Konovalov DA (1995) Natural azulenes. Rastitelnye Resursi. 31: 101 - 132.

Kooten van O and van Howe LWA (1988) Fluorescence as a mean of diagnosing the effect of polutant induced stress in plants. Air pollution and Ecosystems (Proc. Int Symp., 18-22 May 1987, Grenoble, France). (Ed. by P Mathy). Dordrecht. Boston. Lancaster. Tokyo. D. Reidel Publ. Company (Member of the Kluwer Acad. Publ. Group. pp.596-601.

Koppenol WH (1983) Solvation of the superoxide anion. Oxy Radicals and Their Scavenger Systems. (Eds. Cohen G, Greenwald RA). New York. Elsevier Biomedical. 1: 274-277.

Korhonen MS, Pellinen R., Kiiskinen M. And Kangasjärvi JS (1998) Cellular-molecular diagnostics of ozone injury in birch (*Betula pendula* Roth). In: De Kok (ed) Responses of Plant Metabolism to Air Pollution and Global Change,pp. 351-354. SPB Academic Publishing, The Hague.

Kostkarick R and Manning WJ (1992) Effects and interaction of ozone and the antiozonant EDU at different stages of radish (*Raphanus sativus L.*) development. J. Exp. Bot. 43 (257): 1621-1632.

Koyama T (1963) Gaseous metabolism in Lake sediments and paddy soils and production of atmospheric methane and hydrogen. J. Geophys. Res. 63: 3971-3983.

Koziol MJ and Whatley FR (Eds.) (1984) Caseous Air Pollutants and Plant Metabolism. London. Worth. Butterworth. 466 p.

Koziol MJ. Whatley FR and Shelvey JD (1988) An integrated view on the effects of gaseous air pollutants on plant carbohydrate metabolism. Air Pollution and Plant Metabolism. (Eds. Schulte-Hostede S, Darrall NM, Blank LW, Wellburn AR) London. New York: Elsevier Applied Science. pp.148-168.

Kraft M, Weigel HJ, Mejer GJ and Brandes F (1996) Reflectance measurements of leaves for detecting visible and non-visible ozone damage to crops J. Plant Physiol. 148 (1-2): 148-154.

Krause CR (1994) Recent advances using electron beam analysis to detect cuticular changes induced by air pollution. Air Pollutants and the Leaf Cuticle. (Eds. Percy KE, Cape JN, Jagels R, Simpson CJ) NATO ASI Ser. Ser.G: Ecological Sciences. Berlin. Heidelberg. Springer. 36: 329-339.

Krinsky NJ (1979) Biological roles of singlet oxygen. Singlet Oxygen. (Eds. HH Wasserman, RW Murray) pp.597-641. New York. Academic Press.

Krinsky NJ (1966) The role of carotenoid pigments as protective agents against photosensitized oxidation in chloroplast. In: Biochemistry of Chloroplast. (Ed. TW Goodwin) pp.423-430. New York. London. Academic Press.

Krippeit-Drews P, Haberland C, Fingerle J, Drews G and Lang F (1995) Effects of H_2O_2 on membrane potential and $[Ca^{2+}]i$ of cultured rat arterial smooth muscle cells Biochem. Biophys. Res. Cmmun. 209: 139-145.

Krivopishin IP (1973) Effect of ozone on shell microflora and result Chicken egg incubation. Tr. Vses. Nauchno-issled. Tekhnol. inst. Ptisevod. 37: 39-44.

Kronfuss G, Wiesser G, Havranek WM and Polle A (1996) Effects of ozone and mild drought stress on total and apoplastic guaiacol peroxidase and lipid peroxidation in spruce (*Picea abies* L., Karst). J. Plant Physiol. 148 (1-2): 203-206.

Kull O and Moldau H (1994) Absorption of ozone on *Betula pendula* Roth. leaf surface. Water, Air, Soil Pollut. 75 (1-2): 79-86.

Kuo SS, Saad AH, Koong AC, Hahn GM and Giaccia AJ (1993) Potassium-channel activation in response to low doses of γ-irradiation involves reactive oxygen intermediates in nonexcitatory cells. Proc. Natl. Acad. Sci. USA. 9: 908-912.

Kurchii BA (1990) Possible reactions of oxidation and formation of free radicals of ethylene. Fiziologia i Biokhimiya Kulturnikh Rastenii (Physiology and Biochemistry of Cultivated Plants, Russia) 22 (5): 445-454.

Kuzin AM (1995) Ideas of the Radiation Hormesis in Atomic Century. Nauka, Moskow,158 pp.

Kuzin AM (1997) Secondary Biogenic Radiations-Rays of Life. Biological center RAS, Pushchino, 40pp

Labas YuA, Peskin AV, Klebanov GI, Popov Syu and Tikhonov A.V. (1999) Generation of active oxygen species on the surface of water-living organisms. In: Materials of 2-d All-Russian Meeting of Biophysists. ONTI: Moscow. 3: 1046-1047.

Labitzke KG and van Loon H (1999) The Stratosphere. Phenomena, History, and Relevance. Berlin, Heidelberg. 179 pp.

Lacasse NL and Moroz WJ (1969) Handbook of Effects Assessment. Vegetation Damage Center for Air Environment Studies. University Park: Penn. State Univ. 193 p.

Lacasse NL and Treshow M (Eds.) (1978) Diagnosing Vegetation Injury Caused by Air Pollution. Air Pollution Training Institute, North Carolina: US Environmental Protection Agency: Research Triangle Park, Handbook.

Laisk AM, Kull O and Moldau H (1989) Ozone concentration in leaf intercellular air spaces is to close to zero. Plant Physiol. 90 (3): 1163-1167.

Lamb C and Dixon RA (1997) The oxidative burst in plant disease resistance. Annu Rev Plant Physiol Plant Mol Biol 48: 251-275

Landry LG and Pell EJ (1993) Modification of Rubisco and altered proteolytic activity in O_3 -stressed hybrid poplar (*Populus maximowizii x trichocarpa*). Plant Physiol. 101 (4): 1355-1362

Lange OL, Heber U, Schulze ED and Ziegler H (1989) Atmospheric pollutants and plant metabolism. Forest Decline and Air Pollution. A Study of Spruce (*Picea abies*) on Acidic Soils. /Eds. Schulze ED, Lange OL, Oren S (Ecological Studies, vol.77). pp. 238-273. Berlin, Heidelberg: Springer-Verlag.

Langebartels C, Ernst D, Kangasjarvi J and Sandermann H Jr (2000) Ozone effects on plant defense. Methods Enzymol 319: 520-535.

Langebartels C, Heller W, Kerner K, Leonardi S, Rosemann D, Schraunder M, Trost M and Sandermann H (1990) Ozone – induced defense reactions in plant. Environmental Research with Plants in Closed Chambers. (Eds. HD Payer, TR Pfirrmann, P Mathy) Air Pollution Research Reports of the E.C. 26: 358-368.

Langebartels C, Kerner K, Leonardi S, Schrauder M, Trost M, Heller W and Sandermann HJr (1991) Biochemical plant responses to ozone. I. Differential induction of polyamine and ethylene biosynthesis in tobacco. Plant Physiol. 95 (3): 882-889.

Larson RA (1987) Environmental chemistry of reactive oxygen species. CRC Critical Reviews in Environmental Control. 8 (3): 197-246.

Larson RA (1997) Naturally Occurring Antioxidants. Boca Raton, New York: CRC Press. 225 pp.

Laurence JA and Weinstein LH (1981) Effects of air pollutants on plant productivity. Annu Rev. Phytophatol 19: 257-271.

Lavola A, Julkunen Titto R and Paakkonen E (1994) Does ozone stress change the primary metabolites of birch *(Betula pendula Roth*). New Phytol. 126 (4): 637-642.

Le Bras G (Ed) (1997) Chemical Processes in Atmospheric Oxidation. Laboratory Studies of chemistry related to Tropospheric Ozone. Berlin Heidelberg: Springer. 314 pp.

Ledbetter MC, Zimmerman PW and Hitchcock AE (1959) The histopathological effects of ozone on plant foliage. Contrib Boyce Thomson Inst 20 (4): 275-282.

Lee MH and Dawson CR (1973) Ascorbate oxidase. :further studies on the purification of the enzyme. J .Biol.Chem. 248: 6596

Lee TT (1967) Inhibition of oxidative phosphorylation and respiration by ozone in tobacco mitochondria Plant Physiol. 42: 691-696.

Lee TT (1968) Effect of ozone on swelling of tobacco mitochondria. Plant Physiol. 43 (2): 133-139.

Leffler HR and Cherry JH (1974) Destruction of enzymatic activities of corn and soybean leaves exposed to ozone. Can. J. Bot. 43: 677-685.

Legge RL, Thompson JE and Barker JA (1982) Free radicals mediated formation of ethylene from 1-aminocyclopropane-1-carboxylic acid: a spin trap study. Plant Cell Physiol. 23 (2): 171-177.

Lehnherr B, Ggrandjean A, Machler F and Fuhrer J (1987) The effect of ozone in ambient air on ribulose bisphosphate carboxylase/oxygenase activity decreases photosynthesis and grain yield in wheat. J Plant Physiol. 130: 189-200.

Lendzian KJ (1982) Gas permeability of plant cuticles. Planta. 155 (4): 310-315

Lendzian KJ (1984) Permeability of plant cuticles to gaseous air pollutants. Gaseous Air Pollutants and Plant Metabolism. /Eds. Koziol MJ, Whatley FR. pp. 77-81. London, Boston: Butterworths.

Lendzian KJ and Kerstiens J (1988) Interactions between plant cuticles an gaseous air pollutants. Aspects App. Biol. 17 (1): 97-104.

Lendzian KJ and Kerstiens J (1991) Sorption and transport of gases and vapour in plant cuticles. Rev. Environ. Contam. Toxicol. 121 (1): 65-128.

Leshem JJ (1984) Interaction of cytokinins with lipid-associated oxy free radical during; senescence: a prospective mode of cytokinin action. Can. J. Bot. 62 (12): 2943-2949.

Leshem JJ, Halevy AH and Frenkel C (1986) Processes and control of plant senescence. Amsterdam-Oxford-New York. -Tokyo: Elsever. 215 p.

Leshem YY, Shewfelt RL, Willmer CM and Pantoja O (1992) Plant Membranes. A biophysical approach to structure development and senescence. Dordrecht. Boston. London. Kluwer Academic Publishers. 266 p.

Leshem YY, Wurzburger J, Grossman S and Frimer AA (1981) Gytokinin interaction with free radical metabolism and senescence: effects on endogenous lipoxygenase and purine oxidation. Physiologia Plantarum. 53 (1): 9-12.

Leshem YY, Wurzburger Y, Frimer AA, Barness G and Ferqusson IB (1982) Calcium and calmodulin metabolism in senescence: interaction of lipoxygenase and superoxide dismutase with ethylene and cytokinin. Plant Growth Substances. (Ed. Wareing PF). pp. 569-578. London, New York, Paris, San Diego, San Francisco, San Paulo, Sydney, Tokyo, Toronto: Academic Press.

Levitt J (1972) Responses of plants to environmental stress. New York: Acad. Press. 697 p.

Lewis E and Brennan EJ (1977) Of Air Pollution Control Association. 27: 889-891.

Linzon SN Effects of airborne sulphur pollutants of plants. Sulphur in the Environment. Part II: Ecological impacts (Ed. JG Nriaigu) Chichester: John Willey and Sons. pp.110-162.

Linzon SN, Heck WW and MacDowal FDH (1975) Effects of photochemical oxidants on vegetation. Photochemical Air Pollution: Formation, Transport, Effects. Canada: Nat. Res. Coune. pp. 89-142.

Lippmann M (1991) Health effects of tropospheric ozone. Environmental Sci. and Technol. 25 (12): 1954-1962.

Liu DHF and Liptak BG (Eds.) (2000) Air Pollution. Boca Raton: Lewis Publishers, a division of CRC Press. 254 pp.

Longstreth J, de Gruijl FR, Kripke ML, Abseck S, Arnold F, Slaper HI, Velders G, Takizawa Y and van der Leun JC (1998) Health risks. J. Photochemistry and Photo biology. 46 (1-3): 20-39

Losel DM (1978) Lipid metabolism of leaves of *Poa pratensis* during infection by *Puccinia poarum*. New Phytol. 80: 2167-2174.

Loveless AL (1951) Qualitative aspects of the chemistry and biology of radiomimetic (mutagenic) substances. Nature. 167 (4244): 338-342.

Luwe MWF, Takahama U and Heber U (1993) Role of ascorbate in detoxifying ozone in the apoplast of spinach (*Spinacia oleracea* L.) leaves. Plant Physiology. 101 (5): 969-976.

Lichtentaler HK (ed) (1988) Application of Chlorophyll Fluorescence. Dordrecht Boston London: Kluwer Academic Publishers.

Maccarrone M, Veldink GA, Vliegenthart JFG and Finazzi-Argo A (1997) Ozone stress modulates amine oxidase and lipoxygenase expression in lentil (*Lens culinaris*) seedlings. FEBS Letters. 408 (2): 214-244.

MacDowall FDH (1965) Stages of ozone damage to respiration of tobacco leaves. Can. J. Bot. 43 (2): 419-427.
Maggiolo A and Niegowski SJ (1959) Preparation of tertiary amine oxides by ozonation. In: Ozone Chemistry and Technology (Adv Chem Ser, Vol.21) pp.202-204.
Mahmoud GS and Melo TB (1991) The effect of the antioxidant spermine on the photophysical processes and photodegradation of tryptophan in aqueous solution at 293 °K. J Photochem Photobiol 9 (B): 355-367.
Malhotra SS and Khan AA (1985) Biochemical and physiological action of main pollutants. In: Air Pollution and Plant Life (Ed. Treshow M).pp.144-189.
Mandronich S, McKenzie. RL, Bjorn LO and Caldwell MM (1998) Changes in biologically active ultraviolet radiation reaching the Earth surface. J. Photochemistry and Photobiology. 46 (1-3): 5-19
Manes F, Altieri A, Tripodo P, Booth CE and Unsworth MH (1990) Bioindication study of effect of ambient ozone on tobacco and radish plants using a protectant chemical (EDU). Ann Bot. 48 (1): 135-149.
Manning WJ (1975) Interactions between air pollutant and fungal, bacterial, and viral plant pathogenes. Environ Pollut. 9 (1): 87-90.
Manning WJ and Feder WA (1980) Biomonitoring Air Pollutants with Plants. London: Applied Science Publishers. 144 p.
Manning WJ, Feder WA, Papia PM and Perkins I (1971) Influence of foliar zone injury on root development and root surface fungi pinto bean plants. Environ Pollut. 1 (4): 305-311.
Manning WJ, Feder WA, Perkins I an Glickman M (1969) Ozone injury and infection of potato leaves by *Botrytis cinerea*. Plant Dis Rep. 53: 691-693.
Manning WJ and von Tiedemann A (1995) Climate change: Potential effects of increased atmospheric carbon dioxide (CO_2), ozone (O_3), and ultraviolet-BV(UV-B) radiation on plant diseases. Environ Pollut 88: 219-245.
Mansfield TA, Pearson M, Atkinson CJ and Wookey PA (1993) Ozone, sulphur dioxide and nitrogen oxides: some effects on the water relations of herbaceous plants and trees. In: Interacting Stresses on Plants in a Changing Climate Eds. Jacson MB and Black CR (NATO ASI SER. ser. I: Global Environmental Change) Berlin: Springer. 16: 77-88.
Maquire YP and Haard NF (1976) Fluorescent product accumulation in ripening fruit. Nature. (London.) 258 (556): 599-600.
Marclund S (1973) Mechanism of the irreversible inactivation of horseradish peroxidase caused by hydroxymethylhydroperoxide. Arch. Biochem. Biophys. 154 (2): 614-622.
Marklund S (1976) Spectrophotometric study of spontaneous disproportionation of superoxide anion radical and sensitive direct assay for superoxide dismutase. J. Biol. Chem. 251 (23): 7504-7507.
Markowski A and Grzesiak S (1974) Influence of sulphur dioxide and ozone on vegetation of bean and barley plants under different soil moisture conditions. Bul. Acad. Pol. Sci. Biol. 22: 875-888.
Martinez RI, Herron JT and Huie RE (1981) The mechanism of ozone-alkene reactions in the gas phase. A mass spectrometric study of the reactions of eight linear and branched-chain alkenes. J. Am. Chem. Soc. 103 (13): 3807-3820.
Martins EAL and Meneghini R (1990) DNA damage and lethal effects on hydrogen peroxide and menadione in chinese hamster cells: distinct mechanisms are involved. Free radical Biology and Medicine. 8 (5): 433-440.
Maskos Z and Koppenol WH (1991) Oxyradicals and multivitamin tablets. Free Radical Biology and Medicine. 11 (6): 609-610.
Matsunaka S (1973) The utilization of biological indicators in the monitoring systems for photochemical air pollution control. References on photochemical air pollution in Japan. Air Qual Bur. Environ Agency. pp.282-297.
Matushima J (1971) On composite harm to plants by sulphurous acid gas and oxidant. Industrial Public Damage. 7: 218-224.
Matyssek R, Gunthardt-Goerg MS, Keller T and Scheidegger C (1991) Impairrment of the gas exchange and structure in birch leaves (*Betula pendula*) under low ozone concentrations. Trees. 5 (1): 5-13.
Matyssek R, Gunthardt-Goerg MS, Saurer M and Keller T (1992) Seasonal growth, δ^{13} of leaves and stem, and phloem structure in birch (*Betula pendula*) under low ozone concentrations. Tree. 6 (1): 69-76.

Matyssek R, Keller T and Koike T (1993) Branch growth and leaf gas exchange *of Populus tremula* exposed to low ozone concentrations throughout two growing seasons. Swiss Federal Inst. of Forest. (1992) 79 (1): 1-7.

McCool RM, Menge JA and Taylor OC (1979) Effects of ozone and HCl gas on the mycorrhizal fungus *Glomus fasciculatus* and growth of "troyler" Citrange. J. Am. Soc. Hortic Sci. 104: 151-154.

McCool RM, Menge JA and Taylor OC (1977) Influence of ozone and HCl gas on citrus and the vesicular arbuscular mycorrhizae fungus *Glomus fasciculatus*. Proc. Am Phytopathol. Soc. 4: 57.

McCord JM, Beauchamp CO, Goscin S, Misra HP and Fridovich I (1973) Superoxide and Superoxide Dismutase, Oxidases and Related Redox System (Eds. King TE, Mason S and Morrison M). Baltimore: University Press. 1:

McQuattie CJ, Rebbeck J (1994) Effect of ozone and elevated carbon dioxide on cuticular membrane. Ultrastructure of yellow poplar (*Liriodendron tulipifera*) In: Air Pollutants and the Leaf Cuticle. (Eds. Percy KE, Cape JN, Jagels R and Simpson CJ) NATO ASI Ser., Ser. G: Ecological Sciences. Berlin. Heidelberg: Springer. 36: 175-182.

McKee J. (1994) Tropspheric Ozone: Human Health & Agricultural Impacts. Boca Raton, CRC Press Inc.

Mead JF (1984) Free radical mechanisms in lipid peroxidation and prostaglandins. Free radicals in Molecular Biology, Aging and Disease. (Eds. D Armstrong et al.) New York: Raven Press. pp. 53-66.

Mehlhorn H (1990) Ozone/hydrocarbon toxicity and cellular defence mechanisms. Physiologia Plantarum. 79 (2 Part 2): 122.

Mehlhorn H, Cottam DA, Lucas PW and Wellburn AR (1987 a) Induction of ascorbate peroxidase and glutathione reductase activities by interactions of mixture of air pollutants. Free Rad. Res. Commun. 3: 193-197.

Mehlhorn H, O'Shea JM and Wellburn AR (1991) Atmospheric ozone interacts with stress ethylene formation by plants to cause visible plant injury. J. Exp. Bot. 42 (234): 17-24.

Mehlhorn H, Tabner BJ and Wellburn AR (1990) Electron spin resonance evidence for the formation of free radicals in plants exposed to ozone. Physiologia Plantarum. 79 (2): 377-383.

Mehlhorn H and Wellburn AR (1987) Stress ethylene formation determines plant sensitivity to ozone. Nature. 327 (6121): 417-418.

Mehlhorn H and Wellburn AR (1994) Man-induced causes of free radical damage to plants: O_3 and other gaseous pollutants. In: Causes of Photooxidative Stress and Amelioration of Defense Systems in Plants. (Eds. CH Foyer and Mullineaux PM). Boca Raton, Ann Arbor. pp. 155-175. London, Tokyo: CRC Press.

Mehlhorn H and Wenzel AA (1996) Manganese deficiency enhances ozone toxicity in bush beans (*Phaseolus vulgaris* L. cv. Saxa) J. Plant Physiol. 148 (1-2): 155-159.

Mehlhorn H, O'Shea J and, Wellburn AR (1991) Atmospheric ozone interacts with stress ethylene formation by plants to cause visible plant injury. J Exp Bot. 42 (234): 17-24.

Mehlhorn H, Tabner BJ and Wellburn AR (1990) Electron spin resonance evidence for the formation of free radicals in plants exposed to ozone Physiologia Plantarum. 79 (2 part 1): 377-383

Meinander O, Somersalo S, Holopainen T and Strasser RJ (1996) Scots pine after exposure to elevated ozone and carbon dioxide probed by reflectance spectra and chlorophyll a fluorescence transients J. Plant Physiol. 148 (1-2): 226-236.

Menser HA, Chaplin JF, Gheng AL and Sorokin T (1977) Polyphenols, polysterols and reducing sugars in air-cured tobacco leaves injured by ozone air pollution. Tobacco Sci. 21 (1): 35-38.

Menser HA and Heggestad HE (1966) Ozone and sulfur dioxide synergism: Injury to tobacco plants. Science. 153: 424-425.

Menser HK and Street OE (1962) Effects of air pollution, nitrogen levels supplemental irrigation and plant spacing on weather fluck and leaf losses of Maryland tobacco. Tobacco Sci. 6 (2): 165-169.

Men'shchikova EB and Zenkov RK (1993) Antioxidants and inhibitors of radical oxidative processes. Uspekhi Sovremennoi Biologii (Advantages in Modern Biology, Russia) 113: 442-255.

Menzel DB (1971) Oxidation of biologically active reducing substances by ozone. Arch Environ Health. 23 (2): 149-153. Kagan BE, Orlov ON and Prilipko L.L (1986) Problems of Analysis of Endogenic Products of Lipid Peroxidation. ITOGI NAUKI iTEKHNIKI. Ser. Biophysics, Vol 18.. VINITI, Moscow, 135 pp.

Merzlyak MN (1988) Lipo-soluble fluorescent "aging pigments" in plants. Lipofuscin - 1987: State of the Art/Ed. Zs.-Nagy. I./ pp. 185-186. Academiai Kiado/Elsevier: Budapest/Amsterdam..

Merzlyak MN (1989) Active Oxygen and Oxidative Processes in Membranes of Plant Cell. ITOGI NAUKI I TEKHNIKI, Ser. Plant Physiology, Vol.6. VINITI, Moscow. 168 pp.
Merzlyak MN and Zhirov BK (1990) Free radical oxidation in chloroplasts at the ageing of plants. ITOGI NAUKI I TEKHNIKI, Ser.Biophysics, Vol.40.pp.101--125. VINITI, Moscow.
Merzlyak MN and Pogosyan SI (1988) Oxygen Radicals and lipid peroxidation in plant cell. In: OxygenRadicals in Chemistry, Biology and Medicine.(Ed. Afanas'ev IB).pp.232-253. Medicinal Institute, Piga.
Merzlyak MN, Plakunova OV, Gostimsky SA, Rumyantseva VB and Kovak K (1984) Lipid peroxidation in and photodamage to a light-sensitive chlorescence pea mutant. Physiologica Plantarum. 62 (2): 329-334.
Middelton JT, Kendrick JBJr and Schwalm HW (1950) Injury to herbaceous plants by smog or air pollution. Plant Dis. Rep. 34: 245-252.
Mihalatos AM and Calokerinos AC (1995) Ozone chemiluminescence in environmental analysis. Analytica Chimica Acta. 303 (1): 127-135.
Mik Gde and Groot De (1973) The survival of *Escherichia coli* in the open air in open air in different parts of the Netherlands. Airborne Transmission and Airbone Infection. Concepts and methoda. Pres 6th Int. Symp. Aerobiol. Enschede Netherlands. pp.155-158. New York. Toronto. Halsted Press. Wiley.
Milas NA and Nolan JTJr (1959) Some abnormal ozonation reactions. In: Ozone Chemistry and Technology (Adv Chem Ser, Vol. 21) pp.136-139.
Miller JP (1987) Effects of ozone and sulfur dioxide stress on growth and carbon allocation in plants. Phytochemical Effects on Environmental Compounds (Eds. Saunders IA. Kosak-Channing L, Conn EE) Recent Advances is Phytochemistry. New York. London. Plenum press. 21: 55-100.
Miller PR (1973) Oxidant-induced community change in a mixed conifer forest. Air Pollution Damage to Vegetation. (Ed. Maegele JA) Adv. Chem. Ser. 122: 101-117.
Miller PR, Parmeter IE, Flick BH and Martinez CW (1969) Ozone dose response of ponderosa pine seedlings. J. Air Pollut. Cont. Assoc. 19: 435-438.
Miller PR and Rich S (1968) Ozone damage on apples. Plant Dis. Rep. 52: 730-731.
Minibayeva FV, Kolesnikov OP and Gordon LK (1998) Contribution of a plasma membrane redox system to superoxide production by wheat root cells. Protoplasma 205: 101-106.
Minibayeva FV, Rakhmatullina DF, Gordon LK and Vylegzhanina NN (1997) Role of superoxide in the formation of unspecific adaptive syndrome of root cells. Dokl. Akad. Nauk Russ. 355: 554-556
Mitchell A, Ormrod DP and Dietrich HF (1979) Ozone and nickel effects on pea leaf cell ultrastructure. Bull Environ Contamin Toxicol. 22: 379-385.
Miyake H, Matsumura H, Fujinuma Y et al. (l989) Effects of low concentrations of ozone on the fine structure of radish leaves. New Phytol. 111 (2): 187.
Mizun YuG (1993) Ozonic Holes. Myth and Reality. Misl', Moscow. 288 pp.
Moldau H, Sober J, Karolin A and Kallis A (1990) CO_2 uptake and respiration losses in vegetative bean plants due to ozone absorption. Photosynthetica. 25 (3): 341-349.
Montillet JL and Degousee N (1991) Hydroperoxides induce glyceollin accumulation in soybean. Plant. Physiol. Biochem. 29 (6): 689-694.
Mukherjee SP and Choudhuri MA(1983) Implications of water stress-induced changes in the levels of endogenous ascorbic acid and hydrogen peroxide in *Vigna* seedlings. *Physiologia Plantarum*. 58 (2): 166-170
Mori IC, Pinontoan R, Kawano T and Muto S (2001) Involvement of superoxide generation in salicylic acid-induced stomatal closure in *Vicia faba*. Plant Cell Physiol 42 (12): 1383-1388.
Mudd JB (1973) Biochemical effects of some air pollutants on plants. Air pollution Damage to Vegetation. (Er. Naegele JA) (Advances in Chemistry Series. vol. 122.) pp.31-47. Washington: Amer. Chem. Soc.
Mudd JB (1982) Effects of oxidants on metabolic functions. Effects of Gaseous air pollution in Agriculture and Horticulture. (Eds. Unsworth MN and Ormrod DP) pp. 189-203. London: Butterworth Scientific.
Mudd JB (1980) Physiological and biochemical effects of ozone and sulfur doixide. United Nations. Symposium on the Effects of Airborne Pollution on Vegetation. pp. 80-89. Warszawa. Poland.
Mudd JB (1982) Effects of oxidants on matabolic functions. Effects of Gaseous air pollution in Agriculture and Horticulture. /Eds. Unsworth MN, Ormrod DP. pp. 52-208. London: Butterworth Scientific.

Mudd JB (1997) Biochemical basis for the toxicity of ozone. In: Yunus M and Iqbal M (eds) Plants Responce to Air Pollution. Wiley, New York. pp.267-284.
Mudd JB and Freeman BA (1977) Reaction of ozone with biological membrane. Biochemical events of environmental pollutants. /Ed. Lee S. Ann Arbor. Mich. Ann Arbor Science Pub. pp. 97-133.
Mudd JB, Banerjee SK, Doolley MM and Knight KL (1984) Pollutants and plant cells: effects on membranes Gaseous Air Pollutants and Plant Metabolism/Eds. Koziol MJ, Whatley FR. pp. 105-116. London, Boston: Butterworths.
Mudd JB, Leavitt R, Ongun A and McManus TT (1969) Reaction of ozone with amino acids and proteins. Atmos Environ.3: 669-681.
Mudd JB, Leh F and McManus TT (1974) Reaction of ozone with nicotinamide and its derivatives. Arch. Biochem. Biophys. 161 (2): 408-419.
Mudd JB, McManus TT and Ongun A (1971) Inhibition of glycolipid biosynthesis in chloroplasts by ozone and sulfhydryl reagents. Plant Physiol. 48 (3): 335-339.
Mudd JB, McManus TT and Ongun A (1971 a) Inhibition of lipid metabolism in chloroplasts by ozone. Proc. 2nd Int. Clean Air Congr. 1970. (Ed. HM England and WT Beery) pp.256-260. New York: Academic Press.
Müller M, Kohler B, Tausz M, Grill D and Lutz C (1996) The assesment of ozone stress by recording chromosomal abberations in root tips of spruce trees (Picea abies (L.) Karst) J. Plant Physiol. 148 (1-2): 160-165.
Mueller S, Riedel HD and Stremmel W (1997) Determination of catalase activity at physiological hydrogen peroxide concentrations. Anal.Biochem. 245 (1): 55-60.
Mumford RA, Lipke H, Laufer DA and Feder WA (1972) Ozone-induced changes in corn pollen. Environ. Sci. Technol. 6 (3): 427-430.
Mura C and Chung YS (1990) In vitro transcription assay of ozonated T 7 Phage DNA. Enviriomental and Molecular Mutagenesis. 16 (1): 44-47.
Murray RW (1979) Chemical sourses of singlet oxygen. Singlet Oxigen. (Ed. HH Wasserman and RW Murray). pp. 59-114. New York: Academic Press.
Musselman CR, Lefohn SA, and McCool MP (1992) Ozone descriptors for in air quality standard to protect vegetation . Journal of Air & Waste Management Assoc. 44: 1383-1390.

Musselman RC and Hale BA (1997) Methods for controlled and field ozone exposures of forest tree species in North America. Forest Decline and Ozone. A Comparison of Controlled Chamber and Field Experiments /Eds. Sandermann H, Wellburn AR and Heath RL (Ecological Studies, vol. 127) pp. 277-315. Berlin, Heidelberg: Springer-Verlag.
Mustafa M (1990) Biochemical basis of ozone toxicity. Free Radical Biology and Medicine. 9 (3): 245-265.
Murphy TM and Auh Ck (1996) The superoxide synthases of plasma membrane preparations from cultured rose cells. Plant Physiol 110: 621-629.
Nakada M, Fukui S and Kanno S (1976) Effects of exposure to various inurious gaseson germination of lily pollen. Environ Pollution. 11 (3): 181-187.
Nakamura H and Saka H (1978) Photochemical oxidants injury in rice plants. III. Effect of ozone on physiological activities in rice plants. Jpn. J. Crop. Sci. 47: 704-714.
Nash TH and Sigal LL (1979) Gross photosynthetic response of lichens to short-term ozone fumigation. Bryologist 82: 280-285
Nie GG, Tomasevic M and Baker NR (1993) Effect of ozone on the photosynthetic apparatus and leaf development in wheat. Plant. Cell Environment. 16 (6): 543-652.
Nieboer E and Mac Farlane JD (1984) Modification of plant cell buffering capacities by gaseolus air pollutants. Gaseous Air Pollutant and Plant Metabolism. (Ed. Koziol MJ and Whatley FR). London: Butterworth. 466 p.
Nobel PS and Wang CT (1973) Ozone increases the permeability of isolated pea chloroplasts. Arch. Biochem. Biophys. 157 (2): 388-394.
Noll KS, Roush TL, Cruikshank DP, Yohnson RE and Pendleton YJ (1997) Detection of ozone on Saturn's satellites Rhea and Dion. Nature. 388 (6637): 45-47.
Nouchi I and Toyama S (1988) Effects of ozone and peroxyacetylnitrate on polar lipids and fatty acids in leaves of morning glory and kidney bean. Plant Physiol. 87 (3): 638-646.
Noyes WAJr and Leighton PA (1941) The Photochemistry of gases. New York. Dover Publ. Inc. 480 p.

Oberley LW and Oberley TD (1986) Free radicals, cancer, and aging. Free Radicals, Aging, and Degenerative Diseases /Eds. Johnson JE, Jr, Walford, Harman D, Miquel J. pp. 325-371. New York: Alan R. Liss.

Ogier G, Greppin H and Castillo FJ (1991) Biochemical, molecular and physiological aspects of plant peroxidases. (Ed. J Lobarzewski). pp. 391-400. Geneve: Geneve Univ.

Okabe H (1978) Photochemistry of Small Molecules. John Wiley. New York.

Olson JA (1992) Carotenoids and vitamin A - An overview. Lipid-soluble Antioxidants. Biochemistry and Clinical Applications. pp. 178-192. Basel: Birkhauser. Verlag.

Ordin L and Hall MA (1967) Studies on cellulose synthesis by a cell-free oat coleoptile enzyme system: inactivation by airborne oxidants. Plant Physiol. Vol.42. N 2. P.205-212.

Örvar BL and Ellis BE (1997) Transgenic tobacco plants expressing antisense RNA for cytosolic ascorbate peroxidase show increased susceptibility to ozone injury. The Plant Journal 11: 1297-1305.

Oshima RJ, Bennett JP and Braegelmann PK (1978) Effect of ozone on growth and assimilate partitioning in parsley. J. Amer. Soc. Hort. Sci. 103 (3): 348-350.

Osmond DP, Black VJ and Unsworth MH (1981) Depression of net photosynthese in Vicia faba L. Exposed to sulphur dioxide and ozone. Nature. 291: 585-586.

Ostrovskii DN, Martynova MA, Matus VK, Ogrel OD, Lysak EI, Kharat;yan EF, Sibeldina LA and Shchipanova IN (1993) Aspect of the ozone action on microorganisms. Doklady of Russian Academy of Sciences 331: 104-108

Ostrovskaya LK, Trach VV and Mikhailik OM (1990) Superoxide dismutase activation in response to lime-induced chlorosis. New Phytologist. 114 (1): 39-45.

Otto HW and Daines RH (1969) Plant injury by air pollutants: Influence of humidity on stomatal apertures and plant response to ozone. Science. 163 (3872): 1209-1210.

Overmyer K, Tuominen H, Kettunen R, Betz C, Langebartels C, Sandermann HJr and Kangasjarvi J (2000) Ozone-sensitive Arabidopsis rcd1 mutant reveals opposite roles for ethylene and jasmonate signaling pathways in regulating superoxide-dependent cell death. The Plant Cell. 12 (10): 1849-1862.

Pandey J, Agrawal M and Narayan D (1991) An analysis of differential response of woody and herbaceous plants to ozone. Environ Ser. 5 (1): 91-110.

Pauls KJ and Thompson JE (1982) Effects of cytokinins and antioxidants on the susceptibility of membrane to ozone damage. Plant Cell Physiol. 23: 821-832.

Pauls KJ and Thompson JE (1981) Effects of in vitro treatment with ozone on the physical and chemical properties of membrane. Physiol Plant. 53 (3): 255-262.

Pell EJ (1988) Secondary metabolism and air pollutants. Air Pollution and Plant Metabolism. (Eds. Schulte-Hostede S, Darrall NM, Blank LW and Wellburn AR) pp. 222-237. London. New York. Elsevier Applied Science.

Pell EJ and Brennan E (1973) Changes in respiration, photosynthesis, adenosine 5'-triphosphate, and total adenylate content of ozonated pino bean foliage as they relate to symptom expression Plant Physiol. 53 (2): 378-381.

Pell EJ, Eckardt N and Enyedi AJ (1992) Timing of ozone stress and resulting Status of ribulose bis phosphate carboxylase/oxygenase and associated net photosynthesis. New Phytol. 120 (3): 397-405.

Pell EJ and Pearson NS (1983) Ozone-induced reduction in quantity of ribulose-1,5-bisphosphate carboxylase in alfalfa foliage. Plant Physiol. 73 (1): 185-187.

Pell EJ and Puente M (1986) Emission of ethylene by oak plants treated with ozone and simulated acid rains. New Phytol. 103 (3): 709-715.

Pell EJ, Temple PJ, Friend AL, Mooney HA and Winner WE (1994) Compensation as plant response to ozone and associated stresses: an analysis of ROPIS exteriments. J. Environ. Qual. 23 (3): 429-436.

Pell EJ and Weissberger WC (1976) Histopathological characterization of ozone injury to soybean foliage. Phytopathology. 66 (7): 856-861.

Pell EJ, Schlagnhaufe CD and Arteca RN (1997) Ozone-induced oxidative stress: mechanisms of action and reaction. Physiol Plant 100: 264-273.

Pellinen R,Palva T and Kangasjarvi J. (1999) Subcellular localization of ozone-induced hydrogen peroxide production in birch (Betula pendula) leaf cells. Plant J 20: 349-356.

Percy KE, McQuattie CJ and Rebbeck JA (1994) Effects of air pollutants on epicuticular wax chemical composition. In: Air Pollutants and the Leaf Cuticle. (Eds. Percy KE, Cape JN, Jagels R and Simpson CJ) NATO ASI Ser. Ser.G: Ecological Sciences. Berlin. Heidelbeg. Springer. 36: 67-79.

Peregud EA (1976) The Determination of ozone and photooxidants in atmospheric air. In: Chemical Air Analysis (New Improved Methods) pp.270-271. Khimiya, Leningrad

Peregud EA and Bikhovskaya MS and Gernet EB (1962) Fast Methods of the Determination of Hazzard Substances in Air.Goskhimizdat. Moscow, 272 pp.

Peregud EA and Gorelik DO (1981) Instrumental Methods of the Control of Environment. Khimiya, Leningrad, 384 pp.

Perov SP and Khrgian Akh (1980) Modern Problems of Atmospheric Ozone.Hydrometeoizgat, Leningrad, 288 pp.

Pino ME and Mudd J (1995) Ozone-induced alterations in the accumulation of newly synthesized proteins in leaves of maize. Plant Physiol. 108 (2): 777-785.

Pitcher LH, Brennan E, Hurley A, Dunsmuir P, Gepperman JM and Zilinskas BA (1991) Overproduction of *Petunia chloroplastlc* copper/zinc superoxode dimistase does not confer ozone tolerance in transgenic tobacco. Plant Physiol. 97 (1): 452-455.

Podleckis EY, Curtis CR and Heggestad HE (1984) Peroxidase enzyme markers for ozone sensitivity in sweet corn. Phytopathology. 74 (5): 572-577.

Polle A and Rennenberg H (1994) Photooxidative stress in trees. Causes of Photooxidative Stress and Amelioration of Defense Systems in Plants. (Eds. CH Foyer and Mullineaux PM) pp. 199-218. Boca Raton, London, Tokyo. CRC Press: Ann Arbor.

Polle A., Matyssek R., Günthardt-Goerg MS. and Maurer S. (2000) Defence strategies against ozone in trees: the role of nutrition. In: Agrawal S.B., Agrawal M. (eds.), Environmental Pollution and Plant Responses. pp. 223-245. Lewis, Boca Raton, USA.

Posthumus AC (1982) Biological indicators of air pollution. Effect of Gaseous Air Pollution in Agriculture and Horticulture. (Eds. Unsworth MN and Ormrod DP) pp. 27-42. London. Boston. Sydney. Wellington. Toronto: Butterworth Sci.

Posthumus AC (1976) The use or higher plants as indicators air pollution in the Netherlands. Proceedings of the Kuopio meetins on plant damages caused by air pollution. (Ed. Karenlampi L) Kuopio. pp. 115-120.

Priestly DA, Werner BG, Leopold AK and McBride MB (1985) Organic free radical levels in seeds and pollen: the effects of hydration and aging. Physiol. Plantarum. 64 (1): 88-94.

Prokof'eva IA (1951) Atmospheric Ozone. USSR Academy of Sciences, Moscow-Leningrad, 231 pp.

Pryor WA and Church DF (1991) Aldehydes, hydrogen peroxide, and organic radicals as mediators of ozone toxicity. Free Radical Biology and Medicine. 11 (1): 41-46.

Pryor WA, Lightsey JM and Prier DG (1982) Production of free radicals in vivo for the oxidation of xenobiotics: The initiation of autooxidation of polyunsaturated fatty acids by NO_2 and O_3. Lipid Peroxides in Biology and Medicine (Ed. K Yagi). New York: Acad. Press. pp. 1-22.

Puppo A and Halliwell B (1989) On the reaction of plant ferredoxins with hydrogen peroxide what reactive oxidants are generated. Phytochemistry. 28 (12): 3265-3270.

Pye JM (1988) Impact of ozone on the growth and yield of trees: a review. J Environ Quali. 17: 347-360.

Quartacci MF and Navari-Lzzo F (1992) Water stress and free radical mediated in sunflower seedlings. Plant Physiol. 139 (5): 621-625.

Rabinowitch HD and Fridovich I (1983) Superoxide radicals, superoxide dismutases and oxygen toxicity in plants. Photochemistry and Photobiology. 37 (6): 679-690.

Rao MV, Hale BA and Ormrod DP (1995) Amelioration of ozone-induced oxidative damage in wheat plants grown under high carbon dioxide. Role of antioxidant enzymes. Plant Physiology. 109 (2): 421-432.

Rao MV and Davis KR (1999) Ozone-induced cell death occurs via two distinct mechanisms in *Arabidopsis*: The role of salicylic acid. Plant J 17: 603-614.

Rao MV and Davis KR (2001) The physiology of ozone induced cell death. Planta 213: 682-690.

Rao MV, Koch JR and Davis KR (2000) Ozone: A tool for probing programmed cell death in plants. Plant Mol Biol 44: 345-358.

Rao MV, Lee H, Creelman RA, Mullet JE and Davis KR (2000) Jasmonic acid signaling modulates ozone-induced hypersensitive cell death. The Plant Cell 12: 1633-1646.

Rasmussen RA (1972) What do the hydrocarbons from trees contribute to air pollution? J. Air Pollut. Control. Ass. 22 (5): 537-543.

Rasmussen RA and Jones CA (1973) Emission of isoprene from leaf discs of *Hamamelis*. Phytochemistry. 12 (1): 15-19.

Rasmussen RA and Khalil Mak (1988) Isoprene over the Amazon basin. J. Geophys Res. 93 (10): 1417-1421.
Ray DD, Stedman ZH and Wendel GL (1986) Fast chemiluminescent method for measurement of ambient ozone. Analyt. Chem. 58 (3); 594-597.
Razumoskii SD (1979) Oxygen. Elementary Forms and Properties. Khimia, Moskow, 302 pp.
Razumovskii SD and Zaikov GE (1974) Ozone and its Reactions with Organic Compounds (Kinetics and Mechanism). Nauka, Moscow, 322 pp.
Reckendorfer P (1952) Ein Beitrag zur Mikrochemie des Rauchschadens durch Fluor. Die Wanderung des Fluors in pflanzlichem gewebe. 1. Teil: Zie unsichtbaren Schäden. Pflanzenschutzberichte. 9: 33-55.
Reddy GN, Y-R Dai, Negm FB, Flores HE, Arteca RN and Pell EJ. (1991) The effects of ozone stress on the levels of ethylene, polyamines, and rubisco gene expression in potato leaves. In: Active/Oxidative Stress and Plant Metabolism, EJ Pell, K Steffen, eds., The American Society of Plant Physiologists, pp. 52-57.
Reddy GN, Arteca RN, Day YR, Flores HE, Negm FB and Rell EJ (1993) Changes in ethylene and polyamines in relation to messenger RNA levels of the large and small subunits of ribulose bispphosphate carboxylase in ozone stressed potato. Plant Cell and Environment. 16 (7): 819-820.
Regener VH (1964) Measuring of atmospheric ozone with chemiluminescent method. J. Geophys Res V. 69: 3795-3800.
Reich PB (1987) Quantitfying plant response to ozone: a unifying theory. Tree Physiol. 3 (1): 63-91.
Reiling K and Davison AW (1994) Effect of exposure to ozone at different stages in the development of *Plantago major L.* on chlorophyll fluorescence and gas exchange. New Phytol. 128 (3): 509-514.
Reimond P and Farmer EE (1998) Jasmonate and salicylate as global signals for defense gene expression. Curr Opin Plant Biol 1: 404-411.
Reinert RA (1984) Plant response to air pollutant mixtures. Annu Rev Phytopathol 22; 421-442.
Rezchikov VG, Churmasov AA, Gavrilova AA and Sokolova EA (1998) Influence of ozone on the seed germination of *Pisum* and *Hippophae*. Tekhnika in Sel'skom Khozya'stve, Russia (Tecnique in Agriculture) 3: 14-17.
Rich S 919640 Ozone damage to plants. Annu Rev Phytopathol 2: 253-266.
Richards DMC, Dean RT and Jessup W (1988) Membrane proteins are critical targets in free radical mediated cytolysis. Biochimica Biophys. Acta. 946 (2): 281-288.
Richards BL, Middleton JT and Hewitt WB (1958) Air pollution with relation to agronomic crops.V. Oxidant stipple of grape. Agron.J. 50: 559-561.
Richards GA, Mulchi CL and Hall JR (1980) Influence of plant maturity on the sensitivity of turfgrass species to ozone. J .Environ .Qual. 9 (1): 49-53.
Roads AS and Brennan E (1978) The effect of ozone on chloroplast lamellae and isolated mesophyll cell of sensitive and resistant tobacco selections. Phytopathology. 68 (5): 885-886.
Roberts JA and Osborne DJ (1981) Auxin and the control of ethylene production during the development and senescence of leaves and fruits. J. Exp. Bot. 52: 875-887.
Robinson JM and Britz SJ (2000) Tolerance of a field grown soybean cultivar to elevated ozone level is concurrent with higher leafleat ascorbic acid level, higher ascorbate:dehydroascorbate redox status, and long term photosynthetic productivity. Photosynth. Res. 64: 77-87.
Robinson JM and Britz SJ (2001) Ascorbate-dehydroascorbate level and redox status in leaflets of field-grown soybeans exposed to elevated ozone. Int J Plant Sci 162 (1): 119-125.
Rosemann D, Heller W and Sandermann HJr (1991) Biochemical plant responces to ozone. II. Induction of stilbene biosynthesis in Scots Pine (*Pinus sylvestris L.*) seedlings. Plant Physiol. 97 (4): 1280-1286.
Rosen PM, Musselman RC and Kender WJ (1987) Relationship of stomatal resistance to SO_2 and O_3 injury in grapevines. Sci. Hotic. 8 (2): 137-142.
Rosentreter R and Ahmadjian V (1977) Effect of ozone on the lichen *Cladonia arbuscula* and the *Trebouxia* phycobiont of *Cladonia stellaris*. Bryologist 80 (4): 600-605.
Roshchina VD (1971) On composition of transpiration water in plants.Soviet Plant Physiol 18: 433-435.
Roshchina VD (1973) Seasonal dynamics of internal carbon dioxide in shoots of woody plants. Lesnoi Z. 4: 24-31
Roshchina VD and Roshchina VV (1989) The Excretory Function of Higher Plants. Nauka, Moscow
Roshchina VV (1991) Biomediators in plants. Acetylcholine and biogenic amines. Pushchino: Biological Center of USSR Academy of Sciences, 192 pp.

(Forest Journal of USSR) 4: 24-31
Roshchina VV (1996) Volatile plant excretions as natural antiozonants and origin of free radicals. (Eds. Narwal SS and Tauro P) Allelopathy. Field Observation and Methodology. Joghpur: Scientific Publischers. pp. 233-241.
Roshchina VV (1999a) Mechanisms of cell-cell communications Allelopathy. Basic and Applied Aspects. (Ed. Narwal SS) New Delhi, London: Oxford IIBN Publ. 2: 3-25.
Roshchina VV (1999 b) Chemosignalling in Pollen. Advantages in Modern Biology (Russia). 119 (6) : 577-566.
Roshchina VV (2001 a) Neurotransmitters in Plant Life. Science Publisher: Einfield New Hampshire, Plymouth. 284 pp
Roshchina VV (2001 b) Molecular-cellular mechanisms in pollen alllelopathy. Allelopathy Journal. 8 (1): 11-28.
Roshchina VV (2001c) Autofluorescence of plant cells as a sensor for ozone // In: Abstracts of 7 th Conference on Methods and Applications of Fluorescence: Spectroscopy, Imaging and Probes. Amsterdam: 16-19 September 2001. p. 162.
Roshchina VV (2001d) Rutacridone as a fluorescent probe // In: Abstracts of 7 th Conference on Methods and Applications of Fluorescence: Spectroscopy, Imaging and Probes. Amsterdam: 16-19 September 2001. p. 161.
Roshchina VV (2002) Rutacridone as a fluorescent dye for the study of pollen. J.Fluorescence 12 (2): 241-243.
Roshchina VV and Karnaukhov VN (1999) Changes in pollen autofluorescence induced by ozone Biologia Plantarum. 42 (2): 273-278.
Roshchina VV and Mel'nikova EV (1995) Spectral analysis of intact secretory cells and excretions of plants. Allelopathy J. 2 :179-188.
Roshchina VV and Mel'nikova EV (1996) Microspectrofluorimetry: A new technique to study pollen allelopathy. Allelopathy J. 3: 51-58.
Roshchina VV and Mel'nikova EV (1998 a) Chemosensory reactions at the interaction pollen-pistil. Biological Bulleten. 6: 678-685.
Roshchina VV and Melnikova EV (1998 b) Allelopathy and plant generative cells. Participation of acetylcholine and histamine in a signalling at the interactions of pollen and pistil. Allelopathy Journal. 5: 171-182
Roshchina VV and Mel'nikova EV (1999) Microspectrofluorimetry of intact secreting cells applied to allelopathy. In: Principles and Practics in Plant Ecology. Allelochemical Interactions. (Eds. Inderjit KMM and Dakshini CL Foy CRC) Boca Raton: CRS Press. pp.99-126
Roshchina VV and Melnikova EV (2000). Contribution of ozone and active oxygen species in the development of cellular systems of plants. In: Mitochondria, Cells and Active Oxygen Species . (Ed. Skulachev VP and Zinchenko VP). Materials of International Conference, 6-9 June 2000, Pushchino. Pp. 127-120. Biological Center of Russian Academy of Sciences. Pushchino.
Roshchina VV and Melnikova EV (2001) Chemosensitivity of pollen to ozone and peroxides. Russian Plant Physiology (Fiziologia Rastenii) 48 (1): 89-99.
Roshchina VV and Roshchina VD (1993) The Excretory Function of Higher Plants. Berlin, Springer-Verlag. Berlin Heidelberg. 314 pp.
Roshchina VV Mel'nikova EV and Karnaukhov VN (2000) Fluorescing world of plant cells. Science in Russia 6.
Roshchina VV, Mel'nikova EV and Kovaleva LV (1996) Autofluorescence in thge pollen -pistil system of Hippeastrum hybridum. Doklady Biological Sciences 349: 403-405.
Roshchina VV, Mel'nikova EV, Kovaleva LV (1997) The changes in autofluorescence during the male gametophyte development. Russian Journal of Plant Physiology 44: 45-53.
Roshchina VV, Mel'nikova EV, Kovaleva LV and Spiridonov NA (1994) Cholinesterase of pollen grains. Doklady Biological Sciences 337: 424-427.
Roshchina VV, Mel'nikova EV, Spiridonov NA and Kovaleva LV (1995) Azulenes are blue pigments of pollen. Doklady Biological Sciences. 340: 715-718.
Roshchina VV, Mel'nikova EV, Mit'kovskaya LI, Karnaukhov VN (1998 a) Microspectrofluorimetry for the study of intact plant secretory cells. J of General Biology. (Russia) 59: 531-554.
Roshchina VV, Mel'nikova EV, Gordon RYa, Konovalov DA and Kuzin AM (1998 b) A study of the radioprotective activity of proazulenes using a chemosensory model of *Hippeastrum hybridum* pollen. Doklady Biophysics. 358-360: 20-23.

Roshchina VV, Popov VI, Novoselov VI, Melnikova EV, Gordon RYa, Peshenko IV and Fesenko EE (1998 c) Transduction of chemosignal in pollen. Cytology. 40: 964-971.

Roshchina VV, Melnikova EV, Popov VI, Karnaukhov AV, Mit'kovskaya LI, Gorokhov AA and Karnaukhov VN (1998d) Principles of the bank data creation based on the pollen characteristics. Computer analysis of the fluorescence spectra for the cover composition consideration. In: Cryoconservation of Genetic Resurses. Materials of All-Russian Conference 13-15 October 1998, Pushchino. pp.221-225. Biological Center of RAS. Pushchino.

Roshchina VV, Mel'nikova EV, Yashin VA and Karnaukhov VN (2002a) Autofluorescence of intact spores of horsetail *Equisetum arvense* L. during their development. Biophysics (Russia) 47 (2): 318-324.

Roshchina VV, Miller AV and Safronova VG (2002b) Superoxide anionradical on the surface of generative (pollen) and vegetative microspores. In: Materials of 2nd Internationa Symposium on Plant Anatomy and Systematics, 23-25 October 2002, Sankt-Peterburg. Komarov Botanical Institute of RAS: Sankt-Peterburg.

Roshchina VV, Miller AV, Safronova VG and Karnaukhov V.N. (2003a) Active Oxygen Species and Luminescence of Intact Cells of Microspores. Biophysics (Russia) 48: 283-288.

Roshchina VV, Yashin VA and Kononov AV (2003b) Autofluorescence of plant microspores as a probe for their viability and development. In: Abstracts of 8 th Conference on Methods and Applications of Fluorescence: Spectroscopy, Imaging and Probes. Prague: 24-27 August 2003., p. 207

Rossolovskii VYa (1994) Chemical Encyclopedia. Russian Encyclopedia, Moscow. 3: 332-333.

Rousseaux MC, Ballaro CL, Giardano CV, ScopelAL, Zima AM, Szwarzberg -Braccheta M, Searles PS, Caldwell MM and Diaz SB (1999) Ozone depletion and UV-B radiation: Impact on plant DNA damage in Southern South America. Proc Natl Acad Sci USA. 96 (26): 15310-15315.

Rowland AJ, Borland AM and Lea PJ (1988) Changes in amino acids, amines and proteins in response to air pollutants. (Eds. Schulte-Hostede S, Darrall NM, Blank LW and Wellburn AR) Air Pollution and Plant Metabolism. pp. 189-221. London. New York: Elsevier Applied Science.

Rubinstein B and Luster DG (1993) Plasma membrane redox activity: components and role in plant processes. Annu Rev Plant Physiol Plant Mol Biol. 44: 131-155.

Runeckles VC and Vaartnou M (1997) EPR evidence for superoxide anionformation in leaves during exposure to low levels of ozone. Plant Cell Environ 20: 306-314.

Ryerson TB, Trainer M, Holloway JS, Parrish DD, Huey LG, Sueper DT, Frost GJ, Donnelly SG, Schauffler S, Atlas EL, Kuster WS, Goldan PD, Hubler G, Meagher JF and Fehsenfeld FC (2001) Observation of ozone formation in power plant plumes and implications for ozone control strategies. Nature. 292 (5517): 719-723.

Salin ML (1987) Toxic oxygen species and protective systems of the chloroplast. Physiologia Plantarum. 72 (3): 681-689.

Salisbury FB and Marinos NG (1985) The ecological role of plant growth substances. (Eds. Pharis RP and Reid DM) Hormonal Regulation of Development III. Role of environmental factors, Encyclopedia of Plant Physiol. New Ser. pp. 707-766. Berlin, Heidelberg, New York, Tokyo: Springer-Verlag.

Salter I and Hewitt CN (1992) Ozone-hydrocarbons interactions in plants. Phytochemistry. 31 (12): 4045-4050.

Salzer P, Corbiere H and Boller T (1999) Hydrogen peroxide accumulation in *Medicago truncatula* roots colonized by the arbuscular mycorhiza-forming fungus *Glomus intrara* dices.// Planta. 208 (3): 319-325.

Samuilov VD (1999) Role of H_2O_2 in oxygenic photosynthesis. In: Materials of 2nd All-Russian Meeting of Biophysists. Moscow .Biological Center of RAS. Pushchino. 3: 1071-1072.

Samuilov VD (1999) Role of H_2O_2 in oxygen photosynthesis. In: Materials of 2nd All Russian Meeting of Biophysists, Biological Center of Russian Academy of Sciences .Pushchino. 3: 1071-1072.

Sanadze GA (1964) Conditions of diene C_5H_8 release from leaves. Soviet Plant Physiology 2: 49-52

Sanadze GA , Dzhaiani GI and Tevzadze IM (1972) Inclusion of carbon from fixated $C^{13}O_2$ at photosynthesis. Soviet Plant Physiology 19: 24-27.

Sandermann HJr (1996) Ozone and plant health. Annu Rev Phytopathol 34: 347-366.

Sandermann H, Langebartels C and Heller W (1990) Ozonstress bei Pflanzen. Fruhe und "Memory" - Effekte von Ozon bei Nadelbaumen UWSF. Z. Umweltchem Okotox. 2 (1): 14-15.

Sandermann H, Schmitt R, Heller W, Rosenmann D and Langebartels C (1989) Ozone-induced early biochemical reactions in Conifers. Acid Deposition. Sources, Effects and Controls. (Ed. JUS Longhurst). London.: British Library. pp. 243-254.

Sandmann G and Boger P (1980) Copper-mediated lipid peroxidation processes in photosynthetic membranes. Plant Physiol. 66 (5): 797-800.

Sandmann G and Boger P (1982) Volatile hydrocarbons from photosynthetic membranes containing different fatty acids. Lipids. 17 (1): 35-41.

Sandermann H, Wellburn AR, Heath RL (1997) Forest Decline and Ozone. Synopsis. Forest Decline and Ozone. A Comparison of Controlled Chamber and Field Experiments /Eds. Sandermann H, Wellburn AR and Heath RL (Ecological Studies, vol.127). pp. 369-377. Berlin, Heidelberg: Springer-Verlag.

Sandermann H, Wellburn AR and Heath RL (Eds.) (1997) Forest Decline and Ozone. A Comparison of Controlled Chamber and Field Experiments (Ecological Studies, vol.127). Berlin, Heidelberg: Springer-Verlag. 400 pp.

Sandermann HJr, Ernst D, Heller W and Langebartels C (1998) Ozone: an abiotic elicitor of plant defence reactions. Trends in Plant Science 3: 47-50.

Saran M, Michel C and Bors W (1988) Reactivities of free radicals. In: Air Pollution and Plant Metabolism. (Eds. Schulte-Hostede S, Darrall NM, Blank LW and Wellburn AR) pp. 76-93. London. New York: Elsevier Applied Science.

Saran M and Bors W (1990) Radical reactions in vivo - an overview. Radiat. Environ. Biophys. 29 (4): 249-262.

Sasek TW and Richardson CJ (1989) Effects of chronic doses of ozone on loblolly pine: photosynthetic characteristics in the third growing season. Forest Sci. 35: 745-755.

Saunders BK (1975) Peroxidases and catalases. In: Inorganic Biochemistry. (Ed. Eichhorn GB) Vol.2.pp.434-470. Elsevier, Amsterdam, Oxford.

Savich IM (1989) Peroxidases -stressory proteins of plants . Uspekhi Sovremennoi Biologii (Advantages in Modern Biology, Russia) 107 : 406-417.

Sawadaishi K, Miura K, Ohtsuka E, Ueda T, Ishizaki K and Shinriki N (1985) Sequence specificity of ozone degradation of bases in supercoiled plasmid DNA. Nucleit Acids Res. Symp. Ser. 16: 205-208.

Scandalios JG (1994) Regulation and properties of plant catalases. In: Causes of Photooxidative Stress and Amelioration of Defense Systems in Plants. (Eds. CH Foyer and Mullineaux PM) pp. 275-315. Boca Raton, Ann Arbor, London, Tokyo. CRC Press.

Schenone G (1993) Impact of air pollutants on plants in hot dry climates. In: Interacting Stresses on Plants in a Chenging Climate. (Eds. Jacson MB and Black CR) (NATO ASI SER. Ser. I.: Global Environmental Change, vol.16). pp 139-152. Berlin: Springer.

Scherbatskoy T (1994) Leaf cuticles as mediators of environmental infgluences:new developments in the use of isolated cuticles. In: Air Pollutants andthe Leaf Cuticle. (Eds. Percy KE, Cape JN, Jagels R, Simpson CJ) NATO ASISer. Ser.G: Ecological Sciences. Berlin Heidelbeg. Springer. 36: 149-163.

Schmieden U and Wild A (1995) The contribution of ozone to forest decline. Phisiol Plantarum 94; 371-378.

Schneiderbauer A, Back E, Sandermann HJr and Ernst D (1995) Ozone induction of extensin mRNA in Scots pine, Norway spruce and European beech. New Phytol 130: 225-230.

Schraudner M, Ernst D, Langebartels C and Sandermann H (1992) Biochemical plant responses to ozone. III Activation of the defense - related proteins β-1, 3-glucanase and chitinase in tobacco leaves. Plant Physiol. 99: 1321-1328.

Schraudner M, Langebartels C, Negrel J and Sandermann HJr (1993) Plant defence reactions induced in tobacco by the air pollutant ozone. Dev. Plant Phatholog. (Hechanisn of Plant Defense Hesuenses) 2: 286-290.

Schraudner M, Moeder W, Wiese C, van Camp W, Inze D, Langebartels C, and Sandermann HJr (1998) Ozone-induced oxidative burst in the ozone biomonitor plant, tobacco BelW3. Plant J 16: 235-245.

Schreck H and Baeuerle PA (1994) Assessing oxygen radicals as mediators in activation of inducible eukaryotic transcription factor NF-KB. Methods in Enzymology. San Diego, New York, Boston, London, Sydney, Tokio, Toronto. Academic Press. 234 (Part D): 151-163.

Schreck R, Rieber P and Baeurle PA (1991) Reactive oxygen intermediates as apparently widely used messengers in the activation of the NF-Lambda B transcription factor and HIV-1. EMBO Journal. 10 (8): 2247-2258.

Schreiber U, Vidaver W, Runeckles VC and Rosen P (1978) Chlorophyll fluorescence assay for ozone injury in intact plants. Plant Physiol. 61 (1): 80-84.

Schreiber U, Vidaver W, Runeckles VC and Rosen P (1978) Chlorophyll fluorescence assay for ozone injury in intact plants. Plant Physiol. 61 (1): 80-84.

Schultheis AH, Bassett DJP and Fryer AD (1994) Ozone-induced air way hyperresponsivensis and loss of neuronal M2 muscarinic receptor function. J. Appl. Physiol. 76 (3): 1088-1097.

Sellden G, Morre DJ, Sandelius AS, Egger A, Larsson K, Ojanpera K, Carlsson A. and Sutinen S (1990) Effects of ozone on membrane structure, function and biogenesis. Physiolosia Plantarum. 79 (2 Part 2: A122.

Semenov SM., Kounina IM. and Koukhta BA. (1999) Tropospheric ozone and plant growth in Europe. Moscow, Publishing Center " Meteorology and Hydrology", 208 pp.

Setlow BW (1974) The wavellength in sunlight effects in producting skin cancer: a theoretical analysis. Proc. Nat. Acad. Sci. USA. 71 (9): 3363-3366.

Shabala SN and Voinov OA (1994) Dynamics of physiological characteristics of plants as element of the system of ecological monitoring. Uspekhi Sovremennoi Biologii (Advantages in Modern Biology, Russia) 114 : 144-159.

Sharma YK and Davis KR (1997) The effects of ozone on antioxidant responses in plants. Free Rad Biol Med 23; 480-488.

Sharma YK, Leon J, Raskin I and Davis KR (1996)Ozone-induced expression of stress-related genes in Arabidopsis thaliana: the role of salicylic acid in the accumulation of defense-related transcripts and induced resistance. Proc Natl.Acad Sci USA 93: 5099-5104.

Shimazaki K (1988) Thylakoid membrane reactions to air pollutants. (Eds. Schulte-Hostede S, Darrall NM, Blank LW and Wellburn AR. In: Air Pollution and Plant Metabolism. Elsevier Applied Science. pp. 116-133. London. New York.

Shinar E, Navok T, Chevion M (1983) The analogous mechanisms of enzymatic inactivation induced by ascorbate and superoxide in the presence of copper. J. Biol. Chem. 258 (24): 14778-14783.

Shiratori K (1973) Field survey method on damage to plants centering on crops caused by oxidants. In: References on photochemical Air Pollution in Japan. Air Qual Bur. Environ Agency. pp. 246-258.

Shmid C and Ziegler H (1991) Sorption and permeatic properties of cuticles for monoterpene with special regard to the effect of pollutant gases on spruce. GSF-Ber. Proc., Statussemin. PBWU Forschungsschwerpunkt. "Waldschaeden". 2nd. 26 (91): 483-490.

Shobert B and Elstner EF (1980) Production of hexanal and ethane by *Paseodactylum triconutum* and its correlation to fatty acid oxidation and bleoching of photosynthetic pigments. Plant Physiology. 66 (2): 215-219.

Shorning Byu, Smirnova EG, Yaguzhinsky LS and Vanyushin BF (2000) Necessity of the superoxide production for development of etiolated wheat seedlings. Biokhimiya (Biochemistry, Russia) 65 (12): 1612-1617.

Siegel SM (1962) Protection of plants against air born oxidants: *Cucumber* seedlings at extreme ozone levels. Plant Physiol. 37 (1): 35-39.

Sigal LL and Nash TH (1983) Lichen communities on conifers in Southern California mountains: an ecological survey relative to oxidant air pollution. Ecology. 64: 1343-1354.

Simic MG, Bergtold DS and Karam LR (1989) Generation of oxy radicals in biosystems. Mutation Researsh. 214 (1): 3-12.

Singh S.,Suri R and Agrawal CG (1995) Fluorescence properties of oxidised human plasma low-density lipoproteins. Biochim Biophys Acta 1254 (2): 135-139.

Skarby L and Sellden G (1984) The effects on ozone on crops and forests. Ambio. 13 (1): 68-72.

Skulachev VP (1972) Transformation of Energy in Biomembranes. Nauka. Moscow. 203 pp.

Slater GF (1976) Recent Advances in Biochemical Pathology: Toxic Liver Injury. Pion Press. pp. 1-283.

Smith TA (1985) Polyamines. Annu Rev. Plant Physiol. 36: 117-143.

Smith WH (1990) Air Pollution and Forests. In: Interactions between Air Contaminants and Forest Ecosystems. pp. Springer-Verlag, Berlin. 2nd ed.

Sobels FH (1956) Organic peroxides and mutagenic effects in *Drosophila*. Nature. 177 (4517): 979-982.

Sober Anu (1992) Effect of ozone on the rehydration of bean leaves. Изв. АН Эстонии. 41 (1): 35-43.

Soikkeli S and Karenlampi L. (1984) Cellular and ultrastructural effects.In: Air Pollution and Plant Life./Ed. Treshow M) pp. 190-205. Chichester, New York, Wiley .

Staehelin J and Hoigne J (1985) Decomposition of ozone in water in the presence of organic solutes. Environ Sci. Technol. 19 (12)Ж 1206-1213.

Staehelin J and Hoigne Y (1982) Decomposition of ozone in water rate of initiation by hydoxide ion and hydrogen peroxide. Environ Sci/ Technol. 16 (10): 676-681.

Stanley RG and Linskens HF (1974) Pollen. Biology, Biochemistry, Managements. Berlin. Heidelberg. N.Y. Springer. 307 p.

Steinbrecher UP (1988) Role of superoxide in the endothelial cell modification of low density lipoproteins. Biochim. Biophys.Acta. 959 (1): 20-30.

Stern AC (1976) Air Pollution (3rd edition) Mew York: Academic Press. 1-5:3844 pp.

Stich K and Ebermann R (1984) Investigation of hydrogen peroxide formation in plants. Phytochem. 23 (12): 2719-2722.

Stockwell WR, Kramm G, Scheel HE, Mohnen VA and Seiler W (1997) Ozone formation, destraction and exposure in Europe and the United States. Forest Decline and Ozone. A Comparison of Controlled Chamber and Field Experiments /Eds. Sandermann H, Wellburn AR and Heath RL (Ecological Studies, vol.127) pp. 1-38. Berlin, Heidelberg: Springer-Verlag.

Streller S and Wingsle G (1994) *Pinus sylvestris L.* needles contain extracellular Cu-Zn superoxide dismutase. Planta. 192 (2): 195-201.

Stewart CA, Black VJ, Black CR and Roberts JA (1996) Direct effects of ozone on the reproductive development of Brassica species. J. Plant Physiol. 148 (1-2): 172-178.

Subluskey LA, Harris GC, Maggiolo A and Tumolo AL (1959) Improved synthesis of aromatic aldehydes from ozonolysys of olefins . In: Ozone Chemistry and Technology (Adv Chem Ser, vol.21) pp.149-152.

Sukacheva OA, Mareinova OA, Kamburova VN and Gagelgans AI (1999) Influence of hydoperoxide of coumol of energetic metabolism of liver mitochondria. In : Materials of 2nd All-Russian Meeting of Biophysists. 23-27 August 1999, Moscow. pp.209-210. Biological center of RAS. Pushchino

Sutton R and Ting IP (1977 b) Evidence for repair of ozone-induced membrane injury: Alterations in sugar uptake. Atmos Environ. 11 (2): 273-275.

Sutton R and Ting IP (1977) Evidence for the repair of ozone-induced membrane injury. Am: J. Bot. 64: 404-411.

Swanson ES, Thomson WW and Mudd JB (1973) The effect of ozone on leaf cell membranes. Can. J. Botany. 51 (7): 1213-1219.

Takahashi MA and Asada K (1983) Superoxide anion permeability of phospolipid membranes and chloroplast thylakoids. Archives of Biochemistry and Biophysics. 226 (2): 558-566.

Tang X, Madronich S, Wallington T and Calamari D (1998) Changes in tropospheric composition and air quality. J. Photochemistry and Photobiology. 46 (1-3): 83-95.

Tappel AL (1975) Lipid peroxidation and fluorescent molecular damage to membranes. Pathology of Cell Membranes/Ed. BFTrump and AUArstilla. New York: Academic Press. 1: 145-170.

Taylor GE and Hanson PJ (1992) Forest trees and tropospheric ozone-role of canopy deposition and leaf uptake in developing exposure response relationships. Agricultural Ecosystems and Environment. 42 (3-4): 255-274.

Taylor OC (1973) Acute responses of plants to aerial pollutants. Air Pollution Damage to Vegetation. (Ed. Naegele JA (Advances in Chemistry Ser., Vol.122). pp. 9-20. Washigton: Amer. Chem. Soc.

Taylor OC, Cardiff EA, Mersereau JD and Meddleton JT (1958) Effect of air-born reaction products of ozone and 1-N-hexene vapour (synthetic smog) on growth of avocado seedlings. Proc. Am Soc. Hort. Sci. 72: 320-325.

Terry GM and Stoke NJ (1993) The assesment of plant damage by reactive hydrocarbons and their oxidation products. Interacting Stresses on Plants in a Changing Climate. Eds. Jacson MB and Black CR, (NATO ASI SER. Ser.I.:Global Environmental Change). Berlin. Springer. 16: 171-183.

Thomson WW, Nagahashi J and Platt K (1974) Further observation on the effects of ozone on the ultrastructure of leaf tissue. Air Pollution Effects on Plant Growth (Ed. WM Dugger, Jr.) (ACS symposium series 3) pp. 83-93. Washington. Am Chem. Soc.

Thalmair M, Bauw G, Thiel S, Dohring T, Langebartels C and Sandermann HJr (1996) Ozone and ultraviolet B effects on the defense-related protein β-1,3-glucanase and chitinase in tobacco. J. Plant Physiol. 148 (1-2): 222-228.

Ticha I and Catsky J (1977) Ontogenetic changes in the internal limitations to bean leaf photosynthesis. J. Photosynthesis Research. 11 (2): 361-366.

Ting IP and Dugger WM (1968) Factors affecting ozone sensitivity and susceptibility of cotton plants. J. Air Pollut. Control Assoc. 18: 810-813.

Ting IP and Mukerji SKE (1971) Leaf ontogeny as a factor in susceptibility to ozone: Amino-acid and carbohydrate changes during expansion. Amer. J. Bot. 58 (2): 497-504.

Ting IP, Perchorowicz J and Evans L (1974) Effect of ozone on plant cell membrane permiability. Air Pollution Effects on Plant Growth. (Ed. Dugger M) (Symp of Div. Agrcult and Food Chem. at the 167 Meeting of Am Chem. Soc. Los Angeles,California, April 1., 1974), ACS Symposium Ser., 3) Washington. D.C.: Am Chem. Soc. pp. 8-21.

Tingey DT (1977) Ozone induced alterations in plant growth and metabolism. Proceedings Int Conference on Photochemical Oxidant Pollution and its Control. (Ed. Dimitriades B) (Environ Sci. Res. Lab. Office of Res. Usa. North Carolina., Triangle Park., September 12-17, 1976.) 1: 601-610.

Tingey DT (1981) The effect of environment factors on the emission of biogenic hydrocarbons from live oak and slash pine. Atmospheric Biogenic Hydrocarbons. Emissions.(Eds. Bufalini JJ and Arnts RR) Ann Arbor. MI.: Ann Arbor Science. 1: 53-79.

Tingey DT, Fites RC and Wickliff C (1975) Activity changes in selected enzymes from soybean leaves following ozone exposure. Physiologia Plantarum. 33 (4): 316-320.

Tingey DT. Fites RC and Wickliff C (1976 a) Differential foliar sensitivity of soybean. cultivars to ozone associated with Differential enyme Activities. Physiol. Plant. 37 (1): 69-72.

Tingey DT, Fites RC and Wickliff C (1973 b) Foilar sensitivity of soybeans to ozon as related to several leaf parameters. Environ Pollut. 4 (1): 185-192.

Tingey DT, Fites RC and Wickliff C (1973 c) Ozone alteration of nitrate reduction in Soybean. Physiol. Plant. 29 (1): 33-38.

Tingey DT, Reinert RA, Dunning JA and Heck WW (1973 a) Foliar injury responses of eleven plant species to ozone/sulfur dioxide mixtures. Atmos Environ. 7 (2): 201-208.

Tingey DT and Taylor GEJr (1982) Variation in plant response to ozone: a conceptual model of physiological events. Effects of Gaseous Air Pollution in Agriculture and Horticulture. (Eds. Unsworth MN and Ormrod DP) pp. 111-138. London: Butterworth Scietific.

Tingey DT, Wilhour RG and Standley C (1976) The effect of chronic ozone exposures on the metabolite content of ponderosa pine seedlings. Forest Science. 22 (3): 234-241.

Tomlinson H and Rich S (1973) Anti-senescent compounds reduce injury and steroidchanges in ozonated leaves and their chloroplasts. Phytopathology. 63 (9): 903-906.

Tomlinson H and Rich S (1971) Effect of ozone on sterols and sterol derivatives in bean leaves and their chloroplasts. Phytopathology. 63 (9): 903-906.

Tomlinson H and Rich S (1970) Lipid peroxidation a result of injury in bean leaves exposed to ozone. Phytopathology. 61 (12): 1531-1532.

Tomlinson H and Rich S (1967) Metabolic changes in free amino acids of bean leaves exposed to ozone. Phytopathology. 57 (9): 972-974.

Tomlinson H and Rich S (1969) Relating lipid content and fatty acid synthesis to ozone injury of *Tabacco Leaves*. Phytophathology. 59 (11): 1284-1286.

Tonneijck AEG and Leone G (1993) Changes in susceptibility of bean leaves (*Phaseolus vulgaris* L.) to *Sclerotinia sclerotiorum* and *Botrytis cinerea* by pre-inoculatetive ozone exposures. Neth. J. Plant Pathol. 99 (5-6): 313-322.

Toodd GW (1958) Effect of ozone and ozonated 1-hexene on respiration and photosynthesis leaves. Plant Physiol. 33 (2): 416-420.

Toodd GW (1956) The effect of gaseous ozone, hexene, and their reaction products upon the respiration of lemon fruit. Physiol. Plant. 9 (2): 421-428.

Trainer M, Williams EL, Parrish DD, Buhr MP, Allwine EL, Westberg HH, Tehsenfeld FG and Liw SG (1987) Models and observations of the impact of natural hydrocarbons on rural ozone. Nature. 329 (6141): 705-707.

Treshow M (1984a) Diagnostics of the influence of air pollution and similarity of symptoms. In: Air Pollution and Plant Life./Ed. Treshow M) pp. 126-143. Wiley, Chichester, New York, Brisbane Butterworths.

Treshow M (1984b) Epilogue: a biochemical overview. In: Gaseous Air Pollutants and Plant Metabolism./Eds. Koziol MJ and Whatley FR. pp. 425-437. London, Boston: Butterworths.

Treshow M and Anderson FK (1989) Plant Stress from Air Pollutants. Wiley, Chichester, UK

Trevathan LE, Moor LD and Oreutt DN (1979) Symptom expression and free sterol and fatty acid composition of flue-cured Tobacco plants exposed to ozone. Phytopathology. 69 (3): 582-585.

Tuomainen J, Pellinen R, Roy S, Kiiskinen M, Eloranta T, Karjalainen R and Kangasjarvi J (1996) Ozone affects birch (Betula pendula Roth) phenylpropanoid, polyamine and active oxygen detoxifying pathways at biochemical and gene expansion level. J. Plant Physiol. 148 (1-2): 179-188.

Tuomainen J, Betz C, Kangasjärvi J, Ernst D, Yin Z-H, Langebartels C.and Sandermann H. Jr. (1997) Ozone induction of ethylene emission in tomato plants: Regulation by differential transcript accumulation for the biosynthetic enzymes. Plant J, 12:1151-1162.

Tyszkiewicz E and Roux E (1989) Role of oxygen radicals in phenazine mediated ATP synthesis obtained with spinach chloroplasts. Toxicity of OH radicals. Bioelectrochemistry and Bioenergetics 22: 323-329.

Urbach W, Schmidt W, Kolbowski J, Ruhmele S, Reisberg E, Steigner W and Schreiber U (1989) Wirkung von Umweltschadstoffen auf Photosynthese und Zellmembranen von Pflanzen. Statusseminar der PBWU zum Forschungsschwerpunkt" Waldschaden" GSF Bericht 6/89. Neuerberg. pp.195-206.

Vanin AT (2001) Nitrogen oxide: regulation of cell metabolism withot the involvement of the system of cell receptors. Biochemistry (Russia). 46 (4): 631-641.

Varns JL, Mulik JD, Sather ME, Glen G, Smith L and Stallings C (2001) Passive ozone network of Dallas: A modelling opportunity with community involvement. 1. Tropospheric O_3-format. Environment Science and Technology. 35 (5): 845-855.

Vick BA and Zimmerman DC (1987) Pathways of fatty acid hydroperoxide metabolism in spinach leaf chloroplasts. Plant Physiol. 85 (4): 1073-1078.

Victorin K (1992) Review of the genotoxicity of ozone. Mutation Research. 277 (3): 221-238.

Vladimirov YA (1986) Free Radical Lipid Peroxidation in Biomembranes: Mechanism, Regulation and Biological Consequences. Free Radicals, Aging and Degenerative Diseases (Eds. Johnson JE, Walford R, Harman D and Miquel J). New York: Alan R. Liss. (Ser. Modern Aging Research). pp. 141-195.

Vladimirov YA and Archakov AI. (!972) Lipid Peroxidation in Biological Membranes. Nauka, Moscow. 252 pp.

Vladimirov YA, Azizova OA, Deev AI, Kozlov AV , Osipov AN and Roshchupkin DI (1991) Free Radicals in Living Systems. ITOGI NAUKI I TEKHNIKI, Ser. Biophysics, Vol.29. VINITI. Moscow, 252 pp.

Vladimirov YA, Olenev VI, Suslova TB and Cheremisina ZP (1980) Lipid Peroxidation in Mitochondrial Membrane. Advances in Lipid Research. (Eds. R Paoletti and D Kritchevsky). New York. London: Academic Press. 17: 173-249.

Vollenweider P., Ottiger M. and Günthardt-Goerg M.S (2001) Atmospheric ozone pollution: bioindication tools for calibrated expertise of symptoms in leaves and needles from trees. EURASP Newsletter 41: 17-20.

Wallin G, Skarby L and Sellden G (1990) Long term exposure of Norway spruce, *Picea abies*, to ozone in open two chambers. 1. Effects on the capasity of net photosynthesis, dark respiration and leaf conductance of shoots of different ages. New Phytol. 115 (3): 335-344.

Whalen SC and Reeburgh WS (1990) Consumption of atmospheric methane by tundra soils. Nature. 346 (6280): 160-162

Wang CY (1987) Changes of polyamines and ethylene/chiling stress. Physiol. Plant. 69.

Wang WC and Isaksen IS (Eds) (1995) Atmospheric Ozone as a Climatic Gas. Berlin Heidelberg, Springer-Verlag, 461 pp.

Wang WW, Gorsuch JW and Hughes J (Eds.) (1997) Plants for Environmental Studies./ Boca Raton: CRC Press.

Weber JA, Tingey DT and Andersen CP (1999) Plant response to air pollution. Plant-Environment Interactions/ Wilkinson RE Ed. New York: Marcel Dekker. pp.

Weber JA, Tingey DT and Andersen CP (1999) Plant response to air pollution// Plant-Environment Interactions/ Ed. Wilkinson RE New York: Marcel Dekker. pp.

Wegener A, Gimbel W, Werner T, Hani J, Ernst D and Sandermann HJr (1997) Molecular cloning of ozone-inducible protein from *Pinus sylvestris* L. with high sequence similarity to vertebrate 3-hydroxy-3-methylglutaryl-CoA-synthase. Biochim. Biophys. Acta. 1350: 247-252.

Weiner LM (1994) Oxygen radicals generation and DNA scission by antocancer and synthetic quinones. Methods in Enzymol. 233 (part C): 92-105.

Weis J (1935) Investigations on the radical HO_2 in solutions. Trans Farad. Soc. 31: 668-681.

Wellburn AR and Chen Y (1992) Air pollution injury and stress ethylene formation. Physiol. Plantarum. 79 (2 Part 2): A122.

Wellburn AR, Barnes JD, Lucas PW, Mcleod AR and Mansfield TA (1997) Controlled O_3 exposures and field observations of O_3 effects in the UK. In: Forest Decline and Ozone. A Comparison of

Controlled Chamber and Field Experiments /Eds. Sandermann H, Wellburn AR and Heath RL (Ecological Studies, vol. 127). pp. 201-247. Springer-Verlag Berlin, Heidelberg.
Wellburn AR, Robinson DC, Thomson A and Leith ID (1994) Influence of episodes of summer O_3 on delta 5 and delta 9 fatty acids in autumnal lipids of Norway spruce *Picea abies* (L.) Karst. New Phytologist.127 (2): 355-361.
Wellburn FAM and Wellburn AR (1994) Atmospheric ozone affects carbohydrate allocation and winter hardiness of *Pinus halepensis* Mill. J. Exp. Bot. 45: 607-614.
Went FM (1960) Organic matter in the atmosphere, and its possible relation to petroleum formation. Proc. Natl. Acad. Sci. USA. 46 (2): 212-221.
Westermark U (1982) Calcium promoted phenolic coupling by superoxide radical - a possible lignification reaction in Wood. Wood Sci. Technol. 16 (2): 71-78.
Whalen SC and Reeburgh WS (1990) Consumption of atmospheric methane by tundra soils. Nature. 346 (6280): 160-162.
Whatley FR (1984) Gaseous air pollutant and plant metabolism. Gaseus Air Pollutant and Plant Metabolism. (Eds. Koziol MJ and Whatley FR) pp. 77-81. London: Butterworths.
Wieser G and Havranek WM. (1993) Ozone uptake in the sun and shade crown of spruce-quantifying the physiological effects of ozone exposure. Trees. 7: 227-232.
Wilhelm J and Wilhelmova N (1981) Accumulation of lipofiscin-like pigments in chloroplasts from scenescing leaves of *Phaseolus vulgaris*. Photosynthetica. 15 (1): 55-60.
Wilhour RG (1970) The influence of temperature and relative humidity on the respo nse of white ash to ozone. Phytopathology. 60 (4): 576.
Wilkinson TG and Barners RL (1973) Effects of ozone on $^{14}CO_2$ fixation patterns in pine. Can. J. Bot. 51 (9): 1573-1578.
Wingsle G, Mattson A, Ekblad A, Hollgrem JE and Selstam E (1992) Activities of glutathione reductase and superoxide dismutase in relation to changes of lipids and pigments due to ozone in seedlings of *Pinus sylvestris* (L). Plant Science. 82 (2): 167-178.
Winner WE, Gillespie C, Shen WS and Mooney HA (1988) Stomatal responses to SO_2 and O_3. Air Pollution and Plant Metabolism. (Eds. Schulte-Hostede S, Darrall NM, Blank LW and Wellburn AR). pp. 255-271. London. New York: Elsevier Applied Science.
Wojtaszek P (2000) Nitric oxide in plants. To NO or not to NO. Phytochemistry 54: 1-4.
Wolf SP, Garner A and Dean RT (1986) Free radicals, lipids and protein degradation. Trends Biochem. Sci. 11 (1): 27-31.
Wolfenden J, Robinson D, Cape NJ, Paterson IS, Francis BJ, Melhorn H and Wellburn AR (1988) Use of carotenoid ratios ethylene emission and buffer capsacities for the early diagnosis of forest decline. New Phytol. 109 (1): 85-95.
Wolters JH and Martens MJM (1987) Effects of air pollutants on pollen. Botanical Review. 53 (3): 372-414.
Wood FA (1970) The relative sensitivity of sixteen deciduous tree species to ozone. Phytopathology. 60 (4): 579-580.
Yalpani N, Enyedi AJ, Leon J and Raskin J (1994) Ultraviolet light and ozone stimulate accumulation of salicylic acid, pathogenesis-related proteins and virus resistance in tobacco. Planta 193: 372-376.
Yang Sfa and Hoffman NE (1984) Ethylene biosynthesis and its regulation in higher plants. Annu. Rev. Plant Physiol. 35: 155-189.
Year-book "Wastes of harmful substances" (1994) (Ed. ME Berlyand) Geophysical Osevatory of Roshydromet, Sankt-Peterburg.
Yin ZH, Langebartels C and Sandermann H Jr (1994_ Specific induction of ethylene biosynthesis in tobacco plants by the air pollutant, ozone. Proc R Soc Edinburgh B 102: 127-130.
Yoshida M, Nouchi I and Toyama S (1994) Studies on the role of active oxygen in ozone injury to plant cells. I. Generation of active oxygen in rice protoplasts exposed to ozone. Plant Sciences. 95 (2): 197-205.
Yoshida M, Nouchi I and Toyama S (1994 a) Studies on the role of active oxygen in ozone injury to plant cells. I. Generation of active oxygen in rice protoplasts exposed to ozone. Plant Science. 95 (2): 197-205.
Yoshida M, Nouchi I and Toyama S (1994 b) Studies on the role of active oxygen inozone injury to plant cells.II. Effects of antioxidants on rice protoplasts exposed to ozone. Plant Science. 95 (2): 207-212.
Young AJ and Lowe GM (2001) Antioxidant and prooxidant properties of carotenoids. Arch Biochem Biophys. 385 (1): 20-27.

Youngman RJ (1984) Oxigen activation: is the hydroxyl radical always biologically relevant? Trends Biochem. Sci. 9 (3): 280-283.
Youngson C, Nurse C, Yeger H and Cutz E (1993) Oxigen sensing in airway chemoreceptors. Nature. 365 (6442): 153-155.
Zelas RE, Cromroy HL, Bolch WE, Dunavant BG and Bevis HA (1971) Inhaled ozone as a mutagen. 1. Chromosome aberrations induced by Chinese hamster lymphocytes. Environ Res. 4: 262-282.
Zenkov NK, Men'shchikova EB, Shergin and SM (1993) Oxidative Stress. Diagnostics, Therapy, Profilactics. Nauka, Novosibirsk, 181 pp.
Zepp RG, Callaghan TV and Erickson DJ (1998) Effects of enhanced solar ultraviolet radiation on biogeochemical cycles. J. Photochemistry and Photobiology. 46 (1-3): 69-82.
Zhou YC and Zheng RL (1991) Phenolic compound and analogy as superoxide anion scavengers and antioxidants. Biochem. Pharmacol. 42: 1177-1179.
Zimmerman PR, Chatfield RB, Fishman J, Crutzen PJ and Hanst PL (1978) Estimates on the production of CO and H_2 from the oxidation hydrocarbon emissions from vegetation. Geophysical research Letters. 5 (8): 679-682.
Zinger C, Ernst D and Sandermann HJr (1998) Induction of stilbene synthase and cinnamyl alcohol dehydrogenase mRNAs in Scots pine (*Pinus sylvestris* L.) seedlings. Planta. 204: 169-176.
Zinger C, Jungblut T, Heller W, Seidlitz HK, Schnitzler JP Ernst D and Sandermann HJr (2000) The effect of ozone in Scots pine (*Pinus sylvestris* L.): gene expression, biochemical changes and uinteractions with UV-B radiation. Plant, Cell and Environment. 23: 975-982.

SUBJECT INDEX

LATIN INDEX